Autonomous Urban Mobility

This book delves into the complex landscape of autonomous urban mobility, analysing the factors that influence public adoption, stakeholder perspectives, and societal perceptions in this rapidly evolving field. Aimed at scholars, policymakers, urban planners, and industry professionals, this book offers a thorough exploration of the key elements driving the integration of autonomous vehicles in urban settings. Drawing on empirical evidence from diverse case studies, it first investigates public awareness, emphasising the roles of knowledge, exposure, and media in shaping perceptions of autonomous vehicles, and underscores the critical need for targeted awareness campaigns. Subsequent chapters examine pre-trial attitudes towards autonomous shuttles, demonstrating how initial experiences significantly impact adoption willingness, and advocate for pilot programs to cultivate informed, positive perceptions.

This book also explores the potential of smart mobility solutions to bridge first/last mile gaps, presenting data on how autonomous vehicles can improve urban transport efficiency and accessibility. By analysing socio-demographic predictors, it highlights varying perceptions of the benefits and challenges of autonomous demand-responsive transit, underscoring the importance of tailored strategies for diverse urban populations. Stakeholder perspectives, gathered through interviews and case studies, offer practical recommendations for overcoming technological, regulatory, and societal hurdles. Comparative insights from international case studies broaden the understanding of local factors influencing autonomous vehicle acceptance, while advanced modelling techniques identify the key drivers of driver-less car adoption. Finally, this book explores the transformative potential of autonomous vehicles in developing countries, offering visionary insights into their capacity to reshape urban landscapes.

With its data-rich content and forward-thinking analysis, this book is an indispensable guide to the future of urban transportation and the critical role of autonomous mobility.

This volume, alongside its companion—*Autonomous Urban Mobility: Understanding Innovation Principles, Priorities, Policies*—offers a holistic view of autonomous urban mobility. Together, these books provide a comprehensive exploration of the rapidly evolving landscape of autonomous urban mobility, the principles guiding its innovation, the wide-ranging impacts of its adoption on society, policy, and urban environments, and the transformative potential of autonomous vehicles in the future of urban transportation.

"Innovations such as electric mobility, autonomous vehicles, and mobility-as-a-service represent disruptive technologies that drive profound societal transformations. *Autonomous Urban Mobility* provides a comprehensive guide to navigating the complexities of these advancements, helping to avoid common challenges while accelerating the 'Five Transformations'—vehicle transformation, modal-split transformation, lifestyle transformation, economic transformation, and city transformation—essential for achieving sustainable urban development."
Professor Becky Loo, *The University of Hong Kong, China*

"*Autonomous Urban Mobility* dives into the social, technical, and policy landscapes shaping autonomous vehicle adoption in cities. Through empirical research and case studies, the book skilfully examines autonomous vehicles as solutions to urban issues like congestion and access, providing insights for policymakers, planners, and anyone interested in urban mobility's future."
Professor David Levinson, *University of Sydney, Australia*

"A timely work, *Autonomous Urban Mobility* presents a focused collection of Tan Yigitcanlar's research on autonomous driving technology, public acceptance and attitudes towards autonomous vehicles and autonomous transit. This book serves as an invaluable guide for researchers and policymakers, offering insights into public adoption intentions, expectations, and concerns surrounding this transformative mode of mobility."
Professor Sylvia He, *The Chinese University of Hong Kong, China*

"*Autonomous Urban Mobility* by Tan Yigitcanlar provides a timely and essential exploration of the complexities involved in integrating autonomous technologies into urban environments. By addressing foundational principles, actionable policies, and strategic priorities, this book serves as a vital resource for policymakers, urban planners, and professionals in the smart mobility sector."
Professor Jonathan Corcoran, *The University of Queensland, Australia*

"Most discussions about autonomous vehicles focus on the image of a driverless car—a narrow view. In *Autonomous Urban Mobility*, Tan Yigitcanlar presents a thorough analysis of how this transformative technology could reshape the economic, social, and spatial dynamics of cities if understood as a socio-technical system. Through a multidisciplinary lens, he explores its current impacts on urban mobility and its potential to create sustainable and equitable transport systems. Combining theory, critical analysis, and practical insights, this book is an essential guide for urban planners, transportation experts, and anyone shaping the future of cities."
Professor Fábio Duarte, *Massachusetts Institute of Technology, USA*

Autonomous Urban Mobility

Understanding Adoption Parameters, Perceptions, Perspectives

Tan Yigitcanlar

CRC Press
Taylor & Francis Group
Boca Raton London New York

CRC Press is an imprint of the
Taylor & Francis Group, an **informa** business

Designed cover image: Getty Images

First edition published 2026
by CRC Press
2385 NW Executive Center Drive, Suite 320, Boca Raton FL 33431

and by CRC Press
4 Park Square, Milton Park, Abingdon, Oxon, OX14 4RN

CRC Press is an imprint of Taylor & Francis Group, LLC

© 2026 Tan Yigitcanlar

ISBN: 978-1-032-99724-7 (hbk)
ISBN: 978-1-032-99721-6 (pbk)
ISBN: 978-1-003-60567-6 (ebk)

DOI: 10.1201/9781003605676

Typeset in Palatino
by Newgen Publishing UK

This book is dedicated to five remarkable women who have profoundly shaped my life with their love, wisdom, resilience, and an uncanny ability to remind me when I'm being a little too "me":

Selin, Ela, Susan, Cahide, Nahide

Contents

Foreword

In recent years, the potential of autonomous urban mobility has captivated researchers, policymakers, and urban planners alike. The ability to reshape our cities through autonomous vehicles (AVs) is more than a technological vision—it is a catalyst for rethinking how we move, live, and interact in urban spaces. Professor Tan Yigitcanlar's *Autonomous Urban Mobility: Understanding Adoption Parameters, Perceptions, Perspectives* addresses these possibilities with a refreshing depth and thoughtfulness. Tan's expertise in urban planning and transport is unparalleled, and his vision for the future of AVs extends far beyond the mechanics of technology to explore the societal, psychological, and spatial transformations it might enable. This book is a testament to his commitment to forward-thinking research and a shared dedication to advancing knowledge in autonomous mobility.

My relationship with Tan is built on years of collaboration and shared intellectual curiosity. Together, our collaborative research and editorial projects have consistently pushed the boundaries of urban transport research, a commitment that shines in this volume. This book surpasses my expectations on every front; it stands as a unique blend of visionary thinking and practical insight, making it an invaluable resource for anyone seeking to understand the nuances of AV adoption within urban environments. In a world captivated by the future of mobility, this book offers something unique—a comprehensive examination of not just how AVs might function, but how they might transform and be transformed by the people, cities, and cultures they aim to serve.

Tan has structured this book to address the myriad factors influencing AV adoption, but the narrative he weaves here is larger than a simple analysis of parameters and perceptions. Instead, it is an exploration of the human side of technology, presenting autonomous mobility as a reflection of our aspirations, biases, and cultural contexts. Through the empirical studies and insights shared in this work, Tan provides a window into the complex and interwoven dimensions of public awareness, trust, and readiness for AV technology. He introduces readers to the public's varied responses to AVs, reminding us that widespread adoption depends on more than just technological advancement—it requires an informed, engaged, and supportive public that feels both trust and ownership over these innovations.

Throughout the book, Tan also highlights the ways in which AVs could address some of urban mobility's most persistent challenges. The potential of AVs to bridge critical first- and last mile gaps in transit networks, for example, points to a future where these technologies extend beyond convenience to accessibility and inclusivity. The implications are profound, particularly for communities that are underserved by traditional transport networks. Here,

Tan reminds us that autonomous mobility could be an equaliser, reshaping transit systems to cater to diverse urban populations and opening up opportunities for more equitable city planning.

The book further explores the dynamic interactions between people's social and demographic backgrounds and their perceptions of AVs. Tan sheds light on how factors like age, income, and residential location shape public attitudes towards AV adoption. This consideration of socio-demographic diversity enriches our understanding of AV adoption, suggesting that the path to successful integration will be as varied as the urban populations AVs are intended to serve. Rather than a one-size-fits-all approach, Tan advocates for contextually informed strategies that align with the unique characteristics of each community—a perspective that adds a valuable layer of inclusivity to the discourse on autonomous urban mobility.

One of the book's standout contributions is its examination of the AV adoption process from the perspectives of key stakeholders. Drawing from real-world case studies and interviews, Tan goes beyond theory to reveal the experiences, concerns, and aspirations of the policymakers, technology developers, and urban planners at the forefront of this transformation. This collaborative approach invites readers to see AV adoption as a multi-stakeholder endeavour, where overcoming technical and regulatory challenges requires not just innovation but a shared commitment to public benefit. Tan's work underscores the importance of building alliances across sectors, reminding us that a successful transition to autonomous mobility will depend on cooperation and mutual understanding among all involved.

Tan also incorporates a unique cross-cultural perspective, offering comparative insights from diverse urban settings. By analysing how different cities and regions are approaching AV adoption, he illuminates the complex relationship between technology and place. The result is a rich, globally informed narrative that challenges readers to consider how local cultures, policies, and urban structures influence the acceptance and implementation of AVs. This cross-cultural lens is not only enlightening but also necessary, as it encourages us to avoid viewing AV technology through a narrow lens and instead recognise the broader social and cultural contexts that will shape its future.

The book culminates with a forward-looking perspective on how AVs could be deployed in emerging economies, where the potential for transformative change is vast. Here, Tan invites us to imagine a world where AVs are not simply the next step in mobility for advanced economies but a leap towards a more sustainable, inclusive urban future for developing regions. This vision aligns with the most aspirational goals of urban planning, suggesting that autonomous mobility could be a tool for addressing social and environmental challenges on a global scale.

By the end of *Autonomous Urban Mobility*, readers are not only equipped with a deeper understanding of AV technology but also inspired to think critically about how we might collectively shape its future. Tan's work stands out for its ability to connect the granular details of adoption parameters and public perceptions with the broader, transformative possibilities of autonomous mobility. The book is a call to action—an invitation for scholars, practitioners, and policymakers to engage with AVs as more than just machines but as instruments for social change and urban resilience.

Autonomous Urban Mobility: Understanding Adoption Parameters, Perceptions, Perspectives is more than a study of technology; it is a thoughtful, thorough exploration of the human and societal dimensions of urban mobility. Tan's work here will undoubtedly serve as a foundation for future research and inspire ongoing discussions in the field. As we stand at the threshold of a new era in transportation, this book serves as a guide for navigating the challenges and opportunities that lie ahead. It is my pleasure to endorse this work, confident that it will inspire new ideas, spark meaningful dialogues, and contribute substantially to our collective understanding of the future of autonomous urban mobility.

Professor Liton Kamruzzaman
Monash University, Australia

Preface

Imagine a city where cars do not just drive—they think. Picture streets buzzing with sleek vehicles that communicate silently, navigating effortlessly through complex urban landscapes. It is a world reminiscent of the science fiction once confined to the glow of TV screens, where autonomous cars were more fantasy than reality. The futuristic dream of self-driving vehicles, once epitomised by the flying cars of *The Jetsons* zipping effortlessly through utopian skylines, is no longer a far-off possibility. It is a tangible reality taking shape on our streets, fuelled by groundbreaking innovation and a pressing need for smarter, greener, and more inclusive urban mobility. This book dives into the heart of that transformation, dissecting the adoption parameters and societal dynamics that will determine whether this once-fictional future can truly deliver on its promises.

Autonomous Urban Mobility: Understanding Adoption Parameters, Perceptions, Perspectives explores the intricate and rapidly evolving field of autonomous urban mobility by focusing on the adoption side in particular. This book provides an in-depth examination of the adoption parameters that shape public attitudes, the perceptions that influence acceptance, and the diverse perspectives of stakeholders crucial to the successful integration of autonomous vehicles (AVs) in urban environments. As cities confront pressing challenges like traffic congestion, air pollution, accessibility disparities, and the need for sustainable development, AVs emerge as transformative solutions capable of redefining urban transportation systems. This work responds to these challenges by analysing the complex factors driving AV adoption, offering a multidisciplinary view that blends empirical research with theoretical insights. Serving as a vital resource for scholars, policymakers, urban planners, and industry professionals, this book equips readers with the knowledge to understand and guide the future of autonomous mobility in cities worldwide.

The concept of autonomous urban mobility goes beyond technology itself; it encompasses the broader context of urban challenges, societal expectations, and the fundamental goal of improving the quality of urban life. AVs are more than just innovations on wheels; they represent a vision of sustainable, efficient, and accessible transportation. By addressing the unique adoption dynamics for AVs in this book, we hope not only to provide a comprehensive understanding of these dynamics but also to inspire readers to consider AVs as solutions to some of the most pressing issues cities face today.

A Technological Transformation with Human-Centric Goals: In the past decade, AVs have evolved from an experimental technology into a transformative solution for urban mobility, with the potential to redefine transportation in cities worldwide. The book delves into the heart of this transformation, capturing the myriad factors influencing the adoption of AVs in urban settings. This book aims to unravel the complexities of AV adoption by analysing empirical evidence, exploring theoretical frameworks, and investigating real-world case studies to provide a nuanced understanding of public perception and stakeholder perspectives.

The adoption of AVs touches on essential questions about how cities can and should evolve. As we face critical decisions about transportation systems that will serve as the backbone of future urban life, understanding how communities perceive and accept these new technologies becomes increasingly important. AVs bring with them the potential for reducing carbon footprints, lowering transportation costs, and enhancing urban accessibility. However, realising these benefits will require deliberate planning, informed public engagement, and policies that balance the interests of various stakeholders.

This book is a continuation of a journey that began with observing the growing interest in autonomous mobility and recognising its potential to address pressing urban challenges—congestion, accessibility, and environmental sustainability among them. By understanding the public's readiness to embrace autonomous technologies and identifying key socio-demographic and psychographic factors influencing adoption, we can pave the way for smarter, more inclusive, and sustainable urban transportation systems. In this way, the book takes a proactive approach, emphasising the importance of foreseeing and addressing potential barriers to adoption before they become obstacles to the wider integration of AVs in our cities.

Exploring Awareness and Perception: The First Step to Adoption: The journey of exploring autonomous mobility begins in Brisbane, Australia, where the book's authors investigate factors influencing public awareness of AVs, highlighting the importance of knowledge and exposure in shaping public sentiment. Awareness is the first step towards adoption, and this research underscores the need for effective communication strategies to engage urban populations with these transformative technologies. By identifying patterns in how different demographic groups perceive AVs, this book provides actionable insights for policymakers and transportation planners on how to foster awareness and acceptance.

Awareness is essential for acceptance. Studies have shown that people who understand the benefits and risks associated with new technologies are more likely to adopt them. In this book, we discuss the different ways that public awareness can be cultivated and maintained, from educational campaigns to community engagement programs. By examining survey data and drawing insights from public responses, we offer readers an in-depth look at how to build foundational support for AVs within communities, focusing on strategies that go beyond mere marketing to truly educate and inform.

Pilot Programs and Public Demonstrations: Building Trust in Autonomous Mobility: The book then shifts focus to the adoption intention for autonomous shuttles, examining how pre-trial perceptions and early interactions with

AVs can shape long-term acceptance. Here, readers gain insights into the role of pilot programs and public demonstrations in easing people's concerns and building a foundation of trust. The authors propose a conceptual framework for understanding the motivations and reservations of potential AV users, thereby aiding planners in crafting effective policies that encourage engagement with AVs.

Pilot programs play a critical role in the introduction of AVs, as they provide first-hand experiences for the public, allowing them to become familiar with the technology before it becomes mainstream. Early adopters and participants in these programs act as ambassadors for AVs, sharing their experiences and helping to shape public opinion. This book emphasises the importance of carefully planned and well-communicated pilot programs that address common concerns about safety, reliability, and usability. We delve into examples of successful pilot programs and the lessons they offer, including the importance of community involvement and transparent reporting of pilot outcomes.

Bridging the First/Last Mile Gap: AVs as Catalysts for Connectivity: Smart mobility solutions play an essential role in bridging the 'first/last mile' gap, which is a pivotal factor in urban transportation. This book presents compelling evidence from Australian cities, demonstrating the potential of AVs to enhance public transportation networks, especially in areas underserved by traditional transit options. By addressing the first/last mile challenge, autonomous urban mobility has the potential to foster greater connectivity, improving access to essential services and opportunities for all urban residents.

The first/last mile problem remains one of the most significant barriers to public transportation use. Many people rely on personal vehicles simply because public transportation does not reach their neighbourhoods or stops too far from their destination. AVs, particularly autonomous shuttles and ride-sharing models, provide a promising solution to this issue, offering a convenient link between homes and transportation hubs. Through a careful analysis of case studies and statistical data, this book illustrates how AVs can enhance connectivity and reduce dependence on private vehicles, ultimately contributing to more sustainable and accessible urban environments.

Perceptions of Autonomous Demand-Responsive Transit with a Focus on Inclusivity: Next, this book delves into perceptions of autonomous demand-responsive transit. Through a thorough analysis of socio-demographic predictors, it explores how different urban populations view the benefits and limitations of AVs. This section highlights the importance of tailoring deployment strategies to meet the needs of diverse demographic groups, ensuring that AV adoption benefits the entirety of society. The focus on inclusivity extends beyond demographic divides, addressing the unique challenges faced by various communities within urban spaces.

As AVs become more integrated into public transportation systems, it is essential to recognise that different communities will have different needs and concerns. This book examines factors such as age, income, and familiarity with technology, which can all affect how people view AVs. Understanding these differences allows planners to develop policies and programs that are inclusive and responsive to the diverse needs of urban populations. Through survey data and global studies, we explore the various ways AV deployment can be tailored to serve both young professionals and elderly populations, high-income and low-income communities, ensuring that AVs contribute to an equitable urban mobility landscape.

Stakeholder Perspectives and Insights into the Challenges of AV Deployment: Stakeholder perspectives are crucial for successful AV deployment. In this book, a series of interviews and case studies present the viewpoints of industry professionals, policymakers, and academics, shedding light on the barriers and enablers in AV implementation. Through these voices, readers gain insight into the regulatory, technological, and social challenges that may hinder AV adoption and learn practical solutions for overcoming these obstacles.

The transition to autonomous mobility requires the support of a broad coalition of stakeholders, from government regulators to private companies and advocacy groups. Each of these groups has unique perspectives on AV adoption and implementation, often shaped by their distinct goals and priorities. By including these voices, this book provides a balanced view of the opportunities and obstacles associated with AVs. Regulatory hurdles, liability issues, and public safety concerns are among the topics discussed, offering readers a comprehensive understanding of what it takes to create a supportive environment for AVs in urban areas.

Cross-Cultural Comparisons for Understanding Attitudes across Different Contexts: This book also includes an examination of attitudes towards AV deployment in varied urban contexts, drawing on comparative data from Las Vegas, Nevada, US, and major cities in Australia. The cross-cultural analysis illustrates how perceptions vary across different geographies, influenced by local factors such as infrastructure, public policy, and cultural norms. Understanding these differences is essential for developing targeted strategies that resonate with specific urban populations.

While AVs have global appeal, the factors that influence their acceptance are often local. Cultural differences, infrastructure readiness, and regional policies all play a role in shaping public attitudes towards AVs. This book explores these nuances, helping readers appreciate the complex interplay between technology and culture. By analysing cross-cultural data, we highlight how cities can tailor their approach to AV adoption to align with local values and expectations, enhancing the likelihood of success and building community trust.

Adoption in Australian Cities and Insights for Policymakers and Industry Leaders: The analysis of AV adoption within Australian cities provides a detailed look at the influencing factors, from demographic trends to technological familiarity. By identifying the determinants of adoption likelihood, this book offers valuable guidance to policymakers and industry leaders, helping them navigate the complexities of AV integration in urban landscapes. Furthermore, the discussion extends to the adoption of AVs in developing nations, offering a comparative perspective that broadens the global relevance of this work. In many cases, developing nations are uniquely positioned to leapfrog traditional mobility models, embracing AVs as a means of building modern, efficient transportation networks from the ground up.

Through rigorous data analysis and modelling, this book provides insights that will help policymakers make informed decisions about AV integration. The discussion on AV adoption in developing countries is especially relevant as these nations often face limited infrastructure yet hold the potential to implement new mobility solutions from scratch. By examining case studies and policy models, readers will gain a deeper understanding of how to promote AV adoption in various global contexts, making the book a truly international resource for understanding the future of urban mobility.

Social and Psychological Factors: Beyond Technology: In exploring the multifaceted dimensions of autonomous urban mobility, this book goes beyond technical discussions. It emphasises the social and psychological factors that will shape the future of AVs, recognising that public sentiment, stakeholder support, and socio-cultural nuances are just as critical as technological advancements in achieving widespread adoption. The book, therefore, serves not only as a guide for those involved in transportation planning and policy but also as a resource for academics, industry leaders, and anyone interested in the evolving landscape of urban mobility.

The book's multidisciplinary approach encourages readers to consider the human element of autonomous mobility. While technology is a powerful enabler, public trust and social acceptance are equally important for the success of AVs. This work highlights the ways that public perceptions, community engagement, and psychological factors impact the adoption and acceptance of AVs, emphasising the need for a holistic view of urban mobility that accounts for the complex human aspects of technological integration.

An Invitation to Shape the Future of Urban Mobility: As cities grapple with the challenges of modern transportation, this work aims to inform and inspire a new generation of thinkers and doers committed to advancing autonomous urban mobility in ways that are equitable, efficient, and sustainable. It is my hope that this book will contribute to a future where AVs enhance quality of life, reduce environmental impact, and foster greater inclusivity in urban mobility. Together, we can navigate the road ahead, by considering both opportunities and constraints, paving the way for a smarter, safer, and more connected and sustainable world.

As the realm of autonomous urban mobility advances, so too will the dynamic challenges and expansive opportunities it brings. The book equips readers to actively engage with these evolving issues by cultivating a thorough understanding of the factors shaping the future of AV adoption and integration. The book emphasises the risks along with opportunities and offers a forward-thinking approach that anticipates urban needs, champions sustainability, and advocates for equity across diverse communities. By blending theoretical insights with practical frameworks, this work serves as a foundational guide for those dedicated to fostering autonomous mobility solutions that prioritise social responsibility and environmental resilience.

In essence, this book acts as a roadmap for understanding the intricate landscape of autonomous urban mobility. It encourages readers to look beyond the technology itself and to consider the broader societal impacts that AVs and smart mobility systems will have on cities and the lives within them. *Autonomous Urban Mobility: Understanding Adoption Parameters, Perceptions, Perspectives* provides a comprehensive perspective on the interconnected challenges and possibilities that lie ahead, empowering readers to contribute to a future where urban transportation systems are not only smarter but also more inclusive, resilient, and sustainable.

Lastly, this book, *Autonomous Urban Mobility: Understanding Adoption Parameters, Perceptions, Perspectives*, serves as a companion volume to *Autonomous Urban Mobility: Understanding Innovation Principles, Priorities, Policies*. Together, these volumes provide a comprehensive exploration of autonomous urban mobility, with each addressing distinct yet interrelated aspects of AV integration. While *Innovation Principles, Priorities, Policies* delves into the technological and policy frameworks guiding AV innovation, *Adoption Parameters, Perceptions, Perspectives* focuses on the factors driving public acceptance, stakeholder engagement, and the societal dimensions of AV adoption. Together, these works offer a holistic view of the challenges and opportunities presented by AVs, making them essential resources for anyone seeking to understand and shape the future of urban mobility.

May this book spark fresh perspectives and ignite your curiosity as you delve into the transformative world of autonomous urban mobility. Whether you are a scholar, policymaker, urban planner, or simply an enthusiast of futuristic technologies, may the insights within these pages challenge your thinking, broaden your understanding, and inspire innovative solutions for the cities of tomorrow. Let this journey into the complexities of adoption, perceptions, and policies empower you to play a role in shaping smarter, more sustainable, and more inclusive urban transportation systems!

Author

Tan Yigitcanlar is a distinguished Australian researcher known for his significant contributions to urban sustainability, technology, and planning. He is a Professor at the Queensland University of Technology's (QUT) School of Architecture and Built Environment and serves in multiple leadership roles. He leads the QUT Urban AI Hub, the QUT City 4.0 Lab, the QUT Smart City Research Group, and the Australia-Brazil Smart City Research and Practice Network. Additionally, he is a distinguished member of the Australian Research Council College of Experts.

With a career spanning over three decades, Tan's work has encompassed research, teaching, training, and capacity building at prestigious universities in Australia, Brazil, Finland, Japan, and Turkey. His research is dedicated to addressing contemporary challenges in urban planning and development, covering a broad spectrum of economic, societal, spatial, governance, and technology-related issues. At the heart of his work lies a strong commitment to promoting smart and sustainable urbanisation.

Tan has provided research consultancy services to all levels of government, international corporations, and non-governmental organisations both in Australia and internationally. His expertise has been instrumental in helping these entities formulate forward-looking strategies, build resilience, and respond effectively to emerging disruptive challenges. He serves as the Lead Editor-in-Chief of the Elsevier Smart Cities Book Series and holds senior editorial appointments with 12 high-impact academic journals. Tan has held senior leadership roles in more than 40 major international conferences, including serving as chair for 12 of them. He has delivered more than 80 keynote and invited addresses at prominent international academic and industry events.

Tan's research findings have been extensively disseminated, with over 340 articles published in high-impact journals and 33 key reference books published by esteemed international publishing houses. His work has significantly influenced urban policy, practice, and research internationally, with over 30,000 citations and an h-index of 100. He is ranked #1 in Australia and top 10 worldwide in urban and regional planning according to the 2023 Science-wide Author Databases of Standardised Citation Indicators. He was also recognised as an 'Australian Research Superstar' in the Social Sciences Category by *The Australian*'s 2020 Research Special Report and named Australia's Social Sciences 'Research Field Leader' for Urban Studies and Planning in the 2024 edition of *The Australian Research Magazine*. He is acknowledged as one of the top 1% of scientists worldwide by Clarivate in 2023.

1

Factors Affecting Public Awareness of Autonomous Vehicles

1.1 Introduction

Rapid technological changes are sweeping transportation and propelling urban areas towards the fringes of an unknown era in mobility (Faisal et al., 2020). Driven by high-speed communication networks, artificial intelligence, cloud computing, and connected devices and infrastructure, new transport innovations—such as autonomous vehicles (AVs), intelligent transport systems (ITSs), demand responsive transport (DRT), and mobility-as-a-service (MaaS)—are increasingly being investigated as fundamental drivers of smarter, and more equitable transportation systems (Butler et al., 2020b; Yigitcanlar et al., 2020).

At the forefront of the move towards 'smart mobility' is the AV (AV) (Faisal et al., 2019; Yigitcanlar & Kamruzzaman, 2019). While driving automation systems such as advanced cruise control, lane keeping and crash avoidance are standard features in many modern vehicles, over the next two to three decades AVs that sense their environment and drive autonomously start to become increasingly commonplace (Van Brummelen et al., 2018; Butler et al., 2020b).

Parallel with technological changes, urban sprawl—partially a result of unrestrained growth, overreliance on private vehicles, housing affordability, and preference for suburban living—is, in many cities, aggravating existing transport equity issues (Currie, 2009; Dur & Yigitcanlar, 2015). Furthermore, reliance on the Internet and smartphones are creating new risks associated with cyber theft, data privacy, and the digital divide (Seetharaman et al., 2021).

While AVs are expected to revolutionise accessibility for the disabled, they have also been identified for their potential to improve the safety, efficiency, and flexibility of the transportation system, while reducing infrastructure demand, congestion, cost, travel time, and driver stress (Pettigrew et al., 2018; Dai et al., 2021; Huang & Qian, 2021). To harness these social benefits recent trials of shared AVs (SAVs) including shuttles, buses, and smaller lower passenger 'pods' have been implemented throughout the world to provide

DOI: 10.1201/9781003605676-1

alternatives to private vehicle travel and encourage public transport rider-ship by overcoming the 'first-mile/ last-mile' gap (Wen et al., 2018; Vosooghi et al., 2019; Gurumurthy et al., 2020).

Despite this trend, there is a risk that the public for which these programmes aim to benefit, remain unaware, or apathetic, towards the potential advantages of AVs. This is challenging because it is ultimately the individual who will make decisions on how AVs are adopted (Yuen et al., 2020a; Manfreda et al., 2021) and their values, interests, and expectations will influence decision-making and shape future policy development. Without robust policy design guided by foresight and early recognition of opportunities and challenges the diffusion of AVs in urban areas could create greater disparity in transporta-tion equity and increase existing social and economic disadvantage (Dianin et al., 2021).

To date, little research has been completed that aims to understand the extent to which the public understands the benefits of AVs—specifically whether they understand the potential for the technology to ease an issue such as transportation disadvantages. Transport disadvantage is a multidi-mensional concept that includes those who experience physical disabilities as well as those who face challenges associated with the cost of transport, distance, public transport coverage and frequency, overreliance on private vehicles, psychological issues such as perception and safety concerns, ability access to information, and institutional and governance barriers such as policy and regulations (Kamruzzaman et al., 2016; Sochor & Nikitas, 2016; Yigitcanlar et al., 2019b; Butler et al., 2020a). With this in mind the aim of this study is two-fold: (a) To understand the extent to which the public feels that AVs can help ease transportation disadvantages in their neighbourhoods, and (b) To assess the receptiveness to AVs as a way to respond to transportation disadvantage issues by exploring whether individual characteristics, existing travel habits, and challenges impact their awareness of potential benefits of AVs—in the case of Brisbane, Australia.

Following this introduction, Section 1.2 provides a summary and review of relevant literature. Next, Section 1.3 outlines the research design including a description of the case study area, data collection and analysis methods. Then, Section 1.4 discusses the findings and results. Last, Section 1.5 provides some concluding remarks including an overview of the studies limitations and opportunities for future research.

1.2 Literature Background

A few studies have analysed the extent to which the public understands how AV can improve disadvantage. From a disability perspective Bennett et al.

(2019) analysed willingness of people who have mental health disabilities to accept AVs. This research found that despite the potential benefits of the technology in providing increased accessibility for this disadvantaged group there was still scepticism including reservations regarding safety and accessibility of vehicles. Females were shown to be more discouraged from using AV due to fear and less enticed by their potential.

These results correlated with similar research in the field of AV perception (Golbabaei et al., 2021) with some concluding that this fear was likely to influence lower acceptance of AV among females (Dennis et al., 2021), while males are more likely to be motivated by the 'fun' of riding in AVs (Rice & Winter, 2019). Despite other research which shows a correlation between younger age groups (Golbabaei et al., 2021; Huang & Qian, 2021; Othman, 2021), higher income (Wang et al., 2020a; Yuen et al., 2020a, 2020b) and AV acceptance, Bennett et al. (2019) found age and income to be little relevance among those with mental health disabilities. Nonetheless, with regard to the increased freedom offered by AVs, Bennett et al. (2019) found that those who had a disability which impacted their daily lives were more receptive.

Continuing from their research on mental health Bennett et al. (2020) looked at the willingness of people who are blind to accept AV. While this research found no correlation between age and income, it concluded that among the blind, gender was not a good determinant of AV acceptance. A potential reason may be that enthusiasm towards the increased freedom that AV offers may simply outweigh the influence of fear and scepticism among those with disabilities that directly impact their ability to drive or use public transport.

Cordts et al. (2021) further explored the perceptions among those with physical disabilities and found general enthusiasm towards AVs. This correlates with similar observations which discuss how those with difficulty driving were more likely to accept AVs (Gkatrzonikas & Gkritza, 2019) and the notion that mobility impairments are likely to be a significant motivation for accepting AVs (Golbabaei et al., 2020).

Likewise, Faber & van Lierop (2020) conducted a series of focus groups with residents at retirement homes and other care facilities and found that despite previous studies and links between older aged groups and less positive attitudes towards AV, there remains a lot of interest in using AVs to overcome existing barriers associated with accessibility of transport and mobility in the elderly. Older residents highlighted the importance of flexibility and on-demand services as well as the social advantages of being able to travel with friends (Faber & van Lierop, 2020).

Similar results regarding the importance of flexible services and the perceived social benefits were put forward by Sochor and Nikitas (2016) and Zandieh and Achaempong (2021), who also highlight older residents' enthusiasm regarding the potential for AV to create a less stressful driving experience, less parking issues, and improve access to destinations with no public transport coverage—which in turn could improve physical activity such as walking (Zandieh & Achaempong, 2021). However, this research also underlined some concerns regarding increased congestion, reduced safety, loss of driver support, lack of control, fear of new technology, loss of driving culture, and increased costs (Zandieh & Achaempong, 2021).

From a health perspective, Pettigrew et al. (2018) completed an online survey which aimed to understand public awareness of the health benefits of AV. The results from open-ended questions related to the health benefits of AV showed little awareness of the health benefits of AV—the most common benefit mentioned being crash reduction (21%). Nevertheless, when prompted with a list of potential benefits most identified mobility for the elderly and disabled as a potential benefit (73%), with around half of all users identifying reductions in stress, crashes, and emissions as well as improvements in cyclist safety.

Despite the limited research into the awareness of the public benefits of AVs (Dennis et al., 2021), a number of studies have identified the potential benefits of AVs as a way to alleviate transportation disadvantages. Recently, Butler et al. (2020a) completed a systematic literature on innovations in future mobility and categorised their contributions to the easing of transport disadvantage under physical, economic, temporal, physical, psychological, and informational dimensions.

The research highlighted the benefits of AVs to improve accessibility for those unable to drive motor vehicles, increase comfort for those with special needs, reduce operational costs, increase the value of time spent in transit, and reduce the number of crashes (Butler et al., 2020a). Conversely, the research highlighted the risk that AVs could price lower-income populations out of the market, decrease infrastructure capacity (due to increased instances of dead runs, and increased the number of private trips), and increase data and cyber security risk. Furthermore, the research emphasises the importance of shared mobility in overcoming many of these risks (Butler et al., 2020a).

Further to the risks associated with the implementation of future mobility, Dianin et al. (2021) discuss how AVs may transform land use through accessibility polarisation or sprawl. Under the accessibility polarisation model, the diffusion of SAVs could lead to increased density in major urban areas, while potentially improving accessibility in these areas this may come at the cost of accessibility in fringe and rural areas. Similarly, under this model, rising property prices in dense urban areas could inevitably push lower-income residents into areas with lower accessibility and likely higher transport costs.

Conversely, under the accessibility sprawl model, the diffusion of private AV may lead to a new era of suburbanisation due to the increased value of time and comfort associated with daily travel and reduced property prices in fringe and rural areas. This would likely aggravate existing issues associated with urban sprawl (Yigitcanlar et al., 2019a; Dianin et al., 2021).

While AVs can be advantageous in providing a transport option for those with reduced driving abilities there remains potential for many negative impacts associated with how these services are implemented in urban areas (Kovac et al., 2020). This highlights the need for proactive policies that aim to manage future mobility to service the broader public interest rather than facilitating a continuation of the status quo (Kovacs et al., 2020).

Given the importance of SAV and in an effort to optimise these services Papadima et al. (2020) completed a survey and conjoint analysis following a pilot programme of driverless buses in Greece. While the results show a generally positive view of the service with a large percentage of residents wanting the service to be implemented permanently, the results also emphasised the importance of maintaining a cost structure similar to existing public transport services with any cost-benefits gained from reduced personnel used to improve service coverage and frequency. This is reflected by Zandieh and Acheampong (2021), who found older residents with a familiarity with existing public transport systems expressed a preference for the continuation of these services though there was some potential to incorporate SAV as a feeder system which complements the primary public transport system.

1.3 Research Design

1.3.1 Case Study

A case study approach was adopted to investigate public awareness of AVs to ease transportation disadvantages in Brisbane (Queensland, Australia). Despite being the third-largest city in Australia and a claimed smart city (Yigitcanlar et al., 2019c), the urban form of Brisbane is predominately low-density and is typical of many Australian and North American cities where considerable growth occurred parallel with the emergence of the private automobile shaping urban form towards car-dependent suburban neighbourhoods (Li et al., 2015). The greater Brisbane region is the fastest-growing region in Australia, at the most recent census the population was around 2.2 million with this number projected to have grown by 1.9% per year to be at around 2.6 million in 2020. Medium-range population projections expect growth to continue to 3.7 million by 2041 (Queensland Government, 2018).

Due to housing affordability and preference for suburban living, most growth is occurring in low-density fringe urban areas (Li et al., 2015; Willing & Pojani, 2017). This is increasing car dependency, infrastructure demand, and making the provision of equitable public transportation difficult (Li et al., 2015). Brisbane is a highly car-dependent society with 75% of the population travelling to work by private vehicle, 90% of the population owning a private vehicle, and 48% of households owning two or more vehicles (Queensland Government, 2018). Furthermore, lower-income households are typically located in fringe suburban areas where poor public transportation coverage and greater travel distances result in high costs and increased travel time (Dodson & Sipe, 2008; Li et al., 2015). These challenges are further intensified for those residents who are unable to access a private vehicle— e.g., the disabled, elderly, indigenous, and lower-income groups (Rosier & McDonald, 2011).

1.3.2 Data Collection

To better understand public awareness of AVs as a way to ease transportation disadvantages, an online questionnaire was developed and distributed by an external survey company to a sample of residents in Brisbane. A minimum sample size was calculated using Krejcie and Morgan's (1970) sample size formula for populations over 100,000. To achieve a confidence level of 95% and a margin of error of 5% a minimum sample size of 384 was considered appropriate. Rather than focusing on the awareness of those vulnerable to transport-related inequities, the sample was broadened to include any participants who lived within the case study area and were over the age of 18 years. The reasoning behind this approach was that while AVs have the potential to benefit those facing inequities the individual decisions and ultimate policy directions related to the adoption of AV will be likely influenced by the broader society.

An ethical clearance was granted from the host university before the launch of the survey. The survey was conducted online via the Key Survey platform between February and April 2019. Following the removal of cases with missing or incomplete data, a total of 610 valid response were received (27% response rate). An approximate spatial representation of the received responses based on street name and postcode is shown on Figure 1.1.

1.3.3 Descriptive Analysis

1.3.3.1 Dependent Variable

For the purposes of this study, a single dependent variable was used based on the following question: To what extent do you think driverless cars would help in easing the transport disadvantage issue in your neighbourhood? This question aims to analyse the public awareness of AVs to improve transportation equity. As discussed in the literature review, AVs are commonly identified for their potential to improve transportation equity by increasing the accessibility, affordability, and efficiency of transportation systems (Butler et al., 2020a; Dianin et al., 2021).

This research question was placed last in a questionnaire with responses sought for a range of factors including disability, distance to transportation, access to public transport, cost, access to information, and safety concerns. Furthermore, the final question was immediately preceded by six priming questions which directly asked participants to rank to their own abilities to

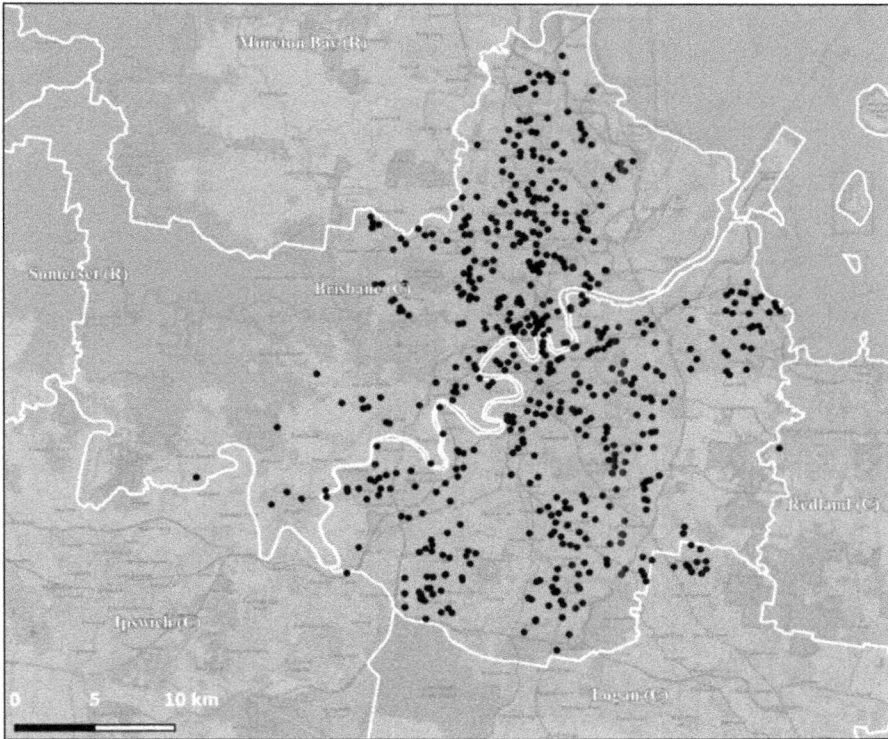

FIGURE 1.1
Distribution of survey data.

participate in transportation activities. The purpose of ordering the questionnaire in this manner was to give participants exposure to the concept of transport disadvantage without explicitly attempting to define the various complexities associated with it.

In order to answer this final question participants were provided with a five-point Likert scale with the following options: 'Not at all', 'A little', 'A moderate amount', 'A lot', and 'A great deal'—coded one to five, respectively, so that lower expectations had lower values and higher expectations has higher values (Table 1.1). The results from this question were analysed two-fold: (a) Original categories were maintained, and (b) the first two categories were merged, the third category was kept, and the last two categories were merged.

The first option shows that a large portion of users have no expectation (28.4%) for the potential for AVs to alleviate transportation disadvantages. Conversely, those with a high level of expectation had the fewest responses (12.3%) (Table 1.1). The second option helps to focus on broader groupings

TABLE 1.1

Survey Results: Original Categories

	To What Extent Do You Think Driverless Cars Would Help in Easing the Transport Disadvantaged Issue in Your Neighbourhood?			
	Frequency	Percent	Valid Percent	Cumulative Percent
Not at all	173	28.4	28.4	28.4
A little	99	16.2	16.2	44.6
A moderate amount	173	28.4	28.4	73.0
A lot	90	14.8	14.8	87.7
A great deal	75	12.3	12.3	100.0
Total	610	100.0	100.0	

by aggregating the data into three distinct categories 'low expectation', 'moderate expectation', and 'high expectation'. Following this step, the distribution of responses was more clearly delineated with those who felt that AVs offer little to no help in easing transportation disadvantage representing 44.6% of all respondents, and those with moderate to high expectations representing 28.4% and 27%, respectively. These results indicate a general indifference within the city regarding the potential for AVs to improve transportation equity.

This research identified a total of 18 variables with the potential to predict user awareness of the transportation equity advantages of AVs. The first four variables relate to personal characteristics and include: (a) gender; (b) age; (c) income; and (d) disability. The next six relate to current transportation habits and include: (a) valid driver's licence; (b) number of vehicles in household; (c) weekday transport mode; (d) weekday transport time; (e) average cost of private vehicle; (f) distance to nearest public transportation station/stop; (g) ability to access information for public transportation; and (h) feeling of safety on public transportation. A description of each of these variables is outlined in Table 1.2.

Table 1.3 shows a descriptive analysis of the first 12 variables which relate to the personal characteristics and transportation habits of participants. Based on a descriptive analysis of the results shown in Table 1.3 only a small percentage of participants (6.4%) have an identified disability that inhibits them from engaging in transportation-related activities. Notwithstanding, other identified variables such as no driver's licence (11.6%), travel time greater than 30 min (24.9%), transport cost in excess of 15% of income (39.3%), and distance to a bus or train greater than 800 m–1000 m, respectively, (28%) are also likely contributors to transportation-related accessibility issues.

Figure 1.2 shows the remaining five variables related to the perception of transportation disadvantages among participants. Data for these variables were collected using a five-point Likert scale ranking from 1 (no issue) to 5 (difficult). Most users in all categories identified no issue with the ability to

TABLE 1.2

Description of Independent Variables

Variable	Question	Description
1. Gender	What gender are you?	The contribution of gender towards positive attitudes and AV is mixed (Bennett et al., 2019, 2020; Rice & Winter, 2019; Golbabaei et al., 2020). The inclusion of this variable will test the hypothesis that males are more likely to be positive towards AV.
2. Age	Which age group best describes you?	Age categories have been divided into four groups (18–29, 30–49, 49–69, and 70+) represent young adults, middle-aged adults, older adults, and the elderly. This variable will test the hypothesis that despite the accessibility benefits offered to older residents they are less likely to have positive attitudes towards AV (Golbabaei et al., 2020; Huang & Qian, 2021; Othman, 2021).
3. Income	Which income group does your household belong to?	Middle- and higher-income residents have been shown to have more positive attitudes towards AV (Wang et al., 2020b; Yuen et al., 2020a; 2020b; Golbabaei et al., 2020). Participants were asked their income level based on categories defined in the Australian census (i.e., no income, less than $300, $300–$649, $650–$799, $800–$999, $1000–$1249, and $1250–$1499). Results were recoded into two distinct categories. As the low-income threshold for Australia is $721 per week, the first three groups were categorised as 'low income' with the remaining groups categorised as middle- to high-income. The inclusion of this variable tests the hypothesis that low-income residents are less likely to have positive attitudes.
4. Disability	Do you regularly need help when travelling because of long-term illness or disability?	The inclusion of this variable will test the hypothesis that those with a disability are more likely to have positive attitudes towards AV due to the increased accessibility offered (Golbabaei et al., 2020; Cordts et al., 2021).
5. Licence	Do you have a valid driver's licence?	Brisbane is a car-dependent city with 75% of all residents travelling to work by private vehicle (Queensland Government, 2018), and the city has been identified for its lack of public transport coverage (Dodson & Sipe, 2008; Li et al., 2015). Therefore, not having a driver's licence may be considered a mobility impairment (Golbabaei et al., 2020) in car-dependent societies. This question will test the hypothesis that those without a driver's licence are more enthusiastic about the potential of AV.

(continued)

TABLE 1.2 (Continued)

Description of Independent Variables

Variable	Question	Description
6. Household vehicles	How many motor vehicles (car/ motorbike) does your household have?	Further to Brisbane's car dependency, over 90% of the population own a private vehicle with 48% of households owning two or more vehicles (Queensland Government, 2018). Due to lack of adequate public transport coverage not having access to private vehicles may increase disadvantage. Furthermore, studies show that car owners are more likely to favour private AV—thus negating potential benefits for SAV (Golbabaei et al., 2020). This question will test if those without easy access to a private vehicle are more aware of the potential benefits of AV.
7. Travel mode	On most weekdays what type of transport do you mainly use?	Residents in Brisbane who travel by public transport are more likely to experience accessibility issues due to a lack of public transport coverage (Dodson & Sipe, 2008; Li et al., 2015). Furthermore, AV have been identified for their potential to improve public transport accessibility and cost. This question will test if public transport users are more aware of the advantages of AV.
8. Travel time	On average, how long does it take you to travel (one way) to your most frequently visited places on weekdays?	AV have been identified for their potential to improve the value of time (VOT) spent in transit (Butler et al., 2020a). Therefore, it may be presumed that residents who travel longer distances are more receptive to the potential benefits of AV. This question will test if distance travel is an indicator of awareness of AV benefits.
9. Travel cost	How much does your household spend on transport as a percentage of household income?	AV have been identified for their potential to improve cost of both private vehicles and public transport (Butler et al., 2020a). Therefore, it may be presumed that residents who spend more on transportation costs are more receptive of the potential benefits of AV. This question will test if money spent on transport is an indicator of awareness of AV benefits.
10. Distance to public transport	How far is your home from the nearest public transport stop?	AV have been identified for their potential to improve public transport coverage by bridging the 'first- and last-mile' gap between home/work and public transport (Butler et al., 2020a). This question will analyse if distance from home to public transport impacts on the expectation of AV to bridge the aforementioned gap.

TABLE 1.2 (Continued)

Description of Independent Variables

Variable	Question	Description
11. Safety	Does anyone in your household ever feel unsafe when using public transport?	Unsafe feelings, particularly in females, have been identified as an example of why many prefer private vehicle travel—particularly as a way to avoid dangerous situations and contact with strangers (Coppola & Silvestri, 2021). AV have been identified for their potential to improve safety by providing better door-to-door transport (Butler et al., 2020a). This question will test whether feelings of unsafety on public transport influence the expectation of AV to improve safety by providing door-to-door transport.
12. Information	Do you think the information availability of public transport is sufficient?	The ability to access good quality, up-to-date information on public transport systems has been identified as a crucial component to allow individuals to be able to independently plan and pay for their journey (Yigitcanlar et al., 2019b). AV particularly when enhanced with vehicle-to-vehicle, vehicle-to-infrastructure, and vehicle-to-user communication systems has been identified for its potential to provide real-time information direct to users and improve public transport accessibility (Butler et al., 2020a). This question will test whether those who experience difficulties accessing information are more excited about the future of AV.
13–18. Perception of disadvantage	How would you rate your ability to participate in transport activities based on the following factors?	Finally, based on the six dimensions of transport disadvantage identified by Butler et al. (2020a), and first identified by Xia et al. (2016) and Yigitcanlar et al. (2019b) users were asked to self-identify their ability to participate in transport activities based on their (a) physical ability, (b) economic constraint, (c) public transport service, (d) distance to destination, (e) concerns regarding safety on public transport, and (f) access to information on public transport. The purpose of this final step is to better understand if perceived disadvantage is a positive predictor of attitudes towards AV.

participate in transportation activities (Figure 1.2). Results were recoded into two distinct categories: (a) little to no issue (rank 1–2); and (b) some issues (rank 3–5). The resulting recoding shows that 13.9% of participants experience some difficulty associated with physically accessing transportation, 22.5% due to monetary constraints, 21.5% due to public transportation service, 23.3% due to distance, 22% due to safety concerns, and 17.2% due to access to information.

TABLE 1.3

Descriptive Analysis of Personal Characteristics and Transportation Habits

Variable	Category	Frequency	Survey %
1. Gender	Male	313	51.3
	Female	297	48.7
2. Age	18–29	127	20.8
	30–49	246	40.3
	49–69	172	28.2
	70+	65	10.7
3. Income	Low income (less than $41,550)	170	27.9
	Med to high income (more than $41,550)	440	72.1
4. Disability	Yes	39	6.4
	No	571	93.6
5. Licence	Yes	539	88.4
	No	71	11.6
6. Household vehicles	0	54	8.9
	1	265	43.4
	2	216	35.4
	3+	75	12.3
7. Travel mode	Private vehicle	391	64.1
	Public transport	163	26.7
	Active transport	49	8
	Taxi/rideshare	7	1.1
8. Travel time	Less than 30 min	458	75.1
	More than 30 min	152	24.9
9. Travel cost	Less than 15%	370	60.7
	More than 15%	240	39.3
10. Distance to public transport	Less than 400 m bus or 800 m to train	439	72
	More than 400 m bus or 800 m to train	171	28
11. Safety	No issues with safety on public transport	453	74.3
	Some issues with safety on public transport	157	25.7
12. Information	Public transport information is sufficient	456	74.8
	Public transport information is insufficient	154	25.2

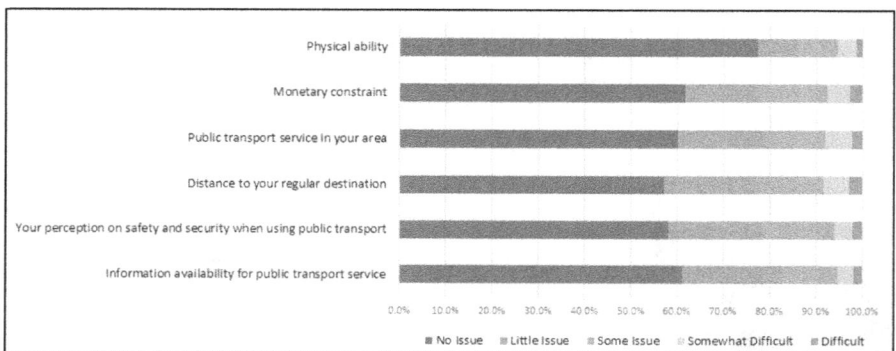

FIGURE 1.2

How would you rate your ability to participate in transportation activities?

1.3.4 Data Analysis

Following descriptive analysis of the survey, an ordinal logistic regression model was used to estimate which of the explanatory variables could help understand awareness regarding AVs and the corresponding opportunity to ease transportation disadvantage. Initially, linear regression was used to assess any potential issues with multicollinearity between independent variables (Daoud, 2017). Table 1.4 shows results and acceptable tolerance (Tolerance > 0.1; VIF <10).

Next to determine the relationship between the identified variables and user awareness of the equity benefits of AV an ordinal logistic regression analysis was completed. A total of 18 independent variables were used in the model. The test of parallel lines was performed to determine whether the model had proportional odds (Table 1.5). The model achieves significance levels >0.05 (p = 0.376) which denotes compliance with the assumption of proportional odds. Table 1.6 provides goodness-of-fit information. Results show that both the Pearson Chi-Square (p = 0.313) and Deviance (p = 1.000) test achieves significant levels >0.05 suggesting the model fits well.

TABLE 1.4

Assessment of Multicollinearity Using Collinearity Statistics

Coefficients[a]		
	Collinearity statistics	
Model	**Tolerance**	**VIF**
01. Gender	0.747	1.338
02. Age	0.666	1.501
03. Income	0.859	1.165
04. Disability	0.808	1.238
05. Licence	0.785	1.273
06. Household vehicles	0.796	1.256
07. Travel mode (public transport user)	0.319	3.134
07. Travel mode (private vehicle user)	0.309	3.232
08. Travel time	0.860	1.162
09. Travel cost	0.884	1.131
10. Distance to public transport	0.900	1.111
11. Safety	0.640	1.562
12. Information	0.875	1.143
13. Disadvantage (physical issues)	0.658	1.521
14. Disadvantage (economic issues)	0.701	1.427
15. Disadvantage (public transport service issues)	0.561	1.781
16. Disadvantage (distance to destination issues)	0.501	1.995
17. Disadvantage (safety issues)	0.453	2.210
18. Disadvantage (information)	0.521	1.918

Note: [a]Dependent variable: Dependent—merged.

TABLE 1.5

Test of Parallel Lines

Test of Parallel Lines[a]				
Model	−2 Log Likelihood	Chi-Square	df	Sig.
Null hypothesis	1657.134			
General	1591.213[b]	65.920[c]	63	0.376

Note: The null hypothesis states that the location parameters (slope coefficients) are the same across response categories.

[a] Link function: Logit.
[b] The log likelihood value cannot be further increased after a maximum number of step-halving.
[c] The chi-square statistic is computed based on the log likelihood value of the last iteration of the general model.

TABLE 1.6

Goodness of Fit

Goodness-of-Fit			
	Chi-Square	df	Sig.
Pearson	2061.596	2031	0.313
Deviance	1572.043	2031	1.000
Link function: Logit			

1.4 Results and Discussion

Table 1.7 provides the estimated model. A total of 18 independent variables and their corresponding likely influence on the dependent variable are shown. In this model, the dependent variable relates to the measurements of the public's awareness of using AVs as a solution to transport disadvantage. Table 1.7 includes an estimate of the regression coefficients, which can be interpreted as the odds of each independent variable being in a higher group in the dependent variable. The dependent variable in our group has five categories the lowest being 'no expectation' and the highest being 'a great deal of expectation'. A reference category in the independent variable was selected so that the hypotheses identified in Table 1.1 could be analysed. Based on this assumption, a positive and negative estimate in the remaining categories represents a predictor of a higher and lower expectation, respectively. Six variables were considered to have a statistically significant relationship to the dependent variable (p-values = <0.05) being (a) number of vehicles in the household, (b) age, (c) income, (d) existing disability, (e) public transport users, and (f) those who feel unsafe on public transport. On the

TABLE 1.7

Parameter Estimates

		Parameter Estimates					95% Confidence interval	
		Estimate	Std. Error	Wald	df	Sig.	Lower Bound	Upper Bound
Threshold	[AV = 1]	-.364	0.335	1.178	1	0.278	-1.020	0.293
	[AV = 2]	0.430	0.335	1.653	1	0.199	-.226	1.086
	[AV = 3]	1.793	0.343	27.358	1	0.000	1.121	2.464
	[AV = 4]	2.859	0.357	64.231	1	0.000	2.160	3.559
Location	Vehicles	-0.185	0.087	4.545	1	0.033	-0.355	-0.015
	[Gender = Female]	0.162	0.172	0.883	1	0.347	-0.175	0.498
	[Gender = Male]	0[a]	.	.	0	.	.	.
	[Age = 18–29]	1.144	0.326	12.344	1	0.000	0.506	1.783
	[Age = 30–49]	1.000	0.290	11.923	1	0.001	0.432	1.568
	[Age = 50–69]	0.180	0.275	0.426	1	0.514	-0.360	0.719
	[Age = 70+]	0[a]	.	.	0	.	.	.
	[Income = Low]	0.370	0.179	4.284	1	0.038	0.020	0.721
	[Income = Mid-High]	0[a]	.	.	0	.	.	.
	[Disability = Yes]	1.091	0.337	10.501	1	0.001	0.431	1.751
	[Disability = No]	0[a]	.	.	0	.	.	.
	[Licence = No]	-0.309	0.258	1.436	1	0.231	-0.814	0.196
	[Licence = Yes]	0[a]	.	.	0	.	.	.
	[Travel Mode PT = Yes]	1.080	0.298	13.183	1	0.000	0.497	1.663
	[Travel Mode PT = No]	0[a]	.	.	0	.	.	.
	[Travel Mode Car = No]	-0.362	0.279	1.682	1	0.195	-0.910	0.185
	[Travel Mode Car = Yes]	0[a]	.	.	0	.	.	.
	[Time => 30 mins]	-0.158	0.183	0.753	1	0.386	-0.516	0.199
	[Time = < 30 mins]	0[a]	.	.	0	.	.	.
	[Cost = > 15%]	-0.235	0.160	2.153	1	0.142	-0.548	0.079

(continued)

TABLE 1.7 (Continued)

Parameter Estimates

Parameter Estimates						95% Confidence interval	
	Estimate	Std. Error	Wald	df	Sig.	Lower Bound	Upper Bound
[Cost = < 15 %]	0[a]	.	.	0	.	.	.
[Distance = > 800/1000]	-0.255	0.173	2.180	1	0.140	-0.593	0.083
[Distance = <800/1000]	0[a]	.	.	0	.	.	.
[Safety = Issues]	0.540	0.209	6.645	1	0.010	0.129	0.950
[Safety = No issues]	0[a]	.	.	0	.	.	.
[Information = Insufficient]	-0.018	0.180	0.010	1	0.922	-0.371	0.336
[Information = Sufficient]	0[a]	.	.	0	.	.	.
[TDA Physical = Yes]	0.038	0.262	0.021	1	0.885	-0.476	0.552
[TDA Physical = No]	0[a]	.	.	0	.	.	.
[TDA Economic = Yes]	-0.043	0.209	0.043	1	0.836	-0.453	0.367
[TDA Economic = No]	0[a]	.	.	0	.	.	.
[TDA Public trans = Yes]	-0.085	0.238	0.127	1	0.722	-0.552	0.382
[TDA Public trans = No]	0[a]	.	.	0	.	.	.
[TDA Distance = Yes]	0.170	0.245	0.483	1	0.487	-0.310	0.651
[TDA Distance = No]	0[a]	.	.	0	.	.	.
[TDA Safety = Yes]	-0.083	0.262	0.101	1	0.751	-0.597	0.431
[TDA Safety = No]	0[a]	.	.	0	.	.	.
[TDA Information = Yes]	-0.155	0.269	0.333	1	0.564	-0.682	0.372
[TDA Information = No]	0[a]	.	.	0	.	.	.

Note: Link function: Logit.

[a] This parameter is set to zero because it is redundant.

other hand, gender, licence, private vehicle use, travel distance, travel cost, distance to public transport, access to information, and perception of disadvantage showed no statistically significant relationship to the dependent variable.

These results were also tested using an ordinal logistic generalised linear regression model (Pearson 1.015; Deviance 0.774) which generated insights similar to those described. Furthermore, a multinomial logistic model (Pearson 0.057; Deviance 1.000) was also generated however results provided less interpretability when describing the ordinal nature of the dependent variable. Nevertheless, this model did identify a statistically significant relationship between those who live close to public transport and those who held 'a great deal' of expectation towards the equity benefits of AV. While not explored in this study this relationship may indicate the usefulness of exploring spatial factors in future research.

1.4.1 Household Vehicle Ownership

When household vehicle ownership is analysed, the results indicate a statistically significant relationship with the dependent variable ($p = 0.033$). Considering Table 1.7, for every one unit increase in vehicles owned per household there is 0.185 times less likely expectation among users that AVs can be used to ease transport disadvantage in their neighbourhood. In other words, households with fewer vehicles are more receptive to the possibility of AVs to ease transport equity issues than those with more vehicles. Given that Brisbane has been identified as a highly car-dependent city, and not having access to a vehicle could be understood as a significant disadvantage, these results suggest that residents without easy access to a vehicle are more receptive to the possibility that AVs can be used to improve transport equity. Results suggest that while there is an opportunity for new AVs to appeal to those more receptive to a product that provides an extension to existing public transport networks, there still may be significant barriers related to overcoming reliance on private vehicles. Furthermore, given the convenience and comfort offered by private vehicle travel, there is a risk that schemes which use AVs as 'first- and last-mile solution' are simply overwhelmed by market forces that promote private AV and entrench the existing culture of car dependency.

1.4.2 Age

When the age variable is analysed a statistically significant relationship between those aged 18–29 ($p = 0.000$) and 30–49 ($p = 0.001$) is identified. Based on the parameter estimates shown in Table 1.7 those aged 18–29 are 1.144 times more likely to have higher expectations of the potential for AV to ease transport disadvantage than those aged 70 or more. Similarly, those aged 30–49 are 1.000 times more likely to have higher expectations. Results substantiate other studies which find younger residents are more likely to have positive attitudes towards AV than the elderly (Golbabaei et al., 2021; Huang & Qian, 2021, Othman, 2021). Given that AV stand to benefit the elderly by providing increased accessibility and more prospects for independent travel later in life these results suggest significant barriers may need to be overcome before older residents are more receptive to the potential for AVs.

Nonetheless, it should be noted that other studies have highlighted increased heterogeneity among older residents with general enthusiasm about AV and its potential to improve accessibility and independence, shown particularly where existing mobility is a challenge (Sochor & Nikitas, 2016; Faber & van Lierop, 2020; Zandieh & Achaempong, 2021). Interestingly, research which is based primarily on quantitative data collection tends to show more negative attitudes among the elderly while research that shows positive attitudes tends to incorporate more qualitative research methods including in-depth interviews (Zandieh & Acheampong, 2021), focus groups (Faber & van Lierop, 2020) and a mixed method incorporating both quantitative and qualitative measures (Sochor & Nikitas, 2016).

1.4.3 Income

When income is analysed, a statistically significant relationship with the dependent variable (p = 0.038) is observed. Table 1.7 shows that residents with lower-income (i.e., $0–$799 per week) are 0.37 times more likely to have positive expectations that AVs will help ease transport disadvantages than their counterparts. Results differ from other studies which indicate that middle to higher-income residents are more likely to have positive attitudes towards AVs (Wang et al., 2020a; Yuen et al., 2020a, 2020b). Given that AVs have been identified for their potential to reduce transport costs and provide greater transport coverage—particularly in fringe urban areas—these results present an opportunity for policymakers to present public AV schemes as a cost-saving transportation option. Furthermore, given that low-income residents in Brisbane are more likely to live in fringe suburban areas it may also highlight a growing need for new transport alternatives and a better understanding of whether spatial or neighbourhood characteristics influence perception.

1.4.4 Disability

When the disability variable is analysed, the results indicate a statistically significant relationship with the dependent variable (p = 0.001). Based on the parameter estimates shown in Table 1.7 residents who regularly need help when travelling because of long-term illness or disability are 1.091 times more likely to have positive expectations that AVs will help ease transport disadvantages than those without disability. Results substantiate other studies which find positive attitude towards AVs among those with physical disabilities (Bennett et al., 2019; Gkatrzonikas & Gkritza, 2019; Cordts et al., 2021; Golbabaei et al., 2021) including declining independence due to age (Sochor & Nikitas, 2016; Faber & van Lierop, 2020; Zandieh & Acheampong, 2021). Similarly, the results highlight opportunities for AVs to provide life-changing benefits to those most in need of transportation alternatives with the potential to provide these residents with increased independence and opportunities to: (a) Access destinations previously unattainable (Faber & van Lierop, 2020); (b) Increase physical activity by giving residents more opportunities to leave the house (Zandieh & Acheampong, 2021); (c) Socialise and meet with friends (Sochor & Nikitas, 2016; Faber & van Lierop, 2020; Zandieh & Acheampong, 2021); and (d) Access services and employment (Bennett et al., 2019).

A key barrier will be to ensure equitable design principles are central to make sure vehicles remain accessible. This is particularly important given that the loss of driver support in vehicles may make those who require help accessing vehicles reluctant to trust the design of vehicles and further increase feelings of exclusion and loss of independence (Zandieh & Acheampong, 2021).

1.4.5 Public Transport and Safety

When the travel mode variable is analysed, the results indicate that there is a statistically significant relationship between those who use public transport as their main form of transportation on weekdays and the dependent variable ($p = 0.000$). That is, residents who regularly travel by public transport are 1.080 times more likely to have positive expectations that AV will help ease transport disadvantages than those who do not. Similarly, for residents who feel safe on public transport, there is a 0.540 predicted increase in the odds of having a positive attitude towards AV (p-value = 0.01) than those who have no issue with safety. These results present opportunities for the implementation of SAVs as an extension of existing public transport networks and to improve safety by providing more flexible, door-to-door transportation options. Notwithstanding, results also highlight potential barriers for residents with existing fears surrounding public transport. SAVs may exacerbate this issue given that drivers will not be able to monitor and discourage anti-social behaviour. A potential solution may be to focus on single, or smaller occupancy SAVs that are used to connect to higher-use transport nodes where increasing passive and active surveillance measures are more practical.

1.5 Conclusion

This chapter investigated how the public perceives the relative advantage of AVs as a solution to overcome transportation disadvantages. In addressing the research question, quantitative analysis reveals that there is a general indifference towards the equity benefits of AVs. These results are disappointing given the potential for AVs, particularly SAVs, to help address issues associated with equity of transportation. This issue highlights challenges for policymakers with respect to the introduction of AV into urban areas. While cities continue to work with industry partners to utilise the benefits of SAV to provide alternatives to private vehicle travel and bridge the 'first-mile/last-mile' gap (Wen et al., 2018; Vosooghi et al., 2019; Gurumurthy et al., 2020), the specific benefits that this technology can have on those vulnerable to transport inequities may go unnoticed. Advertising and education programmes to highlight how these schemes can be used to benefit both individuals and the broader society could be beneficial, as could the continued implementation of pilot programmes in areas with greater risk of vulnerability.

This research also explored how individual characteristics, existing travel behaviour, and challenges affect this perception. Results showed that young and middle-aged adults, those with existing disabilities, and public transport users were the most likely to have positive attitudes towards the potential for AVs to ease transportation disadvantages. To a lesser extent, those who face safety concerns on public transportation and low-income residents were shown to positively influence attitudes, while an increase in the number of household vehicles was shown to have a negative influence. Conversely, gender, access to a driver's licence, walking distance to public transportation, travel time, and cost were found to have no statistically significant relationship with attitude towards AVs. Furthermore, although around 20% of users identified some form of disadvantage in accessing transportation, no relationship was found between this perception and positivity towards the benefits of AVs.

The findings disclose several opportunities and barriers associated with deployment of AVs. Given that individual expectation and experience will influence both the adoption of AVs and the policies that shape transportation systems, understanding these opportunities and barriers are an important step towards introducing SAVs into urban areas.

Nevertheless, it is important to note that AV—as a technology that comes with both the potential to create increased opportunities and risks—represents only one of the many drivers within a complex sociotechnical system where the diffusion of new transport modes will be influenced as much by perception and attitudes as it will by social norms, local economies, available resources, the built environment, and local regulations, policies, and practices (Sochor & Nikitas, 2016). While the introduction of AV may increase accessibility, safety, affordability, and independence for many vulnerable groups and the potential opportunities should be recognised and promoted, there is a risk that AV could lead to greater disadvantages and lead to increased inequities (Butler et al., 2020a). Herein highlights the importance of robust policy, education, and early engagement with developers and users to ensure technological developments in AV, and its continued introduction into urban areas in a way that reflects the values and needs of the broader society and includes recognition of vulnerable groups (Kovac et al., 2020).

This study encountered some limitations: (a) The research did not explore complex factors which may influence user perception and attitudes towards AVs including the influence of psychological factors, social norms, perceived benefit, exposure to technology, motivation, and other behavioural factors; (b) The research assumed a participants had a general understanding of AVs and provided no background knowledge to participants regarding this relatively new technology, or sought information regarding user experience and knowledge of AVs so that this data could be used to analyse variation in results; (c) The question used for the dependent variable may assume participants have a general understanding of the issues surrounding transportation disadvantage which is likely a broader more complex issue than most comprehend; (d) Despite using quantitative data and analysis which was conducted using statistics, there is still likely a bias associated with the researchers' interpretation of this data. This can partially be addressed using an analysis and model estimation framework to support the analyst and remove some of the associated human bias (Paz et al., 2019); and (e) The study did not consider linguistic information regarding public opinions and views which can be collected through narratives and analysed using artificial intelligence (Arteaga et al., 2020). This type of information is likely to reveal aspects that are difficult to capture using scales, categorical, or quantitative variables.

Future research could expand on these findings by: (a) Analysing how existing accessibility influences the perception of a broader range of future mobility strategies including SAV, DRT, and MaaS; (b) Exploring measures to improve the comfort, convenience, and safety of shared travel including analysis of issues associated with sharing the vehicle with strangers and ownership. Including the potential for smaller occupancy vehicles or the provision of designated spaces, seating, and other privacy and health-related measures— such as vulnerability to disease transmission following the extended COVID-19 pandemic (Yigitcanlar et al., 2021); (c) Analysing whether spatial factors such as distance to employment opportunities, recreational facilities, transportation nodes, and neighbourhood characteristics influence user awareness and attitudes towards future mobility—including the potential for decentralisation of workforce through shared spaces, innovation districts, or working from home initiatives; (d) Detail measures to improve equity of accessibility in future vehicles and transit systems; and (e) Investigate methods to optimise SAVs and increase acceptability including the use of statistical methods such as conjoint analysis (Papadima et al., 2020) as well as behaviour-consistent concepts to influence travel choices (Paz & Peeta, 2009).

Acknowledgements

This chapter, with permission from the copyright holder, is a reproduced version of the following journal article: Butler, L., Yigitcanlar, T., & Paz, A. (2021). Factors influencing public awareness of autonomous vehicles: empirical evidence from Brisbane. *Transportation Research Part F: Traffic Psychology and Behaviour*, 82(1), 256–267.

References

Arteaga, C., Paz, A., & Park, J. (2020). Injury severity on traffic crashes. *Safety Science*, 132, 104988.

Bennett, R., Vijaygopal, R., & Kottasz, R. (2019). Willingness of people with mental health disabilities to travel in driverless vehicles. *Journal of Transport & Health*, 12, 1–12.

Bennett, R., Vijaygopal, R., & Kottasz, R. (2020). Willingness of people who are blind to accept autonomous vehicles. *Transportation Research. Part F*, 69, 13–27.

Butler, L., Yigitcanlar, T., & Paz, A. (2020a). How can smart mobility innovations alleviate transportation disadvantage? *Applied Sciences*, 10, 6306.

Butler, L., Yigitcanlar, T., & Paz, A. (2020b). Smart urban mobility innovations. *IEEE Access*, 8, 196034–196049.

Coppola, P., & Silvestri, F. (2021). Gender inequality in safety and security perceptions in railway stations. *Sustainability*, 13, 4007.

Cordts, P., Cotten, S.R., Qu, T., & Bush, T. (2021). Mobility challenges and perceptions of autonomous vehicles for individuals with physical disabilities. *Disability and Health Journal*. https://doi.org/10.1016/j.dhjo.2021.101131

Currie, G. (2009). Australian urban transport and social disadvantage. *Australian Economic Review*, 42, 201–208.

Dai, J., Li, R., & Liu, Z. (2021). Does initial experience affect consumers' intention to use autonomous vehicles? *Accident Analysis and Prevention*, 149, 105778.

Daoud, J. (2017). Multicollinearity and regression analysis. *Journal of Physics*, 949, 12009.

Dennis, S., Paz, A., & Yigitcanlar, T. (2021). Perceptions and attitudes towards the deployment of autonomous and connected vehicles. *Journal of Urban Technology*. https://doi.org/10.1080/10630732.2021.1879606

Dianin, A., Ravazzoli, E., & Hauger, G. (2021). Implications of autonomous vehicles for accessibility and transport equity. *Sustainability*, 13, 4448.

Dodson, J., & Sipe, N. (2008). Shocking the suburbs. *Housing Studies*, 23, 377–401.

Dur, F., & Yigitcanlar, T. (2015). Assessing land-use and transport integration via a spatial composite indexing model. *International Journal of Environmental Science and Technology*, 12, 803–816.

Faber, K., & van Lierop, D. (2020). How will older adults use automated vehicles? *Transportation Research Part A*, 133, 353–363.

Faisal, A., Kamruzzaman, M., Yigitcanlar, T., & Currie, G. (2019). Understanding autonomous vehicles. *Journal of Transport and Land Use*, 12, 45–72.

Faisal, A., Yigitcanlar, T., Kamruzzaman, M., & Paz, A. (2020). Mapping two decades of autonomous vehicle research. *Journal of Urban Technology*. https://doi.org/10.1080/10630732.2020.1780868

Gkartzonikas, C., & Gkritza, K. (2019). What have we learned? A review of stated preference and choice studies on autonomous vehicles. *Transportation Research Part C*, 98, 323–337.

Golbabaei, F., Yigitcanlar, T., & Bunker, J. (2021). The role of shared autonomous vehicle systems in delivering smart urban mobility. *International Journal of Sustainable Transportation*, 15, 731–748.

Golbabaei, F., Yigitcanlar, T., Paz, A., & Bunker, J. (2020). Individual predictors of autonomous vehicle public acceptance and intention to use. *Journal of Open Innovation*, 6, 106.

Gurumurthy, K.M., Kockelman, K.M., & Zuniga-Garcia, N. (2020). First-mile-last-mile collector-distributor system using shared autonomous mobility. *Transportation Research Record*, 2674, 638–647.

Huang, Y., & Qian, L. (2021). Understanding the potential adoption of autonomous vehicles in China. *Psychology & Marketing*, 38, 669–690.

Kamruzzaman, M., Yigitcanlar, T., Yang, J., & Mohamed, M.A. (2016). Measures of transport-related social exclusion. *Sustainability*, 8, 696.

Kovacs, F.S., McLeod, S., & Curtis, C. (2020). Aged mobility in the era of transportation disruption. *Travel Behaviour and Society*, 20, 122–132.

Krejcie, R., & Morgan, D. (1970). Determining sample size for research activities. *Educational and Psychological Measurement*, 30, 607–610.

Li, T., Dodson, J., & Sipe, N. (2015). Differentiating metropolitan transport disadvantage by mode. *Journal of Transport Geography*, 49, 16–25.

Manfreda, A., Ljubi, K., & Groznik, A. (2021). Autonomous vehicles in the smart city era. *International Journal of Information Management*, 58, 102050.

Othman, K. (2021). Public acceptance and perception of autonomous vehicles. *AI and Ethics*, 17, 4419.

Papadima, G., Genitsaris, E., Karagiotas, I., Naniopoulos, A., & Nalmpantis, D. (2020). Investigation of acceptance of driverless buses in the city of Trikala and optimization of the service using conjoint analysis. *Utilities Policy*, 62, 100994.

Paz, A., Arteaga, C., & Cobos, C. (2019). Specification of mixed logit models assisted by an optimization framework. *Journal of Choice Modelling*, 30, 50–60.

Paz, A., & Peeta, S. (2009). Behavior-consistent real-time traffic routing under information provision. *Transportation Research Part C*, 17(6), 642–661.

Pettigrew, S., Talati, Z., & Norman, R. (2018). The health benefits of autonomous vehicles. *Australian and New Zealand Journal of Public Health*, 42, 480–483.

Queensland Government (2018). *Queensland Household Travel Survey*. https://public.tableau.com/profile/qldtravelsurvey#!/vizhome/QueenslandHouseholdTravelSurveyInteractiveReport/QueenslandHouseholdTravelSurvey

Rice, S., & Winter, S.R. (2019). Do gender and age affect willingness to ride in driverless vehicles? *Technology in Society*, 58, 101145.

Rosier, K., & McDonald, M. (2011). *The Relationship Between Transport and Disadvantage in Australia*. Canberra: Australian Institute of Family Studies.

Seetharaman, A., Patwa, N., Jadhav, V., Saravanan, A.S., & Sangeeth, D. (2021). Impact of factors influencing cyber threats on autonomous vehicles. *Applied Artificial Intelligence*, 35, 105–132.

Sochor, J., & Nikitas, A. (2016). Vulnerable users' perceptions of transport technologies. *Proceedings of the Institution of Civil Engineers-Urban Design and Planning*,
169, 154–162.

Van Brummelen, J., O'Brien, M., Gruyer, D., & Najjaran, H. (2018). Autonomous
vehicle perception. *Transportation Research Part C*, 89, 384–406.

Vosooghi, R., Puchinger, J., Jankovic, M., & Vouillon, A. (2019). Shared autonomous
vehicle simulation and service design. *Transportation Research Part C*, 107, 15–33.

Wang, S., Jiang, Z., Noland, R.B., & Mondschein, A.S. (2020a). Attitudes towards
privately-owned and shared autonomous vehicles. *Transportation Research Part
F*, 72, 297–306.

Wang, X., Wong, Y.D., Li, K.X., & Yuen, K.F. (2020b). This is not me! *Transportation
Research Part F*, 74, 345–360.

Wen, J., Chen, Y.X., Nassir, N., & Zhao, J. (2018). Transit-oriented autonomous vehicle
operation with integrated demand-supply interaction. *Transportation Research
Part C*, 97, 216–234.

Willing, R., & Pojani, D. (2017). Is the suburban dream still alive in Australia? *Australian
Planner*, 54, 67–79.

Xia, J.C., Nesbitt, J., Daley, R., Najnin, A., Litman, T., & Tiwari, S.P. (2016). A multidimensional view of transport-related social exclusion. *Transportation Research
Part A*, 94, 205–221.

Yigitcanlar, T., Desouza, K., Butler, L., & Roozkhosh, F. (2020). Contributions and risks
of artificial intelligence (AI) in building smarter cities. *Energies*, 13, 1473.

Yigitcanlar, T., Han, H., Kamruzzaman, M., Ioppolo, G., & Sabatini-Marques, J. (2019c).
The making of smart cities. *Land Use Policy*, 88, 104187.

Yigitcanlar, T., & Kamruzzaman, M. (2019). Smart cities and mobility. *Journal of Urban
Technology*, 26, 21–46.

Yigitcanlar, T., Kankanamge, N., Inkinen, T., Butler, L., Preston, A., Rezayee, M., Gill,
P., Ostadnia, M., Ioppolo, G., & Senevirathne, M. (2021). Pandemic vulnerability
knowledge visualisation for strategic decision-making. *Management Decision*.
https://doi.org/10.1108/MD-11-2020-1527

Yigitcanlar, T., Mohamed, A., Kamruzzaman, M., & Piracha, A. (2019b). Understanding
transport-related social exclusion. *Urban Policy and Research*, 37, 97–110.

Yigitcanlar, T., Wilson, M., & Kamruzzaman, M. (2019a). Disruptive impacts of
automated driving systems on the built environment and land use. *Journal of
Open Innovation*, 5, 24.

Yuen, K.F., Huyen, D.T., Wang, X., & Qi, G. (2020b). Factors influencing the adoption
of shared autonomous vehicles. *International Journal of Environmental Research
and Public Health*, 17, 4868.

Yuen, K.F., Wong, Y.D., Ma, F., & Wang, X. (2020a). The determinants of public
acceptance of autonomous vehicles. *Journal of Cleaner Production*, 270, 121904.

Zandieh, R., & Acheampong, R.A. (2021). Mobility and healthy ageing in the city.
Cities, 112, 103135.

2

Understanding Autonomous Shuttle Adoption Intention

2.1 Introduction

The exponential technological advancements that we are currently experiencing, particularly in artificial intelligence (AI), are providing opportunities for disruption in many sectors (Yigitcanlar et al., 2021). The transport sector is one such sector. With advances in AI, there is a great promise that driverless cars or autonomous vehicles (AVs) will address some of today's transport challenges e.g., facilitating mobility for the transport-disadvantaged population, enhancing mobility efficacy, improving safety, reducing emissions (Yigitcanlar et al., 2019a; Golbabaei et al., 2021), accelerating freight transport (Simpson & Mishra, 2020; Talebian & Mishra, (2022), and enabling underwater transport (Lakhekar & Waghmare, 2022). Emerging business models, like Shared-AVs (SAVs) which are capable of ridesharing in various ways, will make it easier to provide demand-responsive services on non-fixed routes (Peeta et al., 2008; Faisal et al., 2020). Examining the likelihood of adoption and usage intention of AVs is critical, as their integration into the transportation system has the potential to affect passengers' mobility behaviours and lifestyles (Yigitcanlar & Kamruzzaman, 2019; Yigitcanlar et al., 2020). However, as AVs are not yet commercially available in the transport system, with the exception of a few trials, forecasting the precise travel demand is challenging in this domain in terms of, for example, ownership trends, preferred mode choices, and vehicle kilometres travelled (VKT). There might also be reactions to the introduction of this technology in the public domain (Zmud et al., 2016; Golbabaei et al., 2020).

A public transport system's success relies on its ability to attract and retain passengers. The autonomous shuttle bus (ASB) as a form of shared autonomous demand-responsive transit (ADRT) serves multiple passenger trips at a time (Sweet & Laidlaw, 2019) and has the potential to complement other means of travel including regular public transport (RPT), thereby increasing public transport usage overall (Dennis et al., 2021). It is worthwhile exploring

DOI: 10.1201/9781003605676-2

the feasibility of this technology in terms of potential societal advantages. An ASB could have the potential to operate as a feeder to RPT, particularly in peri-urban regions, where first/last mile access to public transit is challenging (Acheampong et al., 2020).

ASBs could also offer personal transport to individuals who do not have a driver's licence due to ageing, or medical ailments—i.e., physical and/or cognitive disabilities (Bradshaw-Martin & Easton, 2014; Musselwhite et al., 2015; Abraham et al., 2017)—or children (Koppel et al., 2021). ASBs are currently at the prototype stage in several projects operating on predetermined routes at restricted speeds. They typically accommodate between eight and ten passengers and need to be supervised by either an on-board steward or an external control room. Many substantial challenges are yet to be overcome for the ASB concept regarding perceptions of the general public, policies, and traffic management. Individuals' perceptions and adoption intentions towards ADRT are important as they influence demand for such technologies, governance policies, and future infrastructure investments (Butler et al., 2021).

Public perception has been defined as "the type of information obtained from a public opinion survey, which is merely the aggregate views of a group of people (usually a randomly selected sample) who are asked directly what they think about particular issues or events" (Dowler et al., 2006). It could be interpreted as "the difference between an absolute truth based on facts and a virtual truth shaped by popular opinion, media coverage and/ or reputation" (Insani, 2013). Responses to "structured questions can be recorded and analyzed in simple, quantitative terms as a sort of snapshot of opinion at a given moment in time" (Dowler et al., 2006). The requirements of different individuals can be identified via their perceptions because public perceptions are affected by their demands. It is essential to recognise our perception because it is the driving force behind our reaction to things (Bansal et al., 2016).

While many surveys have recently been conducted on the public's perception of AVs (Bansal et al., 2016; Xu et al., 2018; Zhang et al., 2020; Butler et al., 2021), limited studies thus far have specifically explored whether the public view ASBs as a viable alternative to their existing modes of travel (Madigan et al., 2017; Nordhoff et al., 2018; Chen & Yan, 2019; Herrenkind et al., 2019a, 2019b; Roche-Cerasi, 2019; Bernhard et al., 2020; Nordhoff et al., 2020; Papadima et al., 2020; Nordhoff et al., 2021; Nordhoff et al., 2021). Acheampong et al. (2020) state that "research on technology and innovation acceptance and diffusion, and choice behavior under volitional control do provide various theoretical models", which can be used as an argument for the investigation of the adoption of such kinds of demand-responsive services. By specifying the interrelationships of the psychological variables of theoretical models, we would be able to provide more systematic predictions about user adoption behaviour and gain deeper insights.

Generally, behavioural theories include a variety of psychological factors that are considered potential predictors of behavioural intention. Additionally, the inclusion of behavioural theories allows researchers to progressively uncover the underlying correlations between psychological factors, ensuring that plausible and profound findings can be presented (Jing et al., 2020). Ultimately, these comprehensive empirical insights would enlighten travel-demand modelling and management strategists.

This research contributes to developing a conceptual framework by using common hypotheses from the Technology Acceptance Model (TAM) (Davis et al., 1989) to explore the psychological factors and their interrelationships, which can affect adult travellers' perceptions and adoption of ADRT options, particularly ASBs as a public transport mode in South East Queensland (SEQ), Australia. The comprehensive insights provided by this research can assist policymakers, transport planners, and engineers in their policy decisions and system plans as well as achieving higher public acknowledgement and potentially wider uptake of ADRT technology solutions where appropriate. The layout of the study design is illustrated in Figure 2.1.

This chapter is structured as follows. Section 2.2 reviews the theoretical and literature background. Section 2.3 explains the research methodology, including the survey design, hypothesis development, and demographics of the survey participants. Section 2.4 focuses on the results of the partial least structural equation modelling (PLS-SEM) analysis including (a) measurement model evaluation, (b) structural model evaluation, and (c) hypotheses testing. Section 2.5 concludes by discussing the main findings and practical implications. Finally, we put forward the research limitations by recommending some further research plans in Section 2.6.

2.2 Literature Background

The ability to predict AV adoption intention is a relatively new concept. Several surveys, however, have been conducted to predict the future of AVs (Herrenkind et al., 2019a, 2019b; Bernhard et al., 2020; Motamedi et al., 2019; Nastjuk et al., 2020; Dirsehan & Can, 2020; Zhang et al., 2020). To explore the span of studies on determinants of public acceptance of AVs, we conducted a systematic review (Golbabaei et al. 2020) and classified the influential factors on willingness to use this technology. In line with our findings, the TAM and its variants are the most commonly utilised underlying theory. TAM was first introduced by Davis and Weber (1985) who adapted the theory of reasoned action (TRA) (Ajzen & Fishbein, 1980) to explore/explain the determinants of usage/non-usage of technological innovations. Behavioural intention is recognised as a reliable predictor of actual usage, particularly in the context

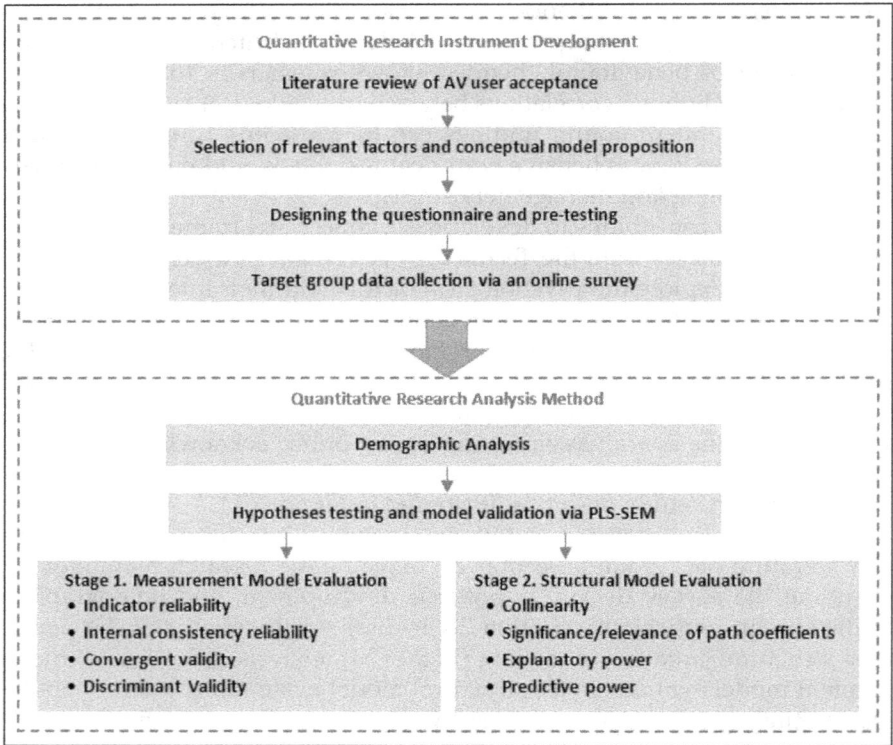

FIGURE 2.1
Study design flowchart.

of studies on current and/or emerging technologies that are not yet commercially available (Högg et al., 2010; Günthner & Proff, 2021; Pai & Huang, 2011). According to Davis and Venkatesh (2004), the

> expectations about a system captured using reliable and valid measures of key expectations, even before hands-on use of the system, are predictive of those that would have been obtained after a brief use of a test prototype, as well as after several weeks of the actual system use.

Davis and Venkatesh (2004) evaluated TAM's capability in the prediction of technology acceptance. Throughout their pre-design and development phases, they discovered that TAM theory is helpful to predict the adoption of technologies. Despite the emergence of other theories in this context, i.e., the unified theory of acceptance and use of technology (UTAUT), TAM continues to be the most robust approach and powerful technique for forecasting technology adoption in various domains (Lee et al., 2003), as it can predict nearly 40% of the variance among individuals' behavioural intention and their actual behaviour (King & He, 2006)]. Therefore, TAM comprising its basic components is considered to be a suitable foundation, while adding constructs to already extended theories may introduce interferences (Herrenkind et al., 2019b).

Rahman et al. (2017) assessed the utility of TAM, TPB, and UTAUT by applying a sample of 430 surveys. The outcomes of Hotelling's T-squared test (Rahman et al., 2017) for non-independent correlations indicated that all models were capable of effectively predicting driver acceptance in terms of behavioural intention; however, the original TAM model was noted to be the most accurate. Their findings support the idea that attitude towards a behaviour is shaped based on relevant beliefs (Fishbein & Ajzen, 1975). Perceived ease of use, identified as the belief in the simplicity of behaviour, has two basic mechanisms to impact attitude: self-efficacy and instrumentality (Davis et al., 1989). If the behaviour is easier to perform using the technology, it will create a sense of efficacy and personal control for the performer.

Additionally, an easier system would contribute to enhanced performance with the same amount of effort. The enhancement of performance corresponds with the belief of usefulness (perceived usefulness (PU)); though, with its self-efficacy mechanism, perceived ease of use influences attitude above and beyond PU. However, in the UTAUT model, the mediation of attitude on how personal beliefs (PU and perceived ease of use) affect behavioural intention was overlooked, and PU and perceived ease of use were only reflected as predicting variables of Behavioural Intention with simple linear regression. Subjective norms showed a positive, though a very small effect, on behavioural intention (Rahman et al., 2017).

As was proven by Rahman et al. (2017), considering the interpretations of these factors and the scales employed in prior research, it is clear that 'performance expectancy' is very similar to PU in TAM, and 'effort expectancy' is very similar to perceived ease of use in TAM, and 'social influence' is very similar to 'subjective norms' in TPB. The high correlation between these pairs of factors and their comparable effects on behavioural intention supports statistical evidence of their similarity. UTAUT was only able to explain 71% of the variance in behavioural intention, the lowest percentage among the evaluated models. Along with this empirical evidence, UTAUT includes a total of eight factors (four components and four moderator variables), which is the highest number of factors among all the models, making the use of this model comparatively challenging. Due to its under-performance, the similarities with TAM and TPB, and the complex nature of the model, UTAUT was not considered to be useful for modelling the acceptance of ASBs in this study.

Regarding the TPB model, the design of both TAM and TPB is very similar, and both proposed three factors, one of which was shared by both: attitude. Hence, researchers had to consider the practical significance of adopting one model over the other. TAM provides a mechanism for explaining the formation of attitude, which was found to be the strongest of the factors in both models, by proposing that PU and perceived ease of use can predict attitude. Of the TAM factors, perceived ease of use has the potential to provide actionable information to the developers of vehicle technologies and is not considered in TPB. TPB provides information on normative beliefs, behavioural control beliefs, and their effects on behavioural intention. The practical implications of these factors are less obvious (Rahman et al., 2017). Considering all these facts, the use of the TAM model to study the acceptance of ASBs could provide more actionable information and explain more variance in behavioural intention compared to the other models.

Extended TAM models have also been proven useful for assessing the public acceptability of new transport technologies. As depicted in Table 2.1, trust (Cho et al., 2017; Lee et al., 2017; Buckley et al., 2018; Panagiotopoulos & Dimitrakopoulos, 2018; Xu et al., 2018; Motamedi et al., 2019; Zhang et al.,

TABLE 2.1

Prior Studies on AV Adoption Based on TAM and Its Extensions

Author	Focus	Data Collection Method	Analysis Method	Investigated Constructs	R^2 (Variance Explained)
Zhang et al. (2020)	AV	Online survey	PLS-SEM	TR, SI, sensation seeking, Big Five personality, PU, PEU → INT	0.54
Nastjuk et al. (2020)	AV	Qualitative research, online survey	PLS-SEM	SN, LOC, PPR, TR, EA, PI, RA, Co, enjoyment, PrE, PU, PEU, ATT → INT	n/a
Motamedi et al. (2019)	Personally owned/ shared-use AV	Focus groups, online survey	CFA, SEM	TR, Co, PSa, PU, PEU → INT	0.91 0.77
Dirsehan and Can (2020)	AV	Online survey	SEM	TR, sustainability concerns, PU, PEU → INT	0.57
Zhang et al. (2019)	AV	Interview	SEM	PEU, PU, PSR, PPR → TR → ATT → INT	0.56, 0.67, 0.61
Wu et al. (2019)	AV	Online survey	SEM	Environmental concern, green perceived usefulness, PEU → INT	n/a
Lee et al. (2017)	AV	Online survey	PLS-SEM	SE, RA, psychological ownership, PR, PU, PEU → INT	0.52

TABLE 2.1 (Continued)

Prior Studies on AV Adoption Based on TAM and Its Extensions

Author	Focus	Data Collection Method	Analysis Method	Investigated Constructs	R^2 (Variance Explained)
Herrenkind et al. (2019b)	ASB	Online survey	PLS-SEM	EA, openness to shared use, PPR, TR, PEn, RA, PrE, residence, family budget, education, social network → INT	0.52
Herrenkind et al. (2019a)	ASB	Qualitative research, interview; revealed preference	CFA, SEM	TR, LOC, PPR, EA, PI, image, SN, PEn, RA, PrE, PU, PEU, ATT → INT	n/a
Xu et al. (2018)	AV	Field experiment	SEM	TR → PU, PEU, PSa → INT Willingness to re-ride	0.55 0.40
Panagiotopoulos and Dimitrakopoulos (2018)	AV	Online survey	Multiple linear regression	PU, PEU, TR, SI → INT	0.44
Buckley et al. (2018)	AV	Interview, revealed preference	bivariate correlations, hierarchical regression	TR, ATT, SN, PBC → INT TR, PU, PEU → INT	0.49 0.44

Note 1: R^2 indicates exogenous construct combined effect on the endogenous construct, and ranges from 0 to 1, where higher values indicate higher levels of prediction accuracy.
Note 2: AV: Autonomous vehicle; ASB: Autonomous shuttle bus; PU: Perceived usefulness; PEU: Perceived ease of use; ATT: Attitudes; INT: Usage intention; TR: Trust; SI: Social influence; SN: Subjective norm; LOC: Locus of control/Desirability of control; RA: Relative advantage; PI: Personal/Consumer Innovativeness; Co: Compatibility; PEn: Perceived enjoyment; PrE: Price evaluation; PSa: Perceived safety; PSR: perceived safety risk; PPR: Perceived privacy risk/concerns; SE: Self-efficacy; PR: perceived risk; DIT: Diffusion of innovation.

2019; Herrenkind et al., 2019a; Dirsehan & Can, 2020; Nastjuk et al., 2020; Zhang et al., 2020), perceived risk (safety, financial, socio-psychological, performance, privacy) (Lee et al., 2017; Zhang et al., 2019; Herrenkind et al., 2019a, 2019b; Nastjuk et al., 2020), personality factors (Zhang et al., 2020), relative advantage (Herrenkind et al., 2019a), social influence or subjective norms (Cho et al., 2017; Panagiotopoulos & Dimitrakopoulos, 2018; Herrenkind et al., 2019a; Nastjuk et al., 2020; Zhang et al., 2020), sensation seeking (Zhang et al., 2020), locus or desirability of control (Herrenkind et al., 2019a; Zhang et al., 2020), ecological awareness (Herrenkind et al., 2019a, 2019b), personal or consumer innovativeness (Herrenkind et al., 2019a; Nastjuk et al., 2020), compatibility (Motamedi et al., 2019; Nastjuk et al., 2020), perceived enjoyment (Herrenkind et al., 2019a, 2019b), price evaluation (Herrenkind et al., 2019a, 2019b; Nastjuk et al., 2020), and self-efficacy (Cho et al., 2017; Lee et al., 2017) are found to be influential factors on AV adoption intention along with the basic TAM constructs.

Whereas prior research acknowledged influential factors on AV usage intention, some interrelationships between TAM constructs remained unclear. For instance, TAM suggests that the perceived ease of use influences the PU and usage intentions, although some prior studies failed to support such cause–effect relationships (Buckley et al., 2018; Motamedi et al., 2019; Herrenkind et al., 2019b). In a similar vein, the misconception of AV systems caused inconsistencies in the previous studies of perceived risks. Xu et al. (2018), Herrenkind et al. (2019a), and Lee et al. (2017) conveyed that perceived risks negatively impacted AV usage intention, while Nastjuk et al. (2020) stated that it did not. Without more investigation of the perceived risks of AVs, specifically ASBs, only a limited understanding of their prospective usage could be drawn.

Similarly, very few studies investigated the impacts of perceived relative advantages on the PU and usage intention simultaneously (Herrenkind et al., 2019a, 2019b; Nastjuk et al., 2020). Prior research was not also able to reveal the perceived relative advantages of ASBs compared to conventional counterparts. Since prospective users of AVs are customers of currently available modes, the impacts of the perceived relative advantages should be studied comprehensively. Some debates have arisen about whether perceived relative advantages and PU need to be considered the same constructs (Lee et al., 2017). Kulviwat et al. (2007) reviewed the prior research and concluded that the two notions are contextually distinct and that perceived relative advantages are a predictor of PU. Correspondingly, prior research has failed to explore the influence of perceived service quality in terms of ADRT acceptance, specifically ASBs. Even though this factor has been a focus in public transport acceptance research (Enoch et al., 2006; de Oña, 2020), research may have underestimated its impact on users' perception, attitude, and usage intention.

Travellers are only starting to consider using ASBs if they perceive this new demand-responsive mode would offer enhanced service quality. To address this issue, and given that the belief and motivation towards using new technology may be influenced by more stimuli (Davis & Weber, 1985), this study proposes an effective conceptual research model for public acceptance of ASBs based on public passenger transport characteristics and focusing on constructs that are most associated with public transport usage, namely perceived service quality, perceived relative advantage, and perceived risks, in addition to the four basic TAM constructs

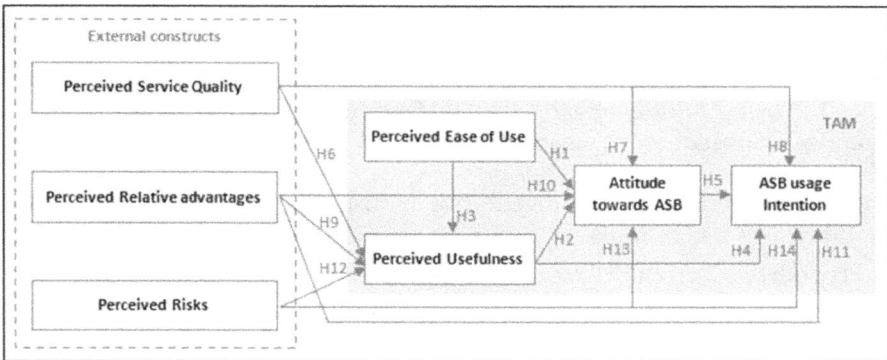

FIGURE 2.2
The conceptual research model for the ASB acceptance investigation.

for the sake of investigating the interrelationships between these key factors in the concept of ASBs, and increasing the context-specific clarity of prediction of adoption intention, and thereby actual usage behaviour, accordingly gaining greater insight for policymaking, as "identifying the strongest concerns relating to ASBs can assist in the planning of proactive efforts to address these issues while building on any perceived positive attributes" (Insani, 2013; Pettigrew & Sophie, 2019). The proposed conceptual model is illustrated in Figure 2.2.

2.3 Research Design

2.3.1 Technology Acceptance Model

As AVs are not yet commercially available in the market to use, and as our conceptual model is founded on TAM, we chose usage intention instead of actual use as the primary endogenous (dependent) factor to study end-users' opinions of ASBs (Venkatesh et al., 2003; Wu & Wang, 2005). Usage intention can be interpreted as the extent to which an individual is willing/ready to utilise an innovation (Davis et al., 1989; Venkatesh & Davis, 2000). In earlier research about AV acceptance, the positive impact of PU on individuals' usage intention was confirmed (Motamedi et al., 2019; Garidis et al., 2020; Zhang et al., 2020). In contrast, literature about AV acceptance has conveyed an unclear connection between perceived ease of use and usage intention (Nordhoff et al., 2017). Many studies reported a positive influence of perceived ease of use on usage intention (Xu et al., 2018; Zhang et al., 2020)], though others could not identify a substantial effect (Madigan et al., 2017; Motamedi et al. 2019).

Following the above-mentioned findings, we assume that individuals who consider ASBs to be useful and effortless to use would have more positive attitudes towards them and, consequently, will be more expected to intend to prioritise using ASBs over cars when this mode becomes available (i.e., usage intention). Subsequently, we postulate the following hypotheses:

Hypothesis 1 (H1): Perceived ease of use positively impacts attitude towards use of ASBs.

Hypothesis 2 (H2): PU positively impacts attitude towards use of ASBs.

Hypothesis 3 (H3): Perceived ease of use of ASBs directly impacts its PU.

Hypothesis 4 (H4): PU of ASBs positively impacts the users' usage intention.

Hypothesis 5 (H5): Attitudes towards ASBs positively impact the users' usage intention.

2.3.2 Perceived Service Quality

In the marketing/management context, service quality is considered to be a key component in investigating customers' behaviour towards a product/ service (Han & Hyun, 2017). Perceived service quality is "the subjective assessment of customers about the overall perfection or superiority of a product/service" (Zeithaml, 1988; Su et al., 2019). Customers' evaluations of services vary according to their expectations or perceptions of service quality (Parasuraman et al., 1988; Asubonteng et al., 1996) concerning technical/ functional service aspects (Grönroos, 1984). Frequently, perceived quality does not correspond to the real quality of the product/service. The perceived quality is influenced by the customer's assessment, while the actual quality is defined by product/service orientation (Garvin, 1983). If customers perceive a higher level of service quality, their intention to use/purchase a product/ service is likely to rise (Su et al., 2019).

In the transport domain, service quality is related to particular service characteristics, such as frequency, cleanliness, comfort, speed, accessibility, timeliness, information, and safety (Oliver, 1980; Ahern et al., 2022). Public transport system service quality has been extensively investigated as a predictor of customer satisfaction or loyalty (Wang et al., 2017). Recently, research has observed the link between perceived service quality and usage intention in the public transport context. Enoch et al. (2006) and de Ona (2020) studied the prospective involvement of DRT systems in sustainable mobility. They indicated that the primary factors determining user adoption of DRT are the range of destination coverage, ease of access to service, availability of comprehensive reliable information, ease of booking, service on-time performance, and affordable fare rates (Tyrinopoulos & Antoniou, 2019).

Perceived quality should be a high-priority concern for both transport policymakers and DRT service providers. Despite the fact that perceived service quality has been the focus of several public transport studies, its impact on the formation of passengers' attitudes towards ADRT services and consequently intention to use ASBs have not received much attention. Therefore, we hypothesise,

Hypothesis 6 (H6): Perceived service quality of ASBs positively impacts their PU.

Hypothesis 7 (H7): Perceived service quality of ASBs positively impacts attitude towards them.

Hypothesis 8 (H8): Perceived service quality of ASBs positively impacts intention to use them.

2.3.3 Perceived Relative Advantages

The perceived relative advantage of such mobility services is another aspect of ADRT research that is worth considering. The perceived relative advantages notion is "a characteristic of innovation, described as how an innovation is evaluated in comparison to its previous manifestation or idea" (Herrenkind et al., 2019b). Kulviwat et al. (2007) identified the link between perceived relative advantages and PU while describing their distinction in the consumer context. Individuals' perception of the usefulness of innovation would be higher when they perceive it as being superior to its precursor. Perceived relative advantage is positively associated with attitude towards use behaviour, as consumers are comparing and contrasting an innovation's attributes, which possibly influence their adoption decision (Herrenkind et al., 2019a; Nastjuk et al., 2020).

In this context, perceived relative advantage is interpreted as how much ASBs are perceived to be superior to conventional shuttle buses (Gkartzonikas & Gkritza, 2019; Nastjuk et al., 2020). This depends on whether the public views ASBs as being advantageous. It has been proven that the perceived relative advantages positively affect public preferences towards AVs (Haboucha et al., 2017). Schoettle and Sivak (2014) discovered that 57% of survey participants had positive attitudes towards AVs, primarily owing to their perceived relative advantage. Talebian and Mishra (2022) also stated that the perceived relative advantages of AVs influence their adoption and diffusion. Following the above-mentioned findings, we hypothesise,

Hypothesis 9 (H9): Perceived relative advantage of ASBs positively impacts their PU.

Hypothesis 10 (H10): Perceived relative advantage of ASBs positively impacts attitude towards them.

Hypothesis 11 (H11): Perceived relative advantage of ASBs positively impacts intention to use them.

2.3.4 Perceived Risks

Within technological innovation research, perceived risks are a major construct in the prediction of intention towards adoption (Yigitcanlar & Kamruzzaman, 2019). Perceived risk can be defined as "a complex feeling of worry, fear, and anxiety that originates from a nervous situation" (Asubonteng et al., 1996; Rittichainuwat, 2011). Individuals accept or reject a technology based on their beliefs and expectations, that is, PU (benefits) and perceived risks. This concept is termed a risk–benefit paradigm in behavioural decision research and is deemed to be the extended bounded rationality notion.

After reviewing the existing literature, the following were found to be the top reasons among potential users being unwilling to use AVs of any type (Golbabaei et al., 2020); perceived risks regarding safety issues, equipment/system breakdown, performance in mixed mode traffic situations and interactions with other road users, cybersecurity and hacking threats, data privacy, and lack of control during accidents. Most of the respondents preferred the vehicle to be supervised by a 'human override' as they thought that human performance is better in the case of instant decision-making.

However, some researchers, e.g., Liu et al. (2019; 2019), reported an insignificant impact of perceived risks on usage intention caused by either respondents' preferential risk tolerance for perceived advantages, or their lack of understanding of AV risks, while Xu et al. (2018) confirmed its negative effects. Acheampong and Cugurullo (2019) reported a positive correlation between perceived risks (fears and anxiety) regarding AV systems and the PU of AVs and advocated that peoples' concern about AV systems' performance or their interactions with other road users does not essentially weaken their PU. The reported contradictions in the findings call for more investigation into the role of perceived risks in the adoption of ASBs. So, we hypothesise,

> **Hypothesis 12 (H12):** Perceived risks and PU of ASBs are positively related.
>
> **Hypothesis 13 (H13):** Perceived risks about ASBs negatively impact attitudes towards them.
>
> **Hypothesis 14 (H14):** Perceived risks about ASBs negatively impact their usage intention.

2.3.5 Survey Design

We followed the measurement development steps outlined in prior research (DeVellis, 2003; Kim, & Eves, 2012). The measurement indicators of each construct of the conceptual model (perceived relative advantage, perceived service quality, perceived risks, perceived ease of use, PU, attitudes, and usage intention) were adapted and confirmed over a three-phase process. Initially, the research in transport/social science literature related to AVs was systematically reviewed to determine the measurement indicators for each research model component. The related indicators were measured by applying a five-level Likert scale. For a clearer interpretation of each shuttle type's typical usage scenario, an introductory paragraph including some photographs of the shuttles was depicted at the beginning of each section of the questionnaire.

In the second step, the list of the measurement items and the introductory paragraphs were refined through consultation with an expert supervisory review panel representing the views of key informants in the field, specifically civil engineering and built-environment academics specialising in transport systems and AVs. An overview of theoretical constructs, their measurement indicators, and adapted references are listed in Table 2.2.

TABLE 2.2

Research Framework Constructs with Measurement Indicators

Construct	Measure	Source
Perceived service quality (PSQ)	Punctuality (on-time performance) [**] Privacy (sharing the shuttle space with other passengers) [**] Comfort (ease of entrance and exit from the vehicle/stations) [**] Affordability (fare price) [**] Safety on-board (regarding accidents) [**] Flexibility (frequency or number of daily services) [**] Convenience (Individual space available inside the vehicle) [**] Speed (getting places quicker) [**]	(de Oña, 2020)
Perceived relative advantages (PRA)	I believe that ASBs will be safer than conventional shuttles. [*] I believe that ASBs will be more efficient than conventional shuttles. [*] ASBs can reduce the need for conventional shuttles. [*] ASBs can reduce traffic congestion and pollutant emissions compared with conventional shuttles. [*] There will be fewer driver errors, in the case of using ASBs. [*] ABSs can allow better access to my intended destinations than other available travel modes. [*]	Self-developed, where items were from (Herrenkind et al., 2019a, 2019b)
Perceived risks (PR)	Unreliable technology (trip interruption) [*] Traffic safety on-board (regarding accidents) [*] AVs won't respond in dangerous situations [*]	Modified from (Liu et al., 2019)

(continued)

TABLE 2.2 (Continued)

Research Framework Constructs with Measurement Indicators

Construct	Measure	Source
Perceived ease of use (PEU)	I believe it would be easy for me to understand/learn how to book a ride. * I believe it would be easy to learn how to interact with ASBs. * I believe it would be easy to learn how to travel in an ASB. *	Modified from (Madigan et al., 2017)
Perceived usefulness (PU)	Riding in ASBs can reduce the stress of driving. * Using ASBs can increase my living and working productivity by reducing the time I spend driving. * I can see more possibilities for my mobility with ASBs. * ASB transport services can serve my travel needs well. * ASB transport service can be a good mobility solution for people who are unable to drive like disabled persons or the elderly. *	Self-developed, where items were from (Liu et al., 2019)
Attitude towards use (ATT)	I believe that ASBs will be more attractive to use than conventional shuttles. * I have a positive attitude towards ASBs. *	Modified from (Jensen et al., 2014)
Intention to use (INT)	If shuttles become available, I will give priority to using them over using a car. * I would be happy to ride in an ASB.*	Modified from (Jing et al., 2019)

Notes:

* Response Anchors: strongly disagree (1)/strongly agree (5).
** Response Anchors: very unlikely (1)/very likely (5).

The remainder of the survey collected respondents' socio-demographic characteristics. The survey's written language was English. In the third step, a pilot study (pre-test) was performed to detect any potential errors or misunderstandings (DeVellis, 2003). A 20–50 respondent sample size was deemed satisfactory with which to obtain feedback to help in identifying possible inconsistencies (Cooper & Schindler, 2008). We targeted higher-degree research (HDR) students and the staff of Queensland University of Technology (QUT) because these people usually have broader knowledge regarding the application of surveys for reliable results.

Accordingly, the survey link was directly sent to the Faculty of Engineering HDR and staff email addresses, which were targeted with the assistance of the university administration. Consequently, a convenience sample of 40 completed answers was collected. The pilot survey enabled us to assess the clarity of question items, to make sure that they were understandable in light of the study objective, and to achieve a satisfactory number of indicator items according to respondents' input. We evaluated the total scale/sub-scale consistency and reliability of the question items by calculating Coefficient Alpha using the pilot survey results. We removed the items that degraded the reliability by analysing the relative contribution of each item to overall scale reliability. After applying this approach, the survey questionnaire was finalised.

2.3.6 Case Study Region

A case study approach was adopted to explore the public adoption of shared DRT services, specifically conventional/autonomous shuttles, by adult residents of the SEQ metropolitan region, which is centred on Brisbane (Australia). SEQ has an adult population of approximately 3,650,000 people. The region was selected mostly because it is a highly car dependent region, which raises transport disadvantage concerns. The region, therefore, "produces unique travel behaviors as a result of its mono-centric physical structure that created a high dependency between suburban areas and the central business district (CBD)" for trades, services, and facilities (Rashid, & Yigitcanlar, 2015).

SEQ has 12 contiguous local government areas (LGAs), where each is a municipality administered by the third, and lowest, tier of government. The survey respondent recruitment methodology involved only individuals living within the urban and peri-urban areas of SEQ. Rural areas were excluded from the study because we expect that offering this service would not be cost-efficient due to its very low-density development and likely low patronage.

2.3.7 Data Collection Procedure

The finalised questionnaire, which was approved by the University Human Research Ethics Committee (UHREC RN: 2000000747), was hosted online and could be self-completed. To respond to the restrictions and risks due to the COVID-19 pandemic, Qualtrics (a professional web-based survey platform provider) was hired to reach out to the target respondents to compile data through a convenient random sampling method for this study. The survey link was passed to the general public by sending an email to each potential respondent. The email explicitly stated a brief description of the project and participant requirements, being adult residents of SEQ. Data were collected during May 2021, with a total of 357 completed questionnaires. According to Krejcie and Morgan (1970), the sample size for a population above 1,000,000 (confidence = 95% and Margin of Error = 6%) is 300. Tabachnick and Fidell (2001) also advise that "it is comforting to have at least 300 cases for factor analysis, however, a smaller sample size (e.g., 150 cases) should be sufficient if solutions have several high loading marker variables (above 0.8)". Stevens (1996) denotes that "the sample size requirements advocated by researchers have been reducing over the years as more research has been done on the topic".

2.3.8 Survey Participants

Ultimately, 300 valid responses with no missing value, invalid observation, or outliers were deemed to be satisfactory for further analysis after screening and cleaning the data. Table 2.3 illustrates the descriptive summary for each demographic group and an assessment of their multicollinearity. Results show that variable inflation factors (VIFs) are all acceptable, at a level of <2.50 (Daoud, 2017).

Of the 300 survey participants, the 18–35 years-old age group was the highest proportion (36%), while 17.7% of the respondents were aged 36–50 years old, the remaining 19.3% and 27% accounted for respondents aged 51–65 and over 66 years-old, respectively. Female respondents were nearly double that of males (65% compared to 35%). Regarding education, 27.4% of participants held tertiary degrees and, of the remainder, 36.3% completed high school, while 36.3% held a vocational certificate. The retired, homemaker

TABLE 2.3

Demographic Attributes and Collinearity Tolerances of Predictor Variables

Predictor Variable	Category	Frequency (n = 300)	Distribution (%)	Collinearity (VIF)
Gender	Male	105	35.0	1.271
	Female: 195	195	65.0	
Age	18–35	108	36.0	1.965
	36–50	53	17.7	
	51–65	58	19.3	
	66 or higher	81	27.0	
Education	High School	109	36.3	1.104
	Vocational	109	36.3	
	Tertiary	82	27.4	
Employment	Retired, homemaker, or not employed	138	46.0	1.531
	Part-time or casual employed	72	24.0	
	Full-time or self-employed	90	30.0	
Household income	Nil to $15,599	37	12.3	1.202
	$15,600–$31,199	43	14.3	
	$31,200–$51,999	55	18.4	
	$52,000–$77,999	65	21.7	
	$78,000–$103,999	54	18.0	
	$104,000 or more	46	15.3	
Residential location	Peri-urban	200	67.0	1.151
	Urban	100	33.0	
Household size	1	66	22.0	1.355
	2	119	39.7	
	3	42	14.0	
	4	43	14.3	
	5 or more	30	10.0	

or not employed group accounted for 46% of respondents while part-time or casual employees accounted for 24% and full-time or self-employed 30% of the remainder. The median and mode annual income bracket was $52,000–$77,999. The survey also showed that two-thirds of respondents (67%) were living in peri-urban areas, and most of these were from two-person households (39.7%).

2.4 Analysis and Results

This research proposed 14 hypotheses exploring causal relationships amongst seven constructs: perceived service quality, perceived relative advantages, perceived risks, perceived ease of use, PU, attitudes towards ASB, and usage intention.

According to Hair et al. (2010), structural equation modelling (SEM) is the best analysis method to "map paths to many dependent (theoretical or observed) variables in the same research model and analyses all the paths simultaneously rather than one at a time" (Gefen & Straub, 2005). Covariance-based SEM (CB-SEM) analytically focuses on shared variance and confirms theoretically assumed relationships, while the partial least squares (PLS-SEM) method is applied for prediction and/or identification of relationships between constructs (Su et al., 2019). PLS-SEM is a variance-based technique that employs total variance in the estimation of parameters (Hair et al., 2011). We considered PLS-SEM to be preferable to CB-SEM for the current study in the following respects, which concur with research in the same context (Herrenkind et al., 2019a, 2019b; Müller, 2019; Su et al., 2019; Nastjuk et al., 2020; Günthner & Proff, 2021).

PLS-SEM is statistically more powerful than CB-SEM and is appropriate for data with non-normal or unknown distributions as it is distribution-free (Lowry & Gaskin, 2014). This indicates that PLS-SEM can identify causal connections in the populations (Astuti & Rukmana, 2021), making it easier to perform exploratory analysis for theory development (Henseler et al., 2009; Hair et al., 2011; Hair Jr et al., 2021) in the case of present work (an extension of an existing structural theory of TAM). Since the proposed conceptual framework is prediction-oriented, which seeks to offer a causal overview of the relationship between pre-trial perceptions/expectations and adoption intention, PLS-SEM is deemed to be the proper approach for studying complex latent interaction effects (Lowry & Gaskin, 2014).

The two-step data analysis procedure recommended by (Hair Jr et al., 2021) using SmartPLS 3.3.3 software (Ringle et al., 2015) involved the following.

Measurement (outer) model evaluation: to check the structural reliability and validity of reflective (highly correlated and interchangeable) indicators

TABLE 2.4

Measurement Model Evaluation Process (Henseler et al., 2009)

Criterion	Description
Composite reliability (ρ_c)	The composite reliability is a measure of internal consistency and must not be lower than 0.6. $\rho_c = \left(\sum \lambda_i\right)^2 / [\left(\sum \lambda_i\right)^2 + \sum \mathrm{Var}\left(\varepsilon_i\right)]$, λ_i: the outer (component) loading to an indicator, and $\mathrm{Var}\left(\varepsilon_i\right) = 1 - \lambda_i^2$ in the case of standardised indicators.
Indicator reliability	Absolute standardised outer (component) loadings should be higher than 0.7.
Average variance extracted (AVE)	$\mathrm{AVE} = \left(\sum \lambda_i\right)^2 / [\left(\sum \lambda_i\right)^2 + \sum \mathrm{Var}\left(\varepsilon_i\right)]$, where λ_i is the component loading to an indicator and $\mathrm{Var}\left(\varepsilon_i\right) = 1 - \lambda_i^2$ in the case of standardised indicators. The AVE should be higher than 0.5.
Fornell–Larcker criterion	To ensure discriminant validity, the AVE of each latent variable should be higher than the squared correlations with all other latent variables. Thereby, each latent variable shares more variance with its block of indicators than with another latent variable representing a different block of indicators.
Cross-loadings	Cross-loadings offer another check for discriminant validity. If an indicator has a higher correlation with another latent variable than with its respective latent variable, the appropriateness of the model should be reconsidered.

and to modify the measurement model to develop a structural model to explain the relationship among the latent (unobserved) variables (Haenlein & Kaplan, 2004; Petter et al., 2007; Hair et al., 2011); see Table 2.4.

Structural (inner) model evaluation: to evaluate the theoretical model and its associated hypotheses and to test whether the hypothesised interrelations among constructs within the measurement model are supported via the survey data (Gefen & Straub, 2005), see Table 2.8.

According to Gefen et al. (2020), applying measurement model and structural model assessment allows a combination of factor analysis and hypotheses testing to be combined in one operation.

2.4.1 Measurement Model Evaluation

In the PLS-SEM approach, measurement model evaluation is conducted by assessing three criteria, internal consistency reliability, convergent validity, and discriminant validity (Fornell & Larcker, 1981; Straub et al., 2004; Hair et al., 2011; Hair Jr et al., 2021). For the model calculation "using the PLS algorithm, the path weighting scheme is selected, with a maximum iteration number of 1000 and a stop criterion of 10-7" (Gefen et al., 2020).

First, as illustrated in Table 2.4, we assessed internal consistency through composite reliability (CR) which should be over 0.7 (Nunnally, 1994), and

TABLE 2.5

Construct Reliability and Validity Results (CA, CR, AVE, and Inter-Construct Correlations)

	Coefficient Alpha (CA > 0.7)	Composite Reliability (CR > 0.7)	Average Variance Extracted (AVE > 0.5)	ATT	INT	PEU	PRA	PR	PSQ	PU
ATT	0.716	0.876	0.779	**0.882**						
INT	0.712	0.874	0.776	0.575	**0.881**					
PEU	0.810	0.887	0.724	0.598	0.504	**0.851**				
PRA	0.891	0.916	0.646	0.490	0.354	0.587	**0.804**			
PR	0.873	0.922	0.797	−0.300	−0.291	−0.149	−0.113	**0.893**		
PSQ	0.886	0.908	0.555	0.522	0.442	0.475	0.352	−0.276	**0.745**	
PU	0.897	0.924	0.709	0.586	0.524	0.688	0.506	−0.132	0.469	**0.842**

Note: * Bolded numbers: square root of AVE.

Coefficient Alpha with a value > 0.7 (Cronbach, 1951). The CR and CA values of all seven constructs are shown in Table 2.5 and range from 0.902 to 0.980, and 0.836 to 0.960, respectively, which are higher than the recommended thresholds, indicating decent internal consistency reliability.

Second, we evaluated the convergent validity by checking the outer loading to be above 0.6 (Hair Jr et al., 2021), the variance inflation factor (VIF) to be less than the adopted threshold of 5 (Hair et al., 2011), and the average variance extracted (AVE) is expected to be over 0.5, as recommended by Fornell and Larcker (1981) and Bagozzi and Yi (1988). Table 2.6 shows loadings of all indicators that exceed 0.6. No item was removed owing to high VIF (see Table 2.6), indicating that there is no collinearity problem. Regarding AVE, all six constructs had AVE values between 0.555 and 0.797 (see Table 2.5), reaching the recommended standard of adequate convergent validity, indicating that all these constructs explain more than half of their indicators' variance (Hair et al., 2011).

Third, discriminant validity, which is identified as "the extent to which a construct is truly distinct from other constructs by empirical standards" (Hair Jr et al., 2021), is assessed by three criteria: the Fornell–Larcker criterion, Cross-loadings, and the heterotrait–monotrait (HTMT) criterion. The Fornell–Larker criterion refers to the "latent construct shares more variance with its assigned indicators than with another latent variable in the structural model" (Fornell & Larcker, 1981). In statistical terms, the AVE of each latent construct should be higher than the construct's highest squared correlation with any other latent construct (Hair et al., 2011).

Table 2.5 illustrates that the AVE value of each construct (ATT, PR, INT, PEU, PRA, PSQ, and PU) is greater than the respective construct's correlation with other constructs in all seven cases. Concerning the cross-loading, "an indicator's loading with its associated latent construct has to be higher than

TABLE 2.6

Factor Loadings (in Bold) and Cross-Loadings

Latent Construct	Indicator	Loadings > 0.6							
		ATT	INT	PEU	PR	PRA	PSQ	PU	VIF
Attitude	ATT1	**0.872**	0.474	0.518	0.479	−0.282	0.409	0.475	1.452
	ATT2	**0.893**	0.538	0.538	0.391	−0.250	0.508	0.556	1.452
Usage intention	INT1	0.565	**0.895**	0.480	0.388	−0.219	0.435	0.462	1.441
	INT2	0.442	**0.867**	0.405	0.227	−0.300	0.340	0.463	1.441
	PEU1	0.492	0.364	**0.788**	0.461	−0.195	0.365	0.465	1.547
Perceived ease of use	PEU2	0.411	0.401	**0.879**	0.442	−0.163	0.393	0.620	2.132
	PEU3	0.608	0.506	**0.883**	0.581	−0.046	0.447	0.653	1.940
Perceived risks	PR1	−0.292	−0.254	−0.156	−0.124	**0.903**	−0.258	−0.127	2.455
	PR2	−0.276	−0.248	−0.144	−0.099	**0.899**	−0.232	−0.116	2.461
	PR3	−0.235	−0.279	−0.097	−0.078	**0.877**	−0.249	−0.110	2.149
Perceived relative advantages	PRA1	0.315	0.219	0.437	**0.822**	−0.066	0.238	0.389	2.709
	PRA2	0.335	0.223	0.411	**0.788**	−0.111	0.267	0.387	2.366
	PRA3	0.392	0.283	0.522	**0.819**	−0.067	0.314	0.415	2.251
	PRA4	0.420	0.316	0.432	**0.772**	−0.105	0.307	0.372	1.923
	PRA5	0.402	0.299	0.475	**0.855**	−0.071	0.298	0.414	2.675
	PRA6	0.468	0.340	0.529	**0.764**	−0.118	0.264	0.446	1.701
Perceived service quality	PSQ1	0.212	0.242	0.207	0.109	−0.124	**0.637**	0.209	1.774
	PSQ2	0.503	0.454	0.407	0.308	−0.271	**0.798**	0.477	2.018
	PSQ3	0.392	0.306	0.376	0.279	−0.263	**0.758**	0.330	2.043
	PSQ4	0.476	0.374	0.425	0.347	−0.262	**0.825**	0.417	2.329
	PSQ5	0.277	0.264	0.243	0.115	−0.145	**0.708**	0.270	1.965
	PSQ6	0.414	0.294	0.356	0.284	−0.168	**0.765**	0.351	2.146
	PSQ7	0.311	0.274	0.354	0.241	−0.152	**0.692**	0.302	1.871
	PSQ8	0.400	0.350	0.383	0.311	−0.195	**0.760**	0.339	1.983
Perceived usefulness	PU1	0.483	0.505	0.605	0.411	−0.089	0.390	**0.860**	2.828
	PU2	0.481	0.398	0.570	0.438	−0.051	0.315	**0.851**	2.760
	PU3	0.508	0.474	0.596	0.469	−0.120	0.417	**0.854**	2.390
	PU4	0.514	0.436	0.536	0.389	−0.163	0.456	**0.825**	2.120
	PU5	0.481	0.385	0.587	0.420	−0.131	0.394	**0.819**	2.098

its loadings with all the remaining constructs" (Daoud, 2017). As depicted in Table 2.6, the outer loadings of all seven constructs are greater than their cross-loadings, and each item has a strong factor loading on its associated construct ($p < 0.01$). Henseler et al. (2016) proposed the HTMT criterion as a value <0.9 in order to test discriminant validity, which shows whether these are the same or different latent factors (Lowry & Gaskin, 2014). In Table 2.7, the HTMT values demonstrate a high level of discriminant validity.

In summary, the measurement model evaluation confirmed a satisfactory degree of internal reliability and validity of all seven constructs (ATT, PR, INT, PEU, PRA, PSQ, and PU) before performing the next stage of the structural model valuation. The standardised root mean square residual (SRMR)

TABLE 2.7

Heterotrait–Monotrait (HTMT) Values

	ATT	INT	PEU	PRA	PR	PSQ	PU
ATT							
INT	0.798						
PEU	0.777	0.652					
PRA	0.608	0.431	0.678				
PR	0.380	0.373	0.187	0.126			
PSQ	0.625	0.537	0.541	0.374	0.301		
PU	0.729	0.654	0.798	0.562	0.149	0.506	

was 0.063, representing a good fit with respect to the cut-off value of 0.08 (Henseler et al., 2009).

2.4.2 Structural Model Evaluation

First, the bootstrapping re-sampling procedure, as suggested by Chin (1998) and Hair et al. (2011) with n = 300 cases and n = 5000 re-samples, was carried out corresponding to Hayes (2009) to evaluate the structural model fit. The evaluation procedure involves examining the predictive power of the con- ceptual framework and analysing interactions among the constructs.

Next, the predictive power of the conceptual framework was assessed by checking two criteria: (a) The model's predictive accuracy, evaluated by the coefficient of determination (R^2) that indicates "the share of the variance of the exogenous variables affecting the endogenous variables. Range of R^2 value is 0–1, but it needs to be at least 0.3 to be considered acceptable" (Günthner & Proff, 2021) also see Table 2.8; (b) the model's predictive relevance is evaluated by the Stone–Geisser Q2 (Geisser, 1974; Stone, 1974), which sig- nifies "an evaluation criterion for the cross-validated predictive relevance of the PLS path model" (Hair Jr et al., 2021). The outcomes demonstrated that PU, ATT, and INT were explained by the other constructs, with R^2 values of 0.51, 0.50, and 0.41, respectively (see Figure 2.3).

According to the rule of thumb advised by Henseler et al. (2009) see Table 2.8, the R^2 value of PU showed a greater level of predictive accuracy than the R^2 value of attitude (ATT) and usage intention (INT), which represented a moderate level power to explain the generated variation amount (Chin, 1998). Along with measuring R^2, the Stone–Geisser Q2 value was estimated by performing the blindfolding technique—"a sample reuse technique that omits singular elements of the data matrix and uses the model estimates to predict the omitted part" (Hair et al., 2011). As indicated by Hair et al. (2014), "a measure of predictive relevance values of 0.02, 0.15, and 0.35 indicated small, medium and large predictive relevance for a certain endogenous con- struct". The results showed that the Q2 of PU, attitude (ATT), and usage

TABLE 2.8

Structural Model Assessment Process (Derived from Henseler et al. (2009))

Criterion	Description
R^2 of endogenous latent variables	R^2 values of 0.67, 0.33, or 0.19 for endogenous latent variables in the inner path model are described as substantial, moderate, or weak (Chin. 1998).
Estimates for path coefficients	The estimated values for path relationships in the structural model should be evaluated in terms of the sign, magnitude, and significance (the latter via bootstrapping).
Prediction relevance (Q^2)	The Q^2 is calculated based on the blindfolding procedure: $Q^2 = 1 - (\sum_D SSE_D) / (\sum_D SSO_D)$, D: the omission distance, SSE: the sum of squares of prediction errors, and SSO: the sum of squares of observations. $Q^2 > 0$: give evidence that the observed values are well reconstructed and that the model has predictive relevance, ($Q^2 < 0$ indicates a lack of predictive relevance).

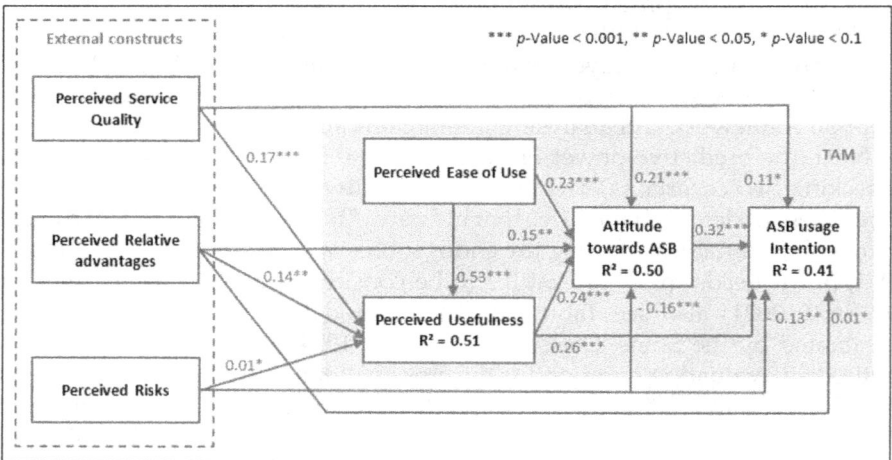

FIGURE 2.3
Structural model.

intention (INT) were 0.355, 0.378, and 0.305 respectively, which are significantly greater than zero, so had large predictive relevance. Thus, the predictive power of the current research conceptual framework was achieved for both R^2 and Q2 values.

2.4.3 Hypothesis Testing

Next, the path relationship between the conceptual framework constructs was assessed by the standardised beta coefficient of ordinary least squares regression (β) that takes the value $\leq |\pm1|$ and its significance. A β value near $|\pm1|$ describes a strong effect on the latent variable, while a value close to zero describes a negligible effect. Values above 0.1 reflect significant influence (Chin, 1998). The recommended values of large, medium, and small loading sizes are 0.5, 0.2, and 0.1, respectively (Harlow, 2014). The path coefficients were deemed to be significant when t-values were larger than $|\pm1.96|$ at significant levels of 5% and confidence levels of 90% (Hair Jr et al., 2014).

Accordingly, Table 2.9 reveals that all paths were significant except for three (Hypotheses H8, H11, and H12), indicating that PSQ and PRA have no direct influence on usage intention. Similarly, PR was shown to have no meaningful direct influence on PU. The outcomes also point out that PSQ has a stronger effect on PU than PRA, with $\beta = 0.17$ compared to 0.14. However, its effect on PU is still weaker than the influence of PEU on PU ($\beta = 0.53$). Out of five predictors of ATT comprising PU, PR, PRA, PSQ, and PEU, the results showed that PU has the strongest impact on ATT with $\beta = 0.24$. This was followed by PEU, PRA, and PSQ with $\beta = 0.23$, $\beta = 0.21$, and $\beta = 0.15$,

TABLE 2.9

Structural Model Evaluation Results

Proposed Hypotheses		Effect	β	T-Value	*p*-Value	Results
H1	Perceived ease of use (PEU) + → attitude (ATT)	+	0.23	2.92	0	Supported
H2	Perceived usefulness (PU) + → attitude (ATT)	+	0.24	3.30	0	Supported
H3	Perceived ease of use (PEU) + → PU	+	0.53	7.94	0	Supported
H4	PU + → usage intention (INT)	+	0.26	3.76	0	Supported
H5	Attitude (ATT) + → usage intention (INT)	+	0.32	4.43	0	Supported
H6	Perceived service quality (PSQ) + → PU	+	0.17	3.09	0	Supported
H7	Perceived service quality (PSQ) + → attitude (ATT)	+	0.21	3.34	0	Supported
H8	Perceived service quality (PSQ) + → usage intention (INT)	+	0.11	1.82	0.07	Not supported
H9	Perceived relative advantage (PRA) + → PU	+	0.14	2.45	0.01	Supported
H10	Perceived relative advantage (PRA) + → attitude (ATT)	+	0.15	2.64	0.01	Supported
H11	Perceived relative advantage (PRA) + → usage intention (INT)	+	0.01	0.19	0.85	Not supported
H12	Perceived risks (PR) + → PU	+	0.01	0.27	0.78	Not supported
H13	Perceived risks (PR) − → attitude (ATT)	−	−0.16	3.65	0	Supported
H14	Perceived risks (PR) − → usage intention (INT)	−	−0.13	2.75	0.01	Supported

Note: β: Path coefficient, t-value: Significance, *p*-value: Significance.

respectively, while PR was the only predictor which influenced ATT negatively, with β = −0.16. Next, regarding the predictors of INT, ATT (β = 0.32),
and PU (β = 0.26) were found to have a positive effect on it, whereas PR had
a significant negative effect (β = −0.13).

2.5 Findings and Discussion

The following sections correspond to prospective policy implications that
can be considered based on the hypothesis findings, which demonstrate how
this methodology was used to identify practical implications.

2.5.1 Technology Acceptance Model

The study findings support TAM's fundamental structure and demonstrate
how ATT might remain to be a statistically substantial predictor of INT in
the ASB context (H5). This is congruent with prior research in the domain,
representing that a generally favourable attitude results in a higher usage
intention and consequently actual use of ASBs (Herrenkind et al., 2019a) and
AV (Nastjuk et al., 2020). The notable influence of PU on ATT (H2), which
was confirmed by Herrenkind et al. (2019a) and Nastjuk et al. (2020), and on
INT (H4), which corresponds with (Xu et al., 2018; Motamedi et al., 2019)
implies that the potential users might recognise ASB as a useful imminent
travel mode.

The positive correlation of PEU with PU (H3), which was also verified
by, e.g., Zhang et al. (2019; 2020), and PEU with ATT (H1), in line with, e.g.,
Nastjuk et al. (2020), indicate that, when individuals can utilise ASBs without
experiencing significant physical or mental exertion, their impression of
usefulness could be enhanced. These results are directly connected to prior
research findings regarding users' pre-trial beliefs about AVs (e.g., Dirsehan
& Can, 2020; Nastjuk et al., 2020).

In contrast, some prior research has shown inconsistencies in the postulated
relationships between the antecedents of TAM. For instance, the direct caus-
ality between PU and ATT could not be supported by Zhang et al. (2019).
Similarly, PU failed to predict INT in the findings of Buckley et al. (2018) and
Herrenkind et al. (2019a). In a similar vein, the association between PEU and
PU was not confirmed. Henceforth, authorities should emphasise the efficacy
and simplicity of ASBs in the technical improvement of impending mobility
services. Policies could also definitely stimulate the public attitude. For
instance, marketing campaigns can highlight the benefits of ASBs, portraying
the mode as having more flexible timetables or consuming less energy than
traditional buses (Herrenkind et al., 2109b).

2.5.2 Perceived Relative Advantage

Furthermore, the results showed a positive effect of PRA on both PU (H9) and ATT (H10), which has been shown in other research in both the ASB (Herrenkind et al., 2109a, 2109b) and AV (Lee et al., 2017; Nastjuk et al., 2020) contexts. This suggests that people who realise the advantages of ASBs over former conventional shuttle buses may find ASBs more useful and accordingly show greater adoption. However, understanding the link between perceived relative advantages and adoption attitude in the ASB setting is not enough and requires further work. Travellers may decide against using ASBs if they do not have a clear and distinct awareness of the benefits that only ASBs could bring. This is endorsed by the findings of Herrenkind et al. (2019a) regarding the insignificant association of PRA with ATT, and Lee et al. (2017), who reported an insignificant association of PRA with INT, showing how individuals choose to adopt and utilise ASBs. They asserted that "the perception of usefulness at the psychological level was not influential on the intention to use but became influential at the system level" (Lee et al., 2017).

Consequently, highlighting the relative advantages could be a useful strategy to encourage travellers to make regular use of ASBs. To increase ASB usage intention, policymakers could accentuate not only the relative advantages over their conventional counterparts but also their inimitable usefulness. ASBs will be ecologically more sustainable than conventional shuttles because automated driving will improve the efficiency, safety, and flexibility of electrically powered ASBs, while also reducing driver errors, traffic congestion, and pollutant emissions (Golbabaei et al., 2020). As ASBs have a larger capacity than private cars, their widespread usage may help balance the growing mobility demand.

2.5.3 Perceived Risks

In alignment with Herrenkind et al. (2019a, 2019b) in the context of ASBs, the results revealed the negative impact of PR on ATT (H13) and INT (H14), implying that perceived risk, as a psychological factor closely related to safety concerns, substantially lowering the potential users' trust and usage intention. However, prior research on the impact of PR has shown contradictory findings in the AV context. For instance, while e.g., Xu et al. (2018) failed to identify its significant effect on the INT, Zhang et al. (2019), and Nastjuk et al. (2020) reported its notable negative influence. Correspondingly, PR is predicted to negatively affect attitudes towards AVs (Zhang et al., 2019) and usage intention indirectly by impacting users' level of trust towards AVs. These inconsistent findings denote respondents' difficulties in evaluating and estimating the implication of implementation of AVs due to the lack of practical evidence in various settings such as several road types, driving situations, or physical/mental conditions (Lee et al., 2017; Keszey,

2020). Therefore, the degree to which people perceive that ASBs have low risks determines whether or not they want to use this mode.

In addition, we could not find any association between PR and PU (H12). This may imply that, although the technical level of AVs is constantly improving, the market requirement for the technical stability and reliability of ASBs is also continuously increasing (Jing et al., 2019). Given that concerns might unfavourably impact potential users' decision-making, "policymakers must work to ensure that the general public understands the technology behind" ASBs (Herrenkind et al., 2019b). The availability of comprehensive accurate information regarding the system's decision-making procedure could be helpful in this regard. Offering trial rides may also assist build or reinforcing trust in ASBs since the initial experience of an ASB ride is probably associated with a feeling of fear/anxiety because control is transferred from the driver to the vehicle (Herrenkind et al., 2019b). Nevertheless, such measures could lead to enhanced perceived trust in ASBs and reduced amount of risk perceptions, hence lessening undesirable impacts on usage intention.

2.5.4 Perceived Service Quality

Turning to the newly added construct, perceived service quality, the examination of the influence of this construct in association with perceptions regarding relative advantages, and risks, along with TAM constructs in the ASB domain has not yet been performed. Thus, our findings are unique and significantly contribute to the ADRT literature. The study revealed that PSQ has a positive direct influence on PU (H6) and ATT (H7). This finding is in alignment with previous public transport literature, e.g., Machado-Leon et al. (2018) indicated that perceived service quality is an antecedent of perceived benefits, and it directly affects involvement. Accordingly, once potential users perceive they can travel in ASBs irrespective of time constraints and that the mode is compatible with their lifestyle or identity, they would find it more beneficial and consequently would show a more positive attitude towards their use.

Nevertheless, INT is not determined by PSQ (H8) according to the statistical results, even though PSQ is found to predict ATT (H7). As such, PU and ATT fully mediated the causal link from PSQ to INT, implying that perceived service quality indirectly affects ASB usage intention. This indirect association between perceived service quality and usage intention is consistent with other research outcomes in the public transport domain (Kuo & Tang, 2011; Koklic et al., 2017), suggesting that while users' attitudes towards ASBs at the beginning of the employment stage will probably be positive, their usage intention may only increase with time and more awareness. However, these relationships are explored in the ASB context for the first time, contributing to extending our understanding of the importance of service quality perceptions in forming both attitudes towards ASBs and intention to use this mode regularly. Further investigation is yet needed to verify these hypotheses.

In order to improve the perceived service quality, authorities could potentially adjust the condition of the ASBs regarding comfort, convenience, and safety, as well as the timeframe of the availability of services to the requirements of the residents, before deploying them into public transport service. The government might also consider marketing and/or the subsidisation of fares to be affordable to complement the existing RPT modes.

2.6 Conclusion

A transition to more sustainable transportation is critical, due to recent ecological trends. The rapid and widespread distribution of shared ADRT may directly contribute to this goal since AV technology promises to deliver ecological benefits as it evolves. Due to the potential for ADRT services, especially ASBs, to complement public transport demand, both researchers and policymakers need to understand how prospective users' behavioural intentions to utilise such services are formed in order to attract and keep consumers in the future. While the interrelationships among perceived relative advantages and perceived risks have been investigated in the autonomous public transit adoption literature, they have attracted less attention in the ASB context. In particular, the potential impacts of perceived service quality as a commonly used notion in both public transport and behaviour research have been overlooked in the ASB domain, indicating a need for more exploration of a robust acceptance concept.

The commercial entities developing ASBs and government agencies responsible for ensuring appropriate physical and social infrastructures are in place are likely to be interested in the extent to which the general public is aware and supportive of the impending roll-out of ASBs. In particular, identifying the strongest concerns relating to ASBs can assist in the planning of proactive efforts to address these issues, while building on any perceived positive attributes (Kulviwat et al., 2007; Bansal et al., 2016; Kelley, 2017; Smith, 2017). Consequently, we developed a conceptual model and included the above-mentioned factors as extra stimulating technology acceptance factors, mainly owing to the nature of the ASB services' potential to be integrated with public transport, to find their interrelationships and weight in ASB adoption (Esterwood et al., 2021).

The findings reported in this study make a noteworthy contribution to the ADRT acceptance literature by demonstrating the usefulness of the method used to underpin public attitude and intention to use ADRT technology. The research findings exemplify managerial implications to assist policymakers, transport planners, and engineers in their policy decisions and system plans as well as achieving higher public acknowledgement and wider uptake of ADRT technology solutions. We also advocate the importance of stimulating technological innovation efforts through government incentives for further advancement of ASB technology to make them smarter and safer (Yigitcanlar et al., 2019). Notwithstanding the comprehensive empirical information presented in the current work, there are certain research limitations worth underlining for future study.

The authors note the potential for bias in the findings due to issues around the representativeness of the small-scale sample size to the SEQ regional population. Future research could adopt the methodology but conduct a more systematic approach befitting a large, cross-sectionally representative sample to lower the possibility of bias, facilitating the development of more robust policy implications.

The data collection was performed by conducting an online survey with potential selection bias. For instance, respondents with higher educational levels are better at using and responding to emails compared to tech-disadvantaged groups. Given that ASBs are not yet available in the market, an online survey is an effective approach to exploring public adoption intention. Nevertheless, further large-scale surveys are required to more broadly understand the public attitude. Particularly, there are other social science theories, such as TPB, for conducting future research on this concept from a different point of view. Some longitudinal surveys could be used to track changing views and behaviours over time.

We have focused on safety risks but the effects of other aspects involving psychological, social, and privacy risks should be investigated in this context as well. Furthermore, trust towards ASBs and the perceived risks of using the mode should be examined jointly. This strategy could provide an in-depth insight into how to improve travellers' adoption intention. Additionally, investigating attitudes towards trust and perceived risks without physically exposing survey participants to riding in ASBs in real traffic systems may lead to biased results as their perceptions may be shaped by information attained from mass media, particular literature, or entertainment venues. Hence, combining mainstream questionnaires with novel methodological techniques like virtual reality, simulation, or gamification (Keszey, 2020) may be useful.

As ASBs facilitate mobility for transport-disadvantaged populations, such as the elderly, those with no driving licence, and physically impaired persons, and help in minimising transport related social exclusion (Kamruzzaman et al., 2015), these groups must be involved in future studies. Moreover, further exploration of the effects of demographic characteristics (e.g., age, gender, education, income, or residential location) that might associate with the adoption of ASBs is desirable. This would result in a more nuanced knowledge of a particular target group (subject to their features) and improve the predictive power of the suggested framework. Lastly, while the current study follows a quantitative method, we recommend using qualitative research approaches, such as interviews to further investigate and validate these study findings.

Acknowledgements

This chapter, with permission from the copyright holder, is a reproduced version of the following journal article: Golbabaei, M., Yigitcanlar, T., Paz, A., & Bunker, J. (2022). Understanding autonomous shuttle adoption intention: Predictive power of pre-trial perceptions and attitudes. *Sensors*, 22(23), 9193.

References

Abraham, H., Lee, C., Brady, S., Fitzgerald, C., Mehler, B., Reimer, B., & Coughlin, J.F. (8–12 January 2017). Autonomous vehicles and alternatives to driving: Trust, preferences, and effects of age. In Proceedings of the Transportation Research Board 96th Annual Meeting (TRB'17), Washington, DC, USA.

Acheampong, R.A., & Cugurullo, F. (2019). Capturing the behavioural determinants behind the adoption of autonomous vehicles: Conceptual frameworks and measurement models to predict public transport, sharing and ownership trends of self-driving cars. *Transportation Research Part F: Traffic Psychology and Behaviour*, 62, 349–375.

Acheampong, R.A., Siiba, A., Okyere, D.K., & Tuffour, J.P. (2020). Mobility-on-demand: An empirical study of internet-based ride-hailing adoption factors, travel characteristics and mode substitution effects. *Transportation Research Part C: Emerging Technologies*, 115, 102638.

Ahern, Z., Paz, A., & Corry, P. (2022). Approximate multi-objective optimisation for integrated bus route design and service frequency setting. *Transportation Research Part B: Methodological*, 155, 1–25.

Ajzen, I., & Fishbein, M. (1980). *Understanding Attitudes and Predicting Social Behavior*; Prentice Hall Inc.: Englewood Cliffs, NJ.

Astuti, S., & Rukmana, D. (2021). Student satisfaction on the implementation of the online undergraduate thesis examination: A PLS-SEM analysis. *Quality Assurance in Education*, 29, 491–508.

Asubonteng, P., McCleary, K.J., & Swan, J.E. (1996). SERVQUAL revisited: A critical review of service quality. *Journal of Services Marketing*, 10, 62–81.

Bagozzi, R.P., & Yi, Y. (1988). On the evaluation of structural equation models. *Journal of the Academy of Marketing Science*, 16, 74–94.

Bansal, P., Kockelman, K.M., & Singh, A. (2016). Assessing public opinions of and interest in new vehicle technologies: An Austin perspective. *Transportation Research Part C: Emerging Technologies*, 67, 1–14.

Bernhard, C., Oberfeld, D., Hoffmann, C., Weismüller, D., & Hecht, H. (2020). User acceptance of automated public transport: Valence of an autonomous minibus experience. *Transportation Research Part F: Traffic Psychology and Behaviour*, 70, 109–123.

Bradshaw-Martin, H., & Easton, C. (2014). Autonomous or 'driverless' cars and disability: A legal and ethical analysis. *European Journal of Current Legal Issues*, 20, 1–10.

Buckley, L., Kaye, S.-A., & Pradhan, A.K. (2018). Psychosocial factors associated with intended use of automated vehicles: A simulated driving study. *Accident Analysis & Prevention*, 115, 202–208.

Butler, L., Yigitcanlar, T., & Paz, A. (2021). Barriers and risks of Mobility-as-a-Service (MaaS) adoption in cities: A systematic review of the literature. *Cities*, 109, 103036.

Butler, L., Yigitcanlar, T., & Paz, A. (2021). Factors influencing public awareness of autonomous vehicles: Empirical evidence from Brisbane. *Transportation Research Part F: Traffic Psychology and Behaviour*, 82, 256–267.

Chen, H.-K., & Yan, D.-W. (2019). Interrelationships between influential factors and behavioral intention with regard to autonomous vehicles. *International Journal of Sustainable Transportation*, 13, 511–527.

Chin, W.W. (1998). The partial least squares approach to structural equation modeling. *Modern Methods for Business Research*, 295, 295–336.

Cho, Y., Park, J., Park, S., & Jung, E.S. (2017). Technology acceptance modeling based on user experience for autonomous vehicles. *Journal of the Ergonomics Society of Korea*, 36, 87–108.

Cooper, C., & Schindler, P. (2008). *Business Research Methods*; McGraw-Hill: Boston, MA.

Cronbach, L.J. (1951). Coefficient alpha and the internal structure of tests. *Psychometrika*, 16, 297–334.

Daoud, J.I. (2017). Multicollinearity and regression analysis. *Journal of Physics: Conference Series*, 949, 012009. https://doi.org/10.1088/1742-6596/949/1/012009

Davis, F.D., & Venkatesh, V. (2004). Toward preprototype user acceptance testing of new information systems: Implications for software project management. *IEEE Transactions on Engineering Management*, 51, 31–46.

Davis, F.D., Bagozzi, R.P., & Warshaw, P.R. (1989). User acceptance of computer technology: A comparison of two theoretical models. *Management Science*, 35, 982–1003.

Davis, J.A., & Weber, R.P. (1985). *The Logic of Causal Order*; Sage: Newbury Park, CA; Volume 55.

de Oña, J. (2020). The role of involvement with public transport in the relationship between service quality, satisfaction and behavioral intentions. *Transportation Research Part A: Policy and Practice*, 142, 296–318.

Dennis, S., Paz, A., & Yigitcanlar, T. (2021). Perceptions and attitudes towards the deployment of autonomous and connected vehicles: Insights from Las Vegas, Nevada. *Journal of Urban Technology*, 28, 75–95.

DeVellis, R. (2003). *Scale Development: Theory and Applications*; Thousand Oaks Publishing: London; New Delhi.

Dirsehan, T., & Can, C. (2020). Examination of trust and sustainability concerns in autonomous vehicle adoption. *Technology in Society*, 63, 101361.

Dowler, E., Green, J., Bauer, M., & Gasperoni, G. (2006). Assessing public perception: Issues and methods. In *Health, Hazard and Public Debate: Lessons for Risk Communication from BSE/CJD Saga*; World Health Organization: Geneva, pp. 39–60.

Enoch, M., Ison, S., Laws, R., & Zhang, L. (2006). Evaluation study of demand responsive transport services in Wiltshire; Final report: Trowbridge.

Esterwood, C., Yang, X.J., & Robert, L.P. (2021). Barriers to AV bus acceptance: A national survey and research agenda. *International Journal of Human–Computer Interaction*, 37, 1391–1403.

Faisal, A., Yigitcanlar, T., Kamruzzaman, M., & Paz, A. (2020). Mapping two decades of autonomous vehicle research: A systematic scientometric analysis. *Journal of Urban Technology*, 28, 45–74.

Fishbein, M., & Ajzen, I. (1975). Belief, attitude, intention, and behavior: An introduction to theory and research. *Philosophy & Rhetoric*, 6, 244–245.

Fornell, C., & Larcker, D.F. (1981). Evaluating structural equation models with unobservable variables and measurement error. *Journal of Marketing Research*, 18, 39.

Garidis, K., Ulbricht, L., Rossmann, A., & Schmäh, M. (7–10 January 2020). Toward a user acceptance model of autonomous driving. In Proceedings of the 53rd Hawaii International Conference on System Sciences, Wailea-Makena, Hawaii.

Garvin, D. (1983). Quality on the line. *Harvard Business Review*, 9, 65–75.

Gefen, D., & Straub, D. (2005). A practical guide to factorial validity using PLS-graph: Tutorial and annotated example. *Communications of the Association for Information Systems*, 16, 5.

Gefen, D., Straub, D., & Boudreau, M.-C. (2020). Structural equation modeling and regression: Guidelines for research practice. *Communications of the Association for Information Systems*, 4, 7.

Geisser, S. (1974). A predictive approach to the random effect model. *Biometrika*, 61, 101–107.

Gkartzonikas, C., & Gkritza, K. (2019). What have we learned? A review of stated preference and choice studies on autonomous vehicles. *Transportation Research Part C: Emerging Technologies*, 98, 323–337.

Golbabaei, F., Yigitcanlar, T., & Bunker, J. (2021). The role of shared autonomous vehicle systems in delivering smart urban mobility: A systematic review of the literature. *International Journal of Sustainable Transportation*, 15, 731–748.

Golbabaei, F., Yigitcanlar, T., Paz, A., & Bunker, J. (2020). Individual predictors of autonomous vehicle public acceptance and intention to use: A systematic review of the literature. *Journal of Open Innovation*, 6, 106.

Grönroos, C. (1984). A service quality model and its marketing implications. *European Journal of Marketing*, 18, 36–44.

Günthner, T., & Proff, H. (2021). On the way to autonomous driving: How age influences the acceptance of driver assistance systems. *Transportation Research Part F: Traffic Psychology and Behaviour*, 81, 586–607.

Haboucha, C.J., Ishaq, R., & Shiftan, Y. (2017). User preferences regarding autonomous vehicles. *Transportation Research Part C: Emerging Technologies*, 78, 37–49.

Haenlein, M., & Kaplan, A.M. (2004). A beginner's guide to partial least squares analysis. *Understanding Statistics*, 3, 283–297.

Hair Jr, J.F., Black, W.C., Babin, B.J, & Anderson, R.E. (2010). *Multivariate Data Analysis: A Global Perspective*, 7th ed.; Pearson Education: Upper Saddle River, NJ, pp. 785–785.

Hair Jr, J.F., Hult, G.T.M., Ringle, C.M., & Sarstedt, M. (2021). *A Primer on Partial Least Squares Structural Equation Modeling (PLS-SEM)*; Sage Publications: Newbury Park, CA.

Hair Jr, J.F., Sarstedt, M., Hopkins, L., & Kuppelwieser, V.G. (2014). Partial least squares structural equation modeling (PLS-SEM): An emerging tool in business research. *Eurasian Business Review*, 26, 106–121.

Hair, J.F., Anderson, R.E., Black, B., Babin, B.J., & Black, W.C. (2013). *Multivariate Data Analysis*; Pearson Education, Limited: Upper Saddle River, NJ.

Hair, J.F., Ringle, C.M., & Sarstedt, M. (2011). PLS-SEM: Indeed a Silver Bullet. *Journal of Marketing Theory and Practice*, 19, 139–152.

Han, H., & Hyun, S.S. (2017). Impact of hotel-restaurant image and quality of physical-environment, service, and food on satisfaction and intention. *International Journal of Hospitality Management*, 63, 82–92.

Harlow, L.L. (2014). *The Essence of Multivariate Thinking: Basic Themes and Methods*, 2nd ed.; Routledge: New York, NY.

Hayes, A.F. (2009). Beyond Baron and Kenny: Statistical mediation analysis in the New Millennium. *Communication Monographs*, 76, 408–420.

Henseler, J., Ringle, C.M., & Sarstedt, M. (2016). Testing measurement invariance of composites using partial least squares. *International Marketing Review*, 33, 405–431.

Henseler, J., Ringle, C.M., & Sinkovics, R.R. (2009). The use of partial least squares path modeling in international marketing. In *New Challenges to International Marketing*; Sinkovics, R.R., and Ghauri, P.N., Eds.; Emerald Group Publishing Limited: Bingley, pp. 277–319.

Herrenkind, B., Brendel, A.B., Nastjuk, I., Greve, M., & Kolbe, L.M. (2019a). Investigating end-user acceptance of autonomous electric buses to accelerate diffusion. *Transportation Research Part D: Transport and Environment*, 74, 255–276.

Herrenkind, B., Nastjuk, I., Brendel, A.B., Trang, S., & Kolbe, L.M. (2019b). Young people's travel behavior – Using the life-oriented approach to understand the acceptance of autonomous driving. *Transportation Research Part D: Transport and Environment*, 74, 214–233.

Högg, R., Schmid, H.P.D.B., & Stanoevska-Slabeva, F.P.D.K. (2010). *Erweiterung und Evaluation des Technologieakzeptanzmodells zur Anwendung bei mobilen Datendiensten*; Universität St. Gallen: St. Gallen.

Insani, P.A. (2013). Public perception towards public service quality. *International Institute for Science, Technology and Education*, 3, 12.

Jensen, A.F., Cherchi, E., & Ortúzar, J.D.D. (2014). A long panel survey to elicit variation in preferences and attitudes in the choice of electric vehicles. *Transportation*, 41, 973–993.

Jing, P., Huang, H., Ran, B., Zhan, F., & Shi, Y. (2019). Exploring the factors affecting mode choice intention of autonomous vehicle based on an extended theory of planned behavior—A case study in China. *Sustainability*, 11, 1155.

Jing, P., Xu, G., Chen, Y., Shi, Y., & Zhan, F. (2020). The determinants behind the acceptance of autonomous vehicles: A systematic review. *Sustainability*, 12, 1719.

Kamruzzaman, M., Hine, J., & Yigitcanlar, T. (2015). Investigating the link between carbon dioxide emissions and transport-related social exclusion in rural

Nort1hern Ireland. *International Journal of Environmental Science and Technology*, 12, 3463–3478.

Kelley, B. (2017). Public health, autonomous automobiles, and the rush to market. *Journal of Public Health Policy*, 38, 167–184.

Keszey, T. (2020). Behavioural intention to use autonomous vehicles: Systematic review and empirical extension. *Transportation Research Part C: Emerging Technologies*, 119, 102732.

Kim, Y.G., & Eves, A. (2012). Construction and validation of a scale to measure tourist motivation to consume local food. *Tourism Management*, 33, 1458–1467.

King, W.R., & He, J. (2006). A meta-analysis of the technology acceptance model. *Information & Management*, 43, 740–755.

Koklic, M.K., Kukar-Kinney, M., & Vegelj, S. (2017). An investigation of customer satisfaction with low-cost and full-service airline companies. *Journal of Business Research*, 80, 188–196.

Koppel, S., Lee, Y.-C., Mirman, J.H., Peiris, S., & Tremoulet, P. (2021). Key factors associated with Australian parents' willingness to use an automated vehicle to transport their unaccompanied children. *Transportation Research Part F: Traffic Psychology and Behaviour*, 78, 137–152.

Krejcie, R.V., & Morgan, D.W. (1970). Determining sample size for research activities. *Educational and Psychological Measurement*, 30(3), 607–610.

Kulviwat, S., Ii, G.C.B., Kumar, A., Nasco, S.A., & Clark, T. (2007). Toward a unified theory of consumer acceptance technology. *Psychology & Marketing*, 24, 1059–1084.

Kuo, C.-W., & Tang, M.-L. (2011). Relationships among service quality, corporate image, customer satisfaction, and behavioral intention for the elderly in high speed rail services. *Journal of Advanced Transportation*, 47, 512–525.

Lakhekar, G.V., & Waghmare, L.M. (2022). Robust self-organising fuzzy sliding mode-based path-following control for autonomous underwater vehicles. *Journal of Marine Engineering & Technology*. https://doi.org/10.1080/20464 177.2022.2120448

Lee, C., Ward, C., Raue, M., D'Ambrosio, L., & Coughlin, J.F. (9–14 July 2017). Age differences in acceptance of self-driving cars: A survey of perceptions and attitudes. In Proceedings of the International Conference on Human Aspects of IT for the Aged Population, Vancouver, BC, Canada.

Lee, Y., Kozar, K.A., & Larsen, K.R.T. (2003). The technology acceptance model: Past, present, and future. *Communications of the Association for Information Systems*, 12, 50.

Liu, H., Yang, R., Wang, L., & Liu, P. (2019). Evaluating initial public acceptance of highly and fully autonomous vehicles. *International Journal of Human–Computer Interaction*, 35, 919–931.

Liu, P., Yang, R., & Xu, Z. (2019). Public acceptance of fully automated driving: Effects of social trust and risk/benefit perceptions: Public acceptance of fully automated driving. *Risk Analysis*, 39, 326–341.

Lowry, P.B., & Gaskin, J. (2014). Partial least squares (PLS) structural equation modeling (SEM) for building and testing behavioral causal theory: When to choose it and how to use it. *IEEE Transactions on Dependable and Secure Computing*, 57, 123–146.

Machado, J.L., de Oña, R., Diez-Mesa, F., & de Oña, J. (2018). Finding service quality improvement opportunities across different typologies of public transit customers. *Transportmetrica A: Transport Science*, 14, 761–783.

Madigan, R., Louw, T., Wilbrink, M., Schieben, A., & Merat, N. (2017). What influences the decision to use automated public transport? Using UTAUT to understand public acceptance of automated road transport systems. *Transportation Research Part F: Traffic Psychology and Behaviour*, 50, 55–64.

Motamedi, S., Wang, P., Zhang, T., & Chan, C.-Y. (2019). Acceptance of full driving automation: Personally owned and shared-use concepts. *Human Factors the Journal of the Human Factors and Ergonomics Society*, 62, 288–309.

Müller, J.M. (2019). Comparing technology acceptance for autonomous vehicles, battery electric vehicles, and car sharing—A study across Europe, China, and North America. *Sustainability*, 11, 4333.

Musselwhite, C., Holland, C., & Walker, I. (2015). The role of transport and mobility in the health of older people. *Journal of Transport & Health*, 2, 1–4.

Nastjuk, I., Herrenkind, B., Marrone, M., Brendel, A.B., & Kolbe, L.M. (2020). What drives the acceptance of autonomous driving? An investigation of acceptance factors from an end-user's perspective. *Technological Forecasting and Social Change*, 161, 1.

Nordhoff, S., de Winter, J., Madigan, R., Merat, N., van Arem, B., Happee, R. (2018). User acceptance of automated shuttles in Berlin-Schöneberg: A questionnaire study. *Transportation Research Part F: Traffic Psychology and Behaviour*, 58, 843–854.

Nordhoff, S., de Winter, J., Payre, W., van Arem, B., & Happee, R. (2019). What impressions do users have after a ride in an automated shuttle? An interview study. *Transportation Research Part F: Traffic Psychology and Behaviour*, 63, 252–269.

Nordhoff, S., Madigan, R., van Arem, B., Merat, N., & Happee, R. (2021). Interrelationships among predictors of automated vehicle acceptance: A structural equation modelling approach. *Theoretical Issues in Ergonomics Science*, 22, 383–408.

Nordhoff, S., Malmsten, V., van Arem, B., Liu, P., & Happee, R. (2021). A structural equation modeling approach for the acceptance of driverless automated shuttles based on constructs from the Unified Theory of Acceptance and Use of Technology and the Diffusion of Innovation Theory. *Transportation Research Part F: Traffic Psychology and Behaviour*, 78, 58–73.

Nordhoff, S., Stapel, J., van Arem, B., & Happee, R. (2020). Passenger opinions of the perceived safety and interaction with automated shuttles: A test ride study with 'hidden' safety steward. *Transportation Research Part A: Policy and Practice*, 138, 508–524.

Nunnally, J.C. (1994). *Psychometric Theory 3E*; Tata McGraw-Hill education: New York, NY.

Oliver, R.L. (1980). A cognitive model of the antecedents and consequences of satisfaction decisions. *Journal of Marketing Research*, 17, 460–469.

Pai, F.-Y., & Huang, K.-I. (2011). Applying the technology acceptance model to the introduction of healthcare information systems. *Technological Forecasting and Social Change*, 78, 650–660.

Panagiotopoulos, I., & Dimitrakopoulos, G. (2018). An empirical investigation on consumers' intentions towards autonomous driving. *Transportation Research Part C: Emerging Technologies*, 95, 773–784.

Papadima, G., Genitsaris, E., Karagiotas, I., Naniopoulos, A., Nalmpantis, D. (2020). Investigation of acceptance of driverless buses in the city of Trikala and optimization of the service using Conjoint Analysis. *Utilities Policy*, 62, 100994.

Parasuraman, A., Zeithaml, V.A., & Berry, L.L. (1988). SERVQUAL: A multiple-item scale for measuring consumer perceptions of service quality. *Journal of Retailing*, 64, 12.

Peeta, S., Paz, A., & DeLaurentis, D. (2008). Stated preference analysis of a new microjet on-demand air service. *Transportation Research Part A*, 42, 629–645.

Petter, S., Straub, D., & Rai, A. (2007). Specifying formative constructs in information systems research. *Management Information Systems Quarterly*, 31, 623–656.

Pettigrew, S., & Sophie L. C. (2019). Stakeholder views on the social issues relating to the introduction of autonomous vehicles. *Transport Policy*, 81, 64–67.

Rahman, M., Lesch, M.F., Horrey, W.J., & Strawderman, L. (2017). Assessing the utility of TAM, TPB, and UTAUT for advanced driver assistance systems. *Accident Analysis & Prevention*, 108, 361–373. https://doi.org/10.1016/j.aap.2017.09.011

Rashid, K., & Yigitcanlar, T. (2015). A methodological exploration to determine transportation disadvantage variables: The partial least square approach. *World Review of Intermodal Transportation Research*, 5, 221–239.

Ringle, C.M., Wende, S., & Becker, J.-M. (2015). *SmartPLS 3. Boenningstedt; SmartPLS GmbH*; SmartPLS GmbH: Oststeinbek, Germany.

Rittichainuwat, B.N. (2011). Tourists' perceived risks toward overt safety measures. *Journal of Hospitality & Tourism Research*, 37, 199–216.

Roche-Cerasi, I. (2019). Public acceptance of driverless shuttles in Norway. *Transportation Research Part F: Traffic Psychology and Behaviour*, 66, 162–183.

Schoettle, B., & Sivak, M. (3–7 November 2014). A survey of public opinion about connected vehicles in the U.S., the U.K., and Australia. In Proceedings of the 2014 International Conference on Connected Vehicles and Expo (ICCVE), Vienna, Austria, pp. 687–692.

Simpson, J.R., & Mishra, S. (2020). Developing a methodology to predict the adoption rate of Connected Autonomous Trucks in transportation organizations using peer effects. *Research in Transportation Economics*, 90, 100866.

Smith, D. (2017). Robocar versus the Pod: A commentary on the state of play in the race for autonomous vehicle commercialisation. *Construction Research and Innovation*, 8, 60–65.

Stevens, J. (1996). *Applied Multivariate Statistics for the Social Sciences*, 3rd ed.; Lawrence Erlbaum: Mahway, NJ.

Stone, M. (1974). Cross-validatory choice and assessment of statistical predictions. *Journal of the Royal Statistical Society*, 36, 111–133.

Straub, D., Boudreau, M.-C., & Gefen, D. (2004). Validation guidelines for IS positivist research. *Communications of the Association for Information Systems*, 13, 24.

Su, D.N., Nguyen-Phuoc, D.Q., & Johnson, L.W. (2019). Effects of perceived safety, involvement and perceived service quality on loyalty intention among ride-sourcing passengers. *Transportation*, 48, 369–393.

Sweet, M.N., & Laidlaw, K. (2019). No longer in the driver's seat: How do affective motivations impact consumer interest in automated vehicles? *Transportation*, 47, 2601–2634.

Tabachnick, B.G., & Fidell, L.S. (2001). *Using Multivariate Statistics*, 4th ed.; HarperCollins: New York, NY.

Talebian, A., & Mishra, S. (2022). Unfolding the state of the adoption of connected autonomous trucks by the commercial fleet owner industry. *Transportation Research Part E: Logistics and Transportation Review*, 158, 102616.

Tyrinopoulos, Y., & Antoniou, C. (2019). Review of factors affecting transportation systems adoption and satisfaction. In *Demand for Emerging Transportation Systems*; Elviser: Amsterdam, pp. 11–36.

Venkatesh, V., & Davis, F.D. (2000). A theoretical extension of the technology acceptance model: Four longitudinal field studies. *Management Science*, 46, 186–204.

Venkatesh, V., Morris, M.G., Davis, G.B., & Davis, F.D. (2003). User acceptance of information technology: Toward a unified view. *Management Information Systems Quarterly*, 27, 425–478.

Wang, T.-L., Tran, P.T.K., & Tran, V.T. (2017). Destination perceived quality, tourist satisfaction and word-of-mouth. *Tourism Review*, 72, 392–410.

Wu, J., Liao, H., Wang, J.-W., & Chen, T. (2019). The role of environmental concern in the public acceptance of autonomous electric vehicles: A survey from China. *Transportation Research. Part F, Traffic Psychology and Behaviour*, 60, 37–46.

Wu, J.-H., & Wang, S.-C. (2005). What drives mobile commerce? An empirical evaluation of the revised technology acceptance model. *Information & Management*, 42, 719–729.

Xu, Z., Zhang, K., Min, H., Wang, Z., Zhao, X., & Liu, P. (2018). What drives people to accept automated vehicles? Findings from a field experiment. *Transportation Research Part C: Emerging Technologies*, 95, 320–334.

Yigitcanlar, T., & Kamruzzaman, M. (2019). Smart cities and mobility: Does the smartness of Australian cities lead to sustainable commuting patterns? *Journal of Urban Technology*, 26, 21–46.

Yigitcanlar, T., Wilson, M., & Kamruzzaman, M. (2019a). Disruptive impacts of automated driving systems on the built environment and land use: An urban planner's perspective. *Journal of Open Innovation: Technology, Market, and Complexity*, 5, 24.

Yigitcanlar, T., Sabatini-Marques, J., da-Costa, E., Kamruzzaman, M., & Ioppolo, G. (2019). Stimulating technological innovation through incentives: perceptions of Australian and Brazilian firms. *Technological Forecasting and Social Change*, 146, 403–412.

Yigitcanlar, T., Corchado, J.M., Mehmood, R., Li, R.Y.M., Mossberger, K., & Desouza, K. (2021). Responsible urban innovation with local government artificial intelligence (AI): A conceptual framework and research agenda. *Journal of Open Innovation: Technology, Market, and Complexity*, 7, 71.

Yigitcanlar, T., Kankanamge, N., Regona, M., Ruiz Maldonado, A., Rowan, B., Ryu, A., & Li, R. (2020). Artificial intelligence technologies and related urban planning and development concepts: How are they perceived and utilized in Australia? *Journal of Open Innovation: Technology, Market, and Complexity*, 6, 187.

Zeithaml, V.A. (1988). Consumer perceptions of price, quality, and value: A means-end model and synthesis of evidence. *Journal of Marketing*, 52, 2–22.

Zhang, T., Tao, D., Qu, X., Zhang, X., Lin, R., & Zhang, W. (2019). The roles of initial trust and perceived risk in public's acceptance of automated vehicles. *Transportation Research Part C: Emerging Technologies*, 98, 207–220.

Zhang, T., Tao, D., Qu, X., Zhang, X., Zeng, J., Zhu, H., & Zhu, H. (2020). Automated vehicle acceptance in China: Social influence and initial trust are key determinants. *Transportation Research Part C: Emerging Technologies*, 112, 220–233.

Zmud, J., Sener, I.N., & Wagner, J. (2016). Self-driving vehicles: Determinants of adoption and conditions of usage. *Transportation Research Record: Journal of the Transportation Research Board*, 2565, 57–64.

3

Bridging the First/Last Mile Gap with Autonomous Vehicles

3.1 Introduction

Bridging the first/last mile gap remains crucial for ensuring reasonable access to public transport (PT) in urban areas (Mohiuddin, 2021). With housing costs in the inner city—close to transport, employment, and cultural hubs—rapidly increasing, many are drawn to fringe suburban areas, where property prices are more affordable (Revington & Townsend, 2016). However, PT is often unable to adequately service these communities. Combined with lower average incomes and longer travel distances, residents are more vulnerable to oil price volatility, and those unable to afford or operate a vehicle risk social and economic exclusion (Scultz et al., 2021).

To promote more affordable, equitable, and sustainable transport many cities have implemented policies aimed at increasing planned density around established centres and transport hubs (Dur & Yigitcanlar, 2015). However, attempts to alter the urban form of cities have been problematic, especially when reliant on market forces (Yigitcanlar & Dur, 2013; Kamruzzaman et al., 2015). Recent challenges have also emerged following the COVID-19 pandemic where working from home (WFH) and a fear of infection in crowded spaces have some predicting PT patronage will never return to normal (Orro et al., 2020). There is a risk that underutilised systems will become redundant, creating greater inequities for those who rely on them (Vickermann, 2021).

Prior to the pandemic and its impact on transport patterns, the evolution of information and communication technologies (ICTs) had influenced the development of a smart city model, which emphasised the role of technology and optimal use of available resources as a fundamental driver of sustainable and knowledge-based planning outcomes (Yigitcanlar et al., 2019; Paiva et al., 2020). In this respect, a focus of smart mobility has been to develop more sustainable transport of which the reduction of private vehicle (PV) use through the establishment of equitable PT services remains fundamental (Jain et al., 2020; Yigitcanlar et al., 2021).

DOI: 10.1201/9781003605676-3

Nonetheless, despite a reduction in PV ownership among younger generations (Schulz et al., 2021), and the potential for technology to affect habitual change (Heyns, 2021), increasing the mode share of PT remains reliant on shifting established behaviours and presenting alternatives that are fundamentally more appealing to the end-user. Furthermore, understanding public attitudes and recognising positive intentions is vital to ensure new mobility solutions do not become an expensive exercise in futility—particularly where the benefits are underappreciated by the broader community (Heyns, 2021; Butler et al., 2022).

This chapter aims to provide insights into user attitudes towards smart mobility including how these attitudes are shaped by location, socio-demographics, and existing travel behaviour. With this aim in mind, the chapter focuses on the question of "how smart mobility can bridge the first/last mile gap". In addition, the purpose of this research is to assist transport and land use planners generating policy and guidelines related to the implementation of smart mobility strategies. The structure of this chapter is as follows: First, a background review of existing literature identifies common strategies, and themes in smart mobility research are undertaken with a focus on bridging the first/last mile gap. Second, our research methodology is outlined including the case study area and survey methods. Third, the results of the survey are analysed including identification of any patterns between attitudes and existing socioeconomic characteristics, travel behaviours, and geographic location. Finally, the chapter concludes with a discussion of the implications for planners and policymakers based on these observations.

3.2 Literature Background

While some research provides a technocentric view of smart mobility—including a focus on the level of integration between ICT and transport systems—others focus on a more outcomes-based approach including the integration of sustainable urban planning theory (So et al., 2020). In this regard, smart mobility has been described as an approach to forming and maintaining a safer and more sustainable transportation system (Maheshwari et al., 2015), with a foundation supported by modern technology and new approaches for improving system efficiency (Paiva et al., 2020; Bıyık et al., 2021; Francini et al., 2021; Scultz et al., 2021). From a sustainability perspective, smart mobility seeks to address transport-related concerns such as emissions, expenditure, time travelled, traffic, safety, accessibility, sprawl, and disadvantage (Tirachini, 2019; Ristvej et al., 2020; Francini et al., 2021; Paz et al., 2013)—including gender equity (Singh, 2020). While technology remains a principal component, retrofitting existing systems remains critical to ensure the development of 'smarter', more efficient networks (Francini et al., 2021). As such, the following strategies aimed at reducing the impact of PV remain central to smart mobility: (a) increasing PT usage; (b) encouraging active and shared transport; (c) cybercommuting; (d) deployment of travel information and incentives; and (e) increasing the capacity of road infrastructure (Ploeger & Oldenziel, 2020; Biyik et al., 2021; Paz & Peeta, 2009).

Given the emphasis on technology, the evolution of intelligent transport systems (ITSs) enhanced by 5G, artificial intelligence (AI), cloud computing, the internet of things (IoT), and sensors are commonly discussed in smart mobility literature (Kelley et al., 2020; Paiva et al., 2020; Biyik et al., 2021; Mahrez et al., 2021). Researchers have highlighted that the increased data collection and communication enabled by ITS, big data modelling, and open information channels (Biyik et al., 2021) will enable a more efficient approach to problem-solving in the face of complex transport systems—including the automation of analytical model building through machine learning (Nikitas et al., 2020; Yigitcanlar & Cugurullo, 2020). Nevertheless, this remains subject to solving data vulnerability and data processing capability concerns (Paiva et al., 2020).

Under this framework, mobility-as-a-service (MaaS)—a service which utilises smartphones to plan, book, and pay for journeys across a range of modes—is a potential platform from which the integration and coordination of services, infrastructure, and real-time data collection and analysis can be implemented and operated (Butler et al., 2021; Field & Jon, 2021). Consequently, MaaS has gained attention for promoting shared multimodality as a means of optimising system efficiency and exploiting a range of developing trends such as AV, passenger drones, ridesharing, and free-floating e-mobility—all while maintaining a strong PT network to form the 'backbone' of the integrated system (Biyik et al., 2021, Santos & Nikolaev, 2021). While the integration of shared transport modes is not a new concept (Ploeger & Oldenziel, 2020), a fully integrated MaaS system would act as a "user-centred architectural framework" (Nikitas et al., 2020) utilising ITS and real-time data collection to provide dynamic, individualised, and context-aware travel plans (Schulz et al., 2021).

3.2.1 Smart Mobility and First/Last Mile Connectivity

The literature identifies the potential to use new mobility technology—including shared autonomous vehicles (SAV) and micro-mobility—to support existing PT networks by acting as a dynamic feeder network, connecting to major PT hubs, and replacing inefficient fixed route bus services (Butler et al., 2020a; Bucchiarone et al., 2021). These 'feeder networks' would be incorporated into an integrated system, such as MaaS, to provide real-time analysis of user and network data—including behavioural changes—to create a journey plan that is more efficient for the user and offers incentives aimed at achieving system optimisation and sustainability goals (Zhang et al., 2020; Bucchiarone et al., 2021; Ahern et al., 2022).

Shared autonomous shuttles (SAS) could be exploited to provide the bulk of these 'feeder networks', essentially replacing lower occupancy bus services, providing greater coverage at reduced costs, and making dynamic route planning and on-demand service provision more viable. Over shorter distances, micro-mobility, including a more widespread implementation of free-floating services (Field & Jon, 2021), ridesharing, and continued development of good quality active transport networks could promote alternatives to PV travel, while directing users towards PT for longer journeys (Butler et al., 2020b; Singh, 2020; Dias et al., 2021). However, these strategies remain subject to good policy and implementation strategies, effective communication among stakeholders (Field & Jon, 2021), user awareness, community support (Butler et al., 2022), and consideration of power grid and smart charging strategies (Vilathgamuwa et al., 2022).

3.2.2 Attitudes and Smart Mobility

Smart mobility feeder systems face many challenges, particularly where user attitudes are likely to influence the uptake of these services. User attitudes can be defined as an individual's positive or negative feelings regarding a particular activity (Acheampong & Cugurullo, 2019). While it is recognised behavioural intention to use is a complex undertaking influenced by a range of factors including social influence, effort, willingness to pay and habit (Haboucha et al., 2017; Golbabaei et al., 2020), understanding attitudes remains of interest in the development of policy and for the assessment of political and social feasibility (Loukopoulos et al., 2005).

To better understand user attitudes, some key themes within smart mobility literature have been identified with a particular focus on where new strategies and technologies have been proposed to improve first/last mile connectivity. Based on this analysis the following themes: (a) technology; (b) smartphones and apps; (c) PT; (d) sharing rides with strangers; (e) multimodality; (f) peer-to-peer (P2) transport; (g) PV use; and (h) environmental consciousness, have been identified as attitudinal factors likely to influence the uptake of smart mobility strategies.

Regarding attitudes towards technology, there remain concerns associated with user trust of new technologies—particularly where online payment is required. Lack of trust is commonly identified as a reason why many are slow to adopt new technology (Golbabaei et al., 2020). Furthermore, trust issues are typically associated with perceived risk, including fear of data breaches and the privacy of personal data, cyber security threats, mass outages, unconscious bias in technology development, familiarity with the technology, and the ethics of automated decision-making (Golbabaei et al., 2020; Nikitas et al., 2020; Rahimi et al., 2020; Dennis et al., 2021).

Continuing from technology, attitudes towards smartphones and apps will also present challenges as many of the new services central to smart mobility planning require access to a smartphone, an internet connection, and the ability to use devices to plan, book, and pay for services requires a certain degree of technical literacy (Alonso-González et al., 2020). Researchers have observed a relationship between 'technological savviness'—including high proficiency in the use of smartphones and the internet—and willingness to use on-demand/ P2P services. Those with low technical literacy, including the elderly, tend to be less likely to embrace new mobility solutions, while younger residents and those high technological affinity are less likely (Haboucha et al., 2017; Alonso-González et al., 2020; Rahimi et al., 2020; Singh, 2020).

Regarding attitudes towards PT, the promotion of shared or active mobility options as a first/last mile feeder network relies on the support and prioritisation of existing PT (Nikitas et al, 2020; Field & John, 2021). Therefore, positive attitudes towards PT remain an important pillar of MaaS and other integrated systems (Alonso-González et al., 2020). However, there continue to be issues associated with maintaining and increasing PT patronage in urban areas— particularly where the weakness of existing systems including coverage inefficiency, overcrowded conditions, and temporal reliability have influenced pre-existing perceptions. In addition, research has shown that individuals who currently avoid PT will be less likely to use the shared services which provide the first/last mile connection—including SAV (Haboucha et al., 2017) while regular users showed increased interest in alternate modes (Rahimi et al., 2020).

Likewise, *attitudes towards sharing rides with strangers* will also be an important challenge for smart mobility planning (Alonso-González et al., 2020). While shared services are often identified as one of the most efficient ways to provide sustainable transport in our cities, these services simply lack mass appeal. In fact, Rahimi et al. (2020) found that around 58.8% of the respondents to their study on shared mobility and autonomous vehicle (AV) expressed a low level of trust related to sharing rides with strangers, with 81.8% feeling that private travel was more convenient (Rahmi et al., 2020). Many users express discomfort over sharing rides with strangers for extended periods (Golbabaei et al., 2020) and when PVs are dominant there remains the need for a major shift in cultural norms (Canitez, 2019). Furthermore, following the COVID-19 pandemic many users have renewed concerns over the safety of shared travel (Scorrano & Daniels, 2021).

In this regard, attitudes towards PV use also present issues. One of the primary goals of smart mobility, and sustainability planning in general, is presenting new alternatives to PV travel (Butler et al., 2021). Notwithstanding the comfort and convenience of PV travel, the habitual and cultural factors associated with owning a vehicle are deeply entrenched in the public consciousness, and loss of 'ownership' rights will present significant challenges particularly when trading off against a more utilitarian perspective (Alonso-González et al., 2020; Nikitas et al., 2020).

Understanding attitudes towards multimodality is an important part of recognising the strengths and weaknesses of an integrated system which provides door-to-door transport while also relying on PT as a central component. The reasoning behind this is that these models typically rely on intermodal options and transferring between modes is central to improving the efficiency, coverage, and flexibility of existing transport networks (Butler et al., 2020a; Kim et al., 2021). Given transferring between modes can be viewed as an inconvenient, unsafe, or stressful step in a trip chain, and car users are less willing to accept multimodality as an option, there remain questions over how user reluctance to transfer between modes will impact the successful implementation of smart mobility strategies (Kim et al., 2021).

Regarding attitudes towards peer-to-peer transport, the recent rise of ridesharing services (such as Uber/OLA/Didi) has been identified as a means of providing a cost-effective first/last mile connection to complement existing PT (Boarnet et al., 2017) while also substituting and providing greater coverage than traditional lower occupancy bus services (Tirachini, 2019). This is evidenced by the reality that P2P ridesharing is becoming one of the most revenue-generating sectors of the sharing economy (Bucchiarone et al., 2021) and those who use these services have been identified as the more likely to adopt new mobility solutions—such as MaaS (Alonso-González et al., 2020). While some advantages of P2P ridesharing include affordability, fare transparency, dynamic routing, responsiveness, ease of payment, productivity, reduced travel time, and global availability, some have expressed concern regarding both personal and data privacy, resource unavailability, and competition across modes (Contreras & Paz, 2018; Neunhoeffer & Teubner, 2018; Lavieri & Bhat, 2019; Hunecke et al., 2021).

Finally, regarding attitudes towards the environment research has shown that those with low sensitivity to environmental concerns are less likely to move away from PV towards new mobility solutions that incorporate shared services (Alonso-González et al., 2020). Conversely, those who express higher levels of concern are more likely to utilise shared services (Haboucha et al., 2017; Li & Kamargianni, 2020). Understanding which population segments have more potential for environmental consciousness can be utilised for targeted advertising campaigns aimed at increasing shared mobility and PT use in urban areas (Li & Kamargianni, 2020).

3.3 Research Design

3.3.1 Case Study Area

This study adopted a case study approach to investigate public attitudes towards smart mobility using the major Australian cities of Brisbane,

Sydney, and Melbourne. Before population growth stagnated because of international travel restrictions during the COVID-19 pandemic, Melbourne (2.3%), Brisbane (2.1%), and Sydney (1.7%) had the highest growth rates of any capital city in Australia. They accounted for 84% of all capital city growth and 66% of all population growth in Australia. Despite policy for infill and transit-oriented development much of the growth in these cities is happening in lower-density fringe suburbs (ABS, 2021). This is primarily due to affordability and preference for living in detached housing. As a result, car dependency and infrastructure demand are increasing. This in turn makes the provision of equitable PT difficult. Furthermore, greater travel distances and inadequate access to PT are resulting in higher fuel costs and increased travel time. These challenges are further intensified by rising fuel costs and recent housing affordability issues with Sydney (26.1%), Melbourne (20.6%), and Brisbane (20.3%) all showing significant growth between September 2018 and 2021 (ABS, 2021).

To differentiate the various zones relative to the central business district (CBD) and thus analyse how attitudes towards smart mobility may be shaped by location, each of the cities was divided into three rings using a method derived from Coffee et al. (2016). Using GIS software, the inner-city (or inner ring) was created based on a 5-km buffer measured from the centroid of the CBD. Similarly, the middle ring was utilised a 5 km–15 km from the CBD and outer ring in any areas that were greater than 15 km.

3.3.2 Survey Questionnaire

An online questionnaire was utilised to analyse attitudes towards smart urban mobility and how personal and spatial factors may influence these results. This was considered an appropriate research method allowing the collection of a substantial amount of data from a large population sample. The survey was divided into three parts and prior to launching the host university granted ethical clearance. Following an ethics approval from the University Ethics Committee, the survey was completed and distributed to respondents using the Qualtrics XM online survey platform.

Services of a professional survey company were acquired to identify and recruit participants who live in the targeted study areas—determined by the postcodes of each case city's inner, middle, and outer rings. The survey link was sent to individuals' email addresses in September 2021 by the selected professional survey company (n = 2,289). The responses were checked for completeness and cases with missing data were excluded from further analysis. This resulted in 618 valid responses, which reduced the response rate to about 27%.

The first section of the questionnaire requested personal details, including respondents' postcode, age, gender, education, income, and dwelling type. The second section requested the respondent's current transport habits, including the number of vehicles per household, whether they have any disabilities that impact mobility, valid driver's licence, their main transport mode, daily commute time, and PT usage.

The third part of the questionnaire focused on attitudes towards smart mobility. A five-point Likert scale was employed with options ranging from strongly agree to strongly disagree. Questions were categorised into eight groups based on the key smart mobility themes identified from the literature, i.e., attitudes towards technology, smartphones and apps, PT, sharing rides with strangers, PVs, multimodality, P2P transport, and the environment. Each group of questions represented a single 'attitudinal variable'.

While the overall research question of "how smart mobility can bridge the first/last mile gap" was kept in mind, specific sub-questions are used for part three of the survey, and the corresponding hypotheses behind the attitudinal variables are shown in Table 3.1. These hypotheses, research, and survey questions have been assembled and inspired following a review of similar research on attitudes to AVs (Haboucha et al., 2017; Acheampong & Cugurullo, 2019; Yuen et al. 2020), shared mobility (Rahimi et al., 2020; Li & Kamargianni, 2020), multimodality (Kim et. al., 2021), and MaaS (Alonso-González et al., 2020).

Prior to undertaking the final analysis, all negatively worded questions (–) under hypotheses 1–7 were reverse coded to ensure a high-value response on all items indicated the same type of response. Similarly, as hypothesis eight is itself a negative indicator of smart mobility attitudes only positive worded questions (+) were reverse coded. By recoding the data in this way, a positive value under each category would be a positive indicator of smart mobility acceptance.

TABLE 3.1

Attitudinal Questions and Research Hypotheses

Question
(H1) A positive attitude regarding technology will positively impact smart mobility acceptance
The possibility offered by modern technologies excites me. (+)
I am sceptical that modern technology will contribute to a better future. (+)
I will often try new products before my friends and family. (+)
When it comes to the latest technology and products, I generally know more than others. (+)
New technology does not interest me. (–)
The problems created by new technology create as many problems as it does solutions. (–)
(H2) A positive attitude regarding public transport will positively impact smart mobility acceptance
Public transport is appealing to me because I can use my time productively (e.g., answering emails, working) or recreationally (e.g., reading, watching videos) (+)
Public transport is unreliable. (–)
I feel safe taking public transport. (–)

TABLE 3.1 (Continued)

Attitudinal Questions and Research Hypotheses

Question

Information regarding public transport services and scheduling is difficult to find and understand. (–)

Public transport is often overcrowded and uncomfortable. (–)

I travel by public transport because it is more affordable. (+)

(H3) A positive attitude regarding shared trips will positively impact smart mobility acceptance

Sharing rides with strangers is unreliable. (–)

Sharing rides with strangers is uncomfortable. (–)

Sharing rides with strangers increases my vulnerability and exposure to crime. (–)

Sharing rides with strangers increases my risk of contracting COVID-19 and other diseases. (–)

I do not mind sharing rides with strangers if it reduces my travel costs. (+)

When travelling privacy is important to me. (–)

(H4) A positive attitude regarding multimodality will positively impact smart mobility acceptance

Combining multiple modes of transportation is an effective way to improve the efficiency of my trips. (+)

Routes that combine multiple modes of transportation are inconvenient. (–)

Transferring between transport modes makes me uncomfortable. (–)

Transferring between transport modes increases the potential for delays. (–)

When planning my commute, I like to compare various options before making my choice. (+)

I find transferring between transport modes stressful. (–)

(H5) A positive attitude regarding peer-to-peer transport will positively impact smart mobility acceptance

I often use peer-to-peer transport apps. (+)

The use of peer-to-peer transport apps is a positive thing. (+)

I feel unsafe using peer-to-peer transport apps. (–)

The use of peer-to-peer transport apps increases my risk of contracting COVID-19. (–)

I do not understand how to access peer-to-peer services like Uber. (–)

Peer-to-peer services such as Uber should be banned (–)

(H6) A positive attitude regarding the smartphones and mobile apps will positively impact smart mobility acceptance

I regularly use smartphone apps. (+)

I would be willing to use a smartphone to book and pay for trips. (+)

It is easy to learn how to use new smartphone apps. (+)

Having internet connectivity everywhere I go is essential to me. (+)

My smartphone makes accessing travelling easier. (+)

(H7) A positive attitude regarding the environment will positively impact smart mobility acceptance

I am concerned about environmental issues such as global warming. (+)

I am consistently looking for new ways to conserve energy in my daily life. (+)

I am consistently looking for new ways to conserve energy and other resources while travelling. (+)

It is important to me to make environmentally friendly transportation choices. (+)

(continued)

TABLE 3.1 (Continued)

Attitudinal Questions and Research Hypotheses

Question
I rarely change my behaviour based on concern for the environment only. (–)
The price of petrol should be raised to combat the negative impacts private vehicles have on the environment. (+)
(H8) A negative attitude regarding private vehicle travel will positively impact smart mobility acceptance
I prefer travelling by private car because it feels safer. (+)
I prefer travelling by private car because I enjoy driving. (+)
I prefer travelling by private car because it is more reliable. (+)
I prefer travelling by private car because it is more comfortable. (+)
I often feel nervous when driving. (–)
I rarely consider travelling in anything but a private vehicle. (+)

3.4 Analysis and Results

3.4.1 Descriptive Analysis

Following the elimination of incomplete data, a total of 618 valid responses were received (27% response rate). The responses were distributed relatively equally among the three cities included in the case study area—Sydney (n = 210), Melbourne (n = 211), and Brisbane (n = 197), and among inner (n = 196), middle (n = 221), and outer rings (n = 201). Salient characteristics of study participants and census info for their locations (cities) are provided in Table 3.2, and a descriptive summary of all responses is shown in Table 3.3.

The responses relating to personal characteristics and transport habits are shown in Table 3.3. Based on these responses inner-city areas had a higher number of residents under the age of 49 (61.7%), compared to middle (59.3%), and outer rings areas (47.2%). Similarly, inner-city areas also have more residents with bachelor and post-graduate degrees (60.2%), and middle-high and high incomes (35.2%). Attached housing (such as units and apartments) is also more common in inner-city areas (53.1%) indicating increased population density around the core of each of the cities which may contribute to the high levels of active transport (21.9%) when compared to middle (7.2%) and outer ring areas (10.9%).

Conversely, when looking at factors that may influence equity and accessibility, we see that disability and lack of driver's licence are higher in the middle ring (10.9%, 10.9%) and outer ring areas (13.4%, 13.9%). However, fears of PT due to safety concerns and ability to access information are higher in the inner (42.3%, 28.1%) and middle ring areas (42.5%, 48.4%, 24%) though this may be due to significantly higher PT use in inner (21.9%) and middle ring areas (17.6%). Interestingly, daily commute is higher in middle ring areas

TABLE 3.2

Salient Characteristics of Study Participants and Their Locations

	All Responses		Sydney Combined			Melbourne Combined			Brisbane Combined		
	Freq (N)	Survey (%)	Freq (N)	Survey (%)	Census (%)	Freq (N)	Survey (%)	Census (%)	Freq (N)	Survey (%)	Census (%)
What is your gender?											
Male	278	45.0	100	47.6	49.4	96	45.5	49.2	82	41.6	49.2
Female	337	54.5	109	51.9	50.6	114	54.0	50.8	114	57.9	50.8
Non-binary	1	0.2	0	0.0	–	0	0.0	–	1	0.5	–
Prefer not to say	2	0.3	1	0.5	–	1	0.5	–	0	0.0	–
What is your age?											
18–29	115	18.6	43	20.5	18.5	37	17.5	19.0	35	17.8	19.1
30–49	232	37.5	80	38.1	38.8	83	39.3	38.9	69	35.0	38.1
50–69	203	32.8	64	30.5	28.6	76	36.0	28.1	63	32.0	29.0
70+	68	11.0	23	11.0	14.1	15	7.1	14.0	30	15.2	13.8
What is the highest level of education you have completed?											
High school or less	154	24.9	52	24.8	–	53	25.1	–	49	24.9	–
Certificate or diploma	183	29.6	61	29.0	–	55	26.1	–	67	34.0	–
Bachelor's degree	177	28.6	57	27.1	–	68	32.2	–	52	26.4	–
Post-graduate	104	16.8	40	19.0	–	35	16.6	–	29	14.7	–
What is your current employment status?											
Unemployed/retired	152	24.6	52	24.8	–	43	20.4	–	57	28.9	–
Unpaid employment	56	9.1	15	7.1	–	19	9.0	–	22	11.2	–
Full-time student	31	5.0	13	6.2	–	8	3.8	–	10	5.1	–
Part-time employment	112	18.1	41	19.5	–	36	17.1	–	35	17.8	–
Full-time/self-employed	267	43.2	89	42.4	–	105	49.8	–	73	37.1	–

(continued)

TABLE 3.2 (Continued)

Salient Characteristics of Study Participants and Their Locations

	All Responses		Sydney Combined			Melbourne Combined			Brisbane Combined		
	Freq (N)	Survey (%)	Freq (N)	Survey (%)	Census (%)	Freq (N)	Survey (%)	Census (%)	Freq (N)	Survey (%)	Census (%)
What is your approximate annual income before tax?											
Low income	263	42.6	93	44.3	54.7	79	37.4	56.8	91	46.2	47.6
Low-mid. income	170	27.5	60	28.6	23.2	53	25.1	23.8	57	28.9	29.7
Mid.-high income	122	19.7	36	17.1	14.5	49	23.2	13.4	37	18.8	16.3
High income	63	10.2	21	10.0	7.5	30	14.2	6.0%	12	6.1	6.3
What best describes your household dwelling?											
Detached	366	59.2	129	61.4	55.8	124	58.8	67.8	113	57.4	76.4
Semi-detached	73	11.8	23	11.0	12.8	28	13.3	16.2	22	11.2	10.0
Attached	175	28.3	56	26.7	30.7	58	27.5	15.6	61	31.0	12.6
Other	4	0.6	2	1.0	0.7	1	0.5	0.4	1	0.5	1.0

Note: All census data has been taken from the 2021 Australian census. Education level and employment data from the 2021 census were not yet publicly available. Increments used for income levels for the survey did not exactly match the increments used in census data. Furthermore, the census data used weekly income not yearly income. The closest approximation was used so that low income which was 0–$50,000 p/a in the survey is shown as 0–$1000 p/w in the census data, low-middle income is $50–$100,000 p/a vs. $1000–$1750 p/w in the census, middle-high income is $90–$140,000 vs. $1,750–$2,999 p/w, and high income $140,000+ p/a vs. $3000+ p/w. Similarly, the age variable for the census is provided in 5-year increments, e.g., 15–19 years and 20–24 years. As the survey did not include respondents below the age of 18 years, the above census (%) column does not include data from the 15- to 19-year age group. This may have resulted in minor differences between the census (%) column and the survey (%) as census respondents aged 18–19 years are not included in the calculated percentage (ABS, 2021).

TABLE 3.3

Descriptive Statistics

	All Responses		Inner Ring Combined		Middle Ring Combined		Outer Ring Combined	
	N	%	N	%	N	%	N	%
What is your gender?								
Male	278	45.0	95	48.5	99	44.8	84	41.8
Female	337	54.5	101	51.5	120	54.3	116	57.7
Non-binary	1	0.2	0	0.0	1	0.5	0	0.0
Prefer not to say	2	0.3	0	0.0	1	0.5	1	0.5
What is your age?								
18–29	115	18.6	40	20.4	45	20.4	30	14.9
30–49	232	37.5	81	41.3	86	38.9	65	32.3
50–69	203	32.8	59	30.1	64	29.0	80	39.8
70+	68	11.0	16	8.2	26	11.8	26	12.9
What is the highest level of education you have completed?								
High school or less	154	24.9	32	16.3	61	27.6	61	30.3
Certificate or diploma	183	29.6	46	23.5	72	32.6	65	32.3
Bachelor's degree	177	28.6	73	37.2	56	25.3	48	23.9
Post-graduate	104	16.8	45	23.0	32	14.5	27	13.4
What is your current employment status?								
Unemployed/retired	152	24.6	41	20.9	55	24.9	56	27.9
Unpaid employment	56	9.1	7	3.6	16	7.2	33	16.4
Full-time student	31	5.0	9	4.6	11	5.0	11	5.5
Part-time employment	112	18.1	38	19.4	43	19.5	31	15.4
Full-time/self-employed	267	43.2	101	51.5	96	43.4	70	34.8
What is your approximate annual income before tax?								
Low income	263	42.6	68	34.7	94	42.5	101	50.2
Low-middle income	170	27.5	59	30.1	65	29.4	46	22.9
Middle-high income	122	19.7	46	23.5	39	17.6	37	18.4
High income	63	10.2	23	11.7	23	10.4	17	8.5
What best describes your household dwelling?								
Detached	366	59.2	65	33.2	150	67.9	151	75.1
Semi-detached	73	11.8	27	13.8	29	13.1	17	8.5
Attached	175	28.3	104	53.1	41	18.6	30	14.9
Other	4	0.6	0	0.0	1	0.5	3	1.5
Do you have any disabilities that impact your mobility?								
Yes	68	11.0	16	8.2	24	10.9	28	13.9
No	550	89.0	180	91.8	197	89.1	173	86.1
Do you have a valid driver's licence?								
Yes	550	89.0	179	91.3	197	89.1	174	86.6
No	68	11.0	17	8.7	24	10.9	27	13.4
How many private vehicles does your household own?								
0	64	10.4	33	16.8	19	8.6	12	6.0
1	346	56.0	121	61.7	115	52.0	110	54.7

(*continued*)

TABLE 3.3 (Continued)

Descriptive Statistics

	All Responses		Inner Ring Combined		Middle Ring Combined		Outer Ring Combined	
	N	%	N	%	N	%	N	%
2	159	25.7	37	18.9	60	27.1	62	30.8
3+	49	7.9	5	2.6	27	12.2	17	8.5
What is the main type of transport you use during the week (Monday to Friday)?								
Private vehicle	417	67.5	103	52.6	160	72.4	154	76.6
Public transport	100	16.2	43	21.9	39	17.6	18	9.0
Taxi/rideshare	12	1.9	6	3.1	3	1.4	3	1.5
Active transport	79	12.8	43	21.9	16	7.2	20	10.0
Electric mobility	7	1.1	1	0.5	3	1.4	3	1.5
Other	3	0.5	0	0.0	0	0.0	3	1.5
Please provide an estimate of your daily commute time (Monday to Friday)								
<15 min	208	33.7	76	38.8	50	22.6	82	40.8
15–30 min	175	28.3	49	25.0	71	32.1	55	27.4
30–60 min	145	23.5	44	22.4	64	29.0	37	18.4%
>60 min	90	14.6	27	13.8	36	16.3	27	13.4
How much on average do you spend on transport as a percentage of your income?								
0–10%	401	64.9	125	63.8	135	61.1	141	70.1
10–30%	164	26.5	49	25.0	69	31.2	46	22.9
>30%	53	8.6	22	11.2	17	7.7	14	7.0
Approximately how far is the nearest bus stop to your home?								
<400 m	423	68.4	145	74.0	153	69.2	125	62.2
400–800 m	111	18.0	35	17.9	35	15.8	41	20.4
800–1200 m	49	7.9	11	5.6	21	9.5	17	8.5
>1200 m	35	5.7	5	2.6	12	5.4	18	9.0
Approximately how far is the nearest train station to your home?								
<400 m	125	20.2	53	27.0	44	19.9	28	13.9
400–800 m	128	20.7	60	30.6	41	18.6	27	13.4
800–1200 m	112	18.1	40	20.4	39	17.6	33	16.4
>1200 m	253	40.9	43	21.9	97	43.9	113	56.2
How many days a week (on average) do you use public transport?								
0	322	52.1	66	33.7	118	53.4	138	68.7
1–3 days	191	30.9	78	39.8	70	31.7	43	21.4
> 4 days	105	17.0	52	26.5	33	14.9	20	10.0
Do you ever feel unsafe when using or waiting for public transport?								
Yes	251	40.6	83	42.3	94	42.5	74	36.8
No	367	59.4	113	57.7	127	57.5	127	63.2
Do you experience difficulties accessing information regarding public transport schedules?								
Yes	151	24.4	55	28.1	53	24.0	43	21.4
No	467	75.6	141	71.9	168	76.0	158	78.6

with over 45.3% of all respondents averaging over 30 min, compared to the inner (36.2%) and outer rings (31.8%).

Congestion within the city core may be a factor, and increased PT use may increase commute time particularly when travelling by bus. Furthermore, spatial proximity to other subregional centres on the fringes of the statistical catchments may result in less travel focus to the city centre for employment, services, and other needs. Transport cost was also higher in middle ring areas with 48.9% of all respondents spending more than 30% of their income on transport compared to inner (36.2%) and outer ring areas (29.9%). The cost disparity may reflect longer commutes in inner-city and outer ring areas due to congestion. However, it may also be a result of respondents underestimating costs associated with PVs while overestimating costs associated with PT (Gardner & Abraham, 2007).

Most respondents in outer (68.7%) and middle ring suburbs (53.4%) use PT less than one day per week. Given outer and middle ring areas have been identified for lower incomes and accessibility issues such as disability and lack of driver's licence, these results are problematic and likely contribute to increased socioeconomic disadvantage as residents are increasingly dependent on PV travel. This issue is exacerbated by the density changes across the city which result in increased travel distances and less attractive environments for active transport, in addition to reduced PT coverage in middle and outer suburbs. While greater than 50% of all respondents are within 400 m walking distance to a bus stop when looking at trains—which typically provide faster, more direct routing—distance increased from 42.3% of inner-city residents living within further than 800 m of a train station to 61.5% and 72.6% of those in the middle and outer rings.

3.4.2 Confirmatory Factor Analysis

Confirmatory factor analysis (CFA) is a particular form of factor analysis, it is a multivariate technique that uses empirical data to confirm a theoretical model. It is part of the larger multivariate technique of structural equation modelling. Individual confirmatory component analyses (CFA) were conducted for each set of questions collected under eight attitudinal categories. In addition, a correlated CFA also has been performed. The purpose of CFA was to verify the factor structure of a set of observed variables (or attitudinal factors) for use in further statistical analysis. As part of CFA, validity and reliability tests were conducted using RMSEA with P-values ≤ 0.5, and the comparative fit index (CFI) was considered very good for values equal to or greater than 0.95, good between 0.9 and 0.95, suffering between 0.8 and 0.9, and bad for values less than 0.8. The Tucker-Lewis Index (TLI) was considered very good if it is equal to or greater than 0.95, good between 0.9 and 0.95, suffering between 0.8 and 0.9, and bad if it is less than 0.8.

TABLE 3.4

The Assessment for Validity

	Technology	Public Trans.	Sharing	Multimodality
(p-value)	0.000	0.000	0.000	0.000
CFI	1.000	0.979	0.978	0.976
TLI	1.000	0.936	0.955	0.952
RMSE	0.000	0.090	0.098	0.013
	Peer-to-Peer	Smartphones	Environment	Private Vehicle
(p-value)	0.000	0.000	0.000	0.000
CFI	0.863	0.972	0.969	0.981
TLI	0.589	0.954	0.938	0.962
RMSE	0.1261	0.102	0.123	0.093

The confirmatory factor analysis was conducted by using the Lavaan package. The R package Foreign was used to access the questionnaire data. For each attitudinal category, a single factor was extracted and only items with significant factor loading variables λ from CFA were (p-value <0.5) considered. The goodness of fit from the previous measurement using the significant loading factors is shown in Table 3.4, while the retained items and final factor loading scores are shown in Table 3.5. In general, the model is appropriate for different mode attributes where all the p-values are significant and the CFI >0.8. Notably, for the technology attribute the model fit perfectly with CFI = 1 and RMSE equivalent to zero.

Following the completion of CFA, a composite score was created for each attitudinal category by calculating the means of the remaining items. To test the suitability of using the mean to calculate the composite score a loading factor analysis based on the loading factors identified in Table 3.5 was performed. This loading factor was tested against the means using Spearman's Correlation. A score greater than 0.989 (p = <0.01) for all variables indicated a high degree of correlation between the two methods. Due to the high level of correlation, it was decided to use the mean to calculate the composite score as it was a more straightforward dimension reduction technique.

Figure 3.1 shows the distribution of the composite scores for each of the attitudinal categories. The boxplot displays variation in the composite score across the survey sample. The results show that among all survey participants, negative attitudes towards PV remain low (M = 2.2, skewness = 0.768). This indicates a high dependency and appreciation of PV and represents the strongest barrier against future smart mobility implementation. Similarly, negative attitudes against sharing rides with strangers (M = 2.23, skewness = 0.566) and multimodality (M = 2.65, skewness = 0.230) also represent significant barriers. Conversely, positive attitudes towards smartphones (M = 3.5426, skewness = –0.667), environment (M = 3.34,

TABLE 3.5

Component Matrix (Loading Variable λ)

	Technology	Public Trans.	Sharing	Multimodality
The possibilities offered by modern technology excite me.	0.521			
I will often try new products before my friends and family.	1.007			
When it comes to the latest technology and products, I generally know more than others.	1.044			
Public transport is appealing to me because I can use my time productively.		0.917		
I feel safe taking public transport.		0.633		
Information regarding public transport services and scheduling is difficult to find and understand.		−0.195		
I travel by public transport because it is more affordable.		0.957		
Sharing rides with strangers is unreliable.			0.771	
Sharing rides with strangers is uncomfortable.			0.818	
Sharing rides with strangers increases my vulnerability and exposure to crime.			0.780	
Sharing rides with strangers increases my risk of contracting COVID-19 and other diseases.			0.727	
When travelling privacy is important to me.			0.573	
Routes that combine multiple modes of transportation are inconvenient.				0.662
Transferring between transport modes makes me uncomfortable.				0.885
Transferring between transport modes increases the potential for delays.				0.536
When planning my commute, I like to compare various options before making my choice.				−0.214
I find transferring between transport modes stressful.				0.903

(continued)

TABLE 3.5 (Continued)

Component Matrix (Loading Variable λ)

	Peer-to-Peer	Smartphones	Environment	Private Vehicle
I feel unsafe using peer-to-peer transport apps.	0.789			
The use of peer-to-peer transport apps increases my risk of contracting COVID-19.	0.482			
I do not understand how to access peer-to-peer services like Uber.	0.804			
Peer-to-peer services such as Uber should be banned.	0.937			
I regularly use smartphone apps.		1.018		
I would be willing to use a smartphone to book and pay for trips.		1.005		
It is easy to learn how to use new smartphone apps.		0.861		
My smartphone makes accessing transport easier.		0.910		
Having internet connectivity everywhere I go is essential to me.		1.030		
I regularly access online journey planners on my smartphone.		0.945		
I am concerned about environmental issues such as global warming.			0.821	
I am consistently looking for new ways to conserve energy in my daily life.			0.779	
I am consistently looking for new ways to conserve energy and other resources while travelling.			0.918	
It is important to me to make environmentally friendly transportation choices.			0.873	
The price of petrol should be raised to combat the negative impacts private vehicles have on the environment.			0.551	
I prefer travelling by private car because it feels safer.				0.800
I prefer travelling by private car because I enjoy driving.				0.678
I prefer travelling by private car because it is more reliable.				0.819
I prefer travelling by private car because it is more comfortable.				0.854
I rarely consider travelling in anything but a private vehicle				0.598

Note: All factors were extracted using the confirmatory factor analysis method.

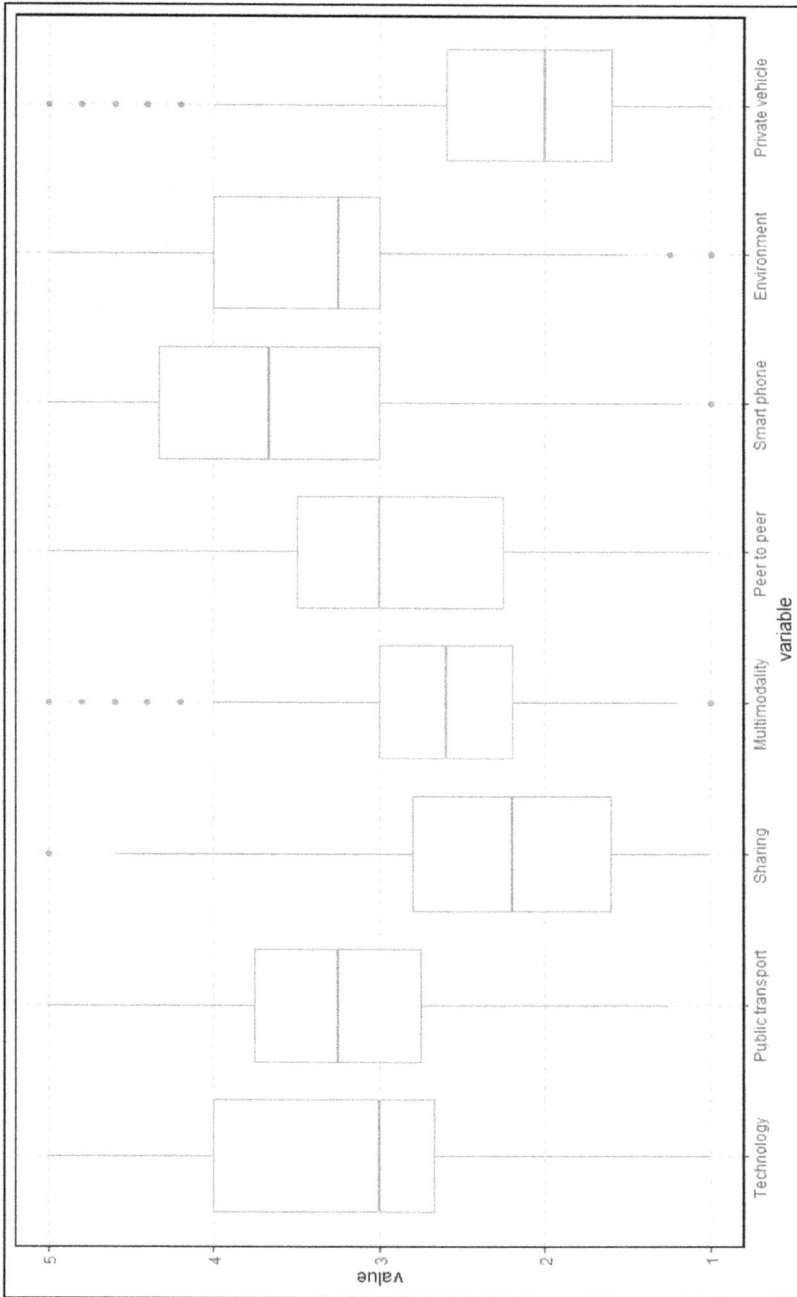

FIGURE 3.1
Composite score box plots.

skewness = –0.355), and to a lesser extent PT (M = 3.21, skewness = –0.255) represent opportunities for smart mobility. Attitudes towards technology and P2P transport are more equally distributed with a negative skew for technology (–0.044) and towards P2P (–0.172).

As a final test the established attitudinal variables were tested for redundancy using Pearson's Correlation removing any factor with a moderate to strong correlation coefficient (i.e., >0.500). As shown in Table 3.6, there are no moderate to strong correlations between the resulting variables.

The low score for negative attitudes to PV across all parts of the case study area highlights respondents' preference for PV as a mode of transport and represents a significant barrier to any mobility plan with a goal to reduce PV use. This conclusion is supported by other research including Scorrano and Danielis (2021) who found the majority preferred PV as a mode of transport. The specific advantages highlighted in the responses including, the perceived safety, the reliability of door-to-door transport, increased comfort, and the enjoyment and habitual nature of PV travel reflect similar results from a series of interviews conducted by Beirão and Sarsfield Cabral (2007) who stress how the personal attachment, freedom, and comfort of PV travel provides a fundamental basis for perception of the mode—in addition to other advantages including speed, flexibility, expectation, and privacy.

Further challenges to smart mobility planning identified in all locations include a general aversion to: (a) sharing rides with strangers including concerns regarding the unreliability of shared services, discomfort, and lack of privacy and vulnerability to crime and infection, and (b) multimodality, including concerns regarding inconvenience, stress, and potential for delays when transferring between modes. Despite negative attitudes towards sharing and multimodality most respondents view PT favourably, especially when the focus is on positive attributes such as affordability, safety, and ability to use time productively. Further similarities are identified in Beirão and Sarsfield Cabral (2007) who also identify additional social, recreational, and psychological benefits of PT.

Similar opportunities highlighted on a city-wide scale include positive attitudes towards the use of smartphones to book and pay for transportation and a generally high environmental consciousness including growing concerns regarding global warming and the importance of environmentally friendly transportation choices. These results highlight potential opportunities for promoting and introducing new transport modes into urban areas and reflect Haboucha et al. (2017) and Rahimi et al. (2020) who identify that groups who express concern for the environment and/or have greater technological savviness—including the use of smartphone apps—are more likely to utilise shared transport.

This study employed confirmatory factor analysis to assess items measured individually in Table 3.5 and the correlation between different mode attributes. Using the results from Table 3.5, the study constructed CFA using

TABLE 3.6

Correlation Matrix: Composite Factor Scores

		Technology	Public Trans.	Sharing	Multimodality	Peer-to-Peer	Smartphones	Environment	Private Vehicle
Technology	Pearson correlation	1							
	N	618							
Public transport	Pearson correlation	0.191**	—						
	Sig. (2-tailed)	0.000							
	N	618	618						
Sharing	Pearson correlation	−00.216**	0.023	—					
	Sig. (2-tailed)	0.000	0.563						
	N	618	618	618					
Multimodality	Pearson correlation	−0.135**	0.271**	0.393**	—				
	Sig. (2-tailed)	0.000	0.000	0.000					
	N	618	618	618	618				
Peer-to-peer	Pearson correlation	−0.149**	0.002**	0.412**	0.325**	—			
	Sig. (2-tailed)	0.000	0.000	0.000	0.002				
	N	618	618	618	618	618			
Smartphones	Pearson correlation	0.567**	0.271**	−0.230**	−0.124**	0.030	—		
	Sig. (2-tailed)	0.000	0.000	0.000	0.002	0.458			
	N	618	618	618	618	618	618		
Environment	Pearson correlation	−0.448	0.309**	−0.267**	−0.116**	−0.270**	0.418**	—	
	Sig. (2-tailed)	0.490	0.000	0.000	0.000	0.000	0.000		
	N	618	618	618	618	618	618	618	
Private vehicle	Pearson correlation	−1.79**	0.157**	0.492**	0.393**	0.340**	−0.200**	−0.182**	—
	Sig. (2-tailed)	0.109	0.070	0.000	0.000	0.000	0.000	0.000	
	N	618	618	618	618	618	618	618	618

** Correlation is significant at the 0.01 level (2-tailed).

* Correlation is significant at the 0.05 level (2-tailed).

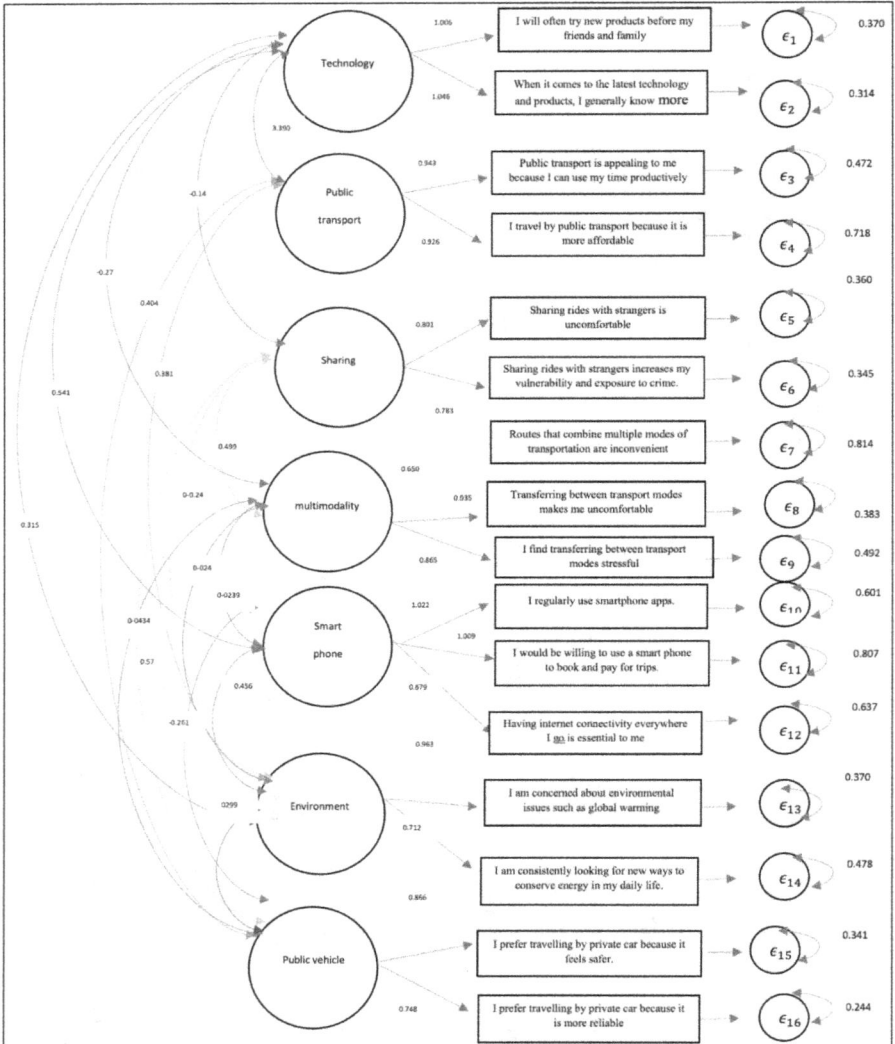

FIGURE 3.2
Correlated CFA.

37 items. By using the same sample of 618 respondents, we validate the measurement model using the components mentioned above. Results from CFA using 37 items presented a model with p <0.0001, CFI = 0.733, TLI = 0.707, and RMSEA=0. 090. This means that the model needs some adjustments. In this sense, variables with less contribution to its factor were removed from the model. Figure 3.2 shows the final correlated CFA measurement model after removing some variables. In this model p <0.0001, CFI = 0.963, TLI = 0.947, and RMSEA = 0. 052. A better fit for the model is noticed when we consider the correlation between mode attributes shown in Table 3.6.

3.4.3 Inferential Analysis

A Kruskal-Wallis H-Test was conducted to analyse the difference between the resulting attitudinal variables created through CFA and the location relative to the city centre. The analysis found statistically significant relationships between location and positive attitudes towards PT ($H = 11.196$, $p = <0.001$), smartphones ($H = 9.929$, $p = 0.007$), and environment ($H = 10.856$, $p = 0.005$) and negative attitudes towards PVs ($H = 14.594$, $p = 0.001$). However, no statistically significant relationship was identified between location and positive attitudes towards technology ($H = 7.87$, $p = 0.2$), sharing rides with strangers ($H = 2.031$, $p = 0.362$), multimodality ($H = 0.606$, $p = 0.739$), and P2P services ($H = 2.662$, $p = 0.264$), (Table 3.7).

Following the completion of the Kruskal-Wallis H-Test and the subsequent identification of all statistically significant differences between attitudinal variation and location relative to the city centre, Dunn's post hoc test was used to calculate multiple pairwise comparisons with Bonferroni correction performed for each relationship. With regard to positive attitudes towards PT, the post hoc analysis results (Table 3.8) reveal statistically significant relationships between the outer ring (mean rank = 284.63) and the inner ring (mean rank = 342.79) ($p = 0.001$), and the middle ring (mean rank = 284.63) and inner ring (mean rank = 342.79) ($p = 0.021$), but not between the middle and outer rings. In both cases, the mean rank of the inner ring is higher than the middle and outer rings. These scores reveal that positive attitudes toward PT are significantly higher in the inner ring when compared to the middle and outer rings.

With regard to positive attitudes towards smartphones, the post hoc analysis results (Table 3.8) reveal statistically significant relationships between the outer ring (mean rank = 284) and the inner ring (mean rank = 339.92) ($p = 0.002$) and the middle (mean rank = 305.71) and inner ring (mean rank = 339.92) ($p = 0.05$) but not between the middle and outer rings. In both cases the mean rank of the inner ring is higher than the middle and outer rings. These scores reveal that positive attitudes towards smartphones are significantly higher in the inner ring when compared to the middle and outer rings.

With regard to positive attitudes towards the environment, the post hoc analysis results (Table 3.8) reveal statistically significant relationships between the middle ring (mean rank = 294.70) and the inner ring (mean rank = 343.43) ($p = 0.004$) and the outer ring (mean rank = 292.69) and inner ring (mean rank 343.43) ($p = 0.005$) but not between the middle and outer rings. In both cases, the mean rank of the inner ring is higher than the middle and outer rings. These scores reveal that positive attitudes towards the environment are significantly higher in the inner ring when compared to the middle and outer rings.

With regard to negative attitudes towards PVs, the post hoc analysis results (Table 3.8) reveal statistically significant relationships between the outer ring

TABLE 3.7

Kruskal-Wallis H Analysis

	H	Sig	N	Mean Rank
Positive attitude towards technology				
Inner ring suburb	7.87	0.2	196	337.48
Middle ring suburb			221	303.62
Outer ring suburb			201	289.17
Total			618	309.5
Positive attitude towards public transport				
Inner ring suburb	11.196	<0.001	196	342.79
Middle ring suburb			221	302.60
Outer ring suburb			201	284.63
Total			618	309.5
Positive attitude towards sharing rides with strangers				
Inner ring suburb	2.031	0.362	196	321.85
Middle ring suburb			221	310.40
Outer ring suburb			201	296.46
Total			618	309.5
Positive attitude towards multimodality				
Inner ring suburb	0.606	0.739	196	302.76
Middle ring suburb			221	309.02
Outer ring suburb			201	316.60
Total			618	309.5
Positive attitude towards peer-to-peer services				
Inner ring suburb	2.662	0.264	196	317.56
Middle ring suburb			221	317.62
Outer ring suburb			201	292.71
Total			618	309.5
Positive attitude towards smartphones				
Inner ring suburb	9.929	0.007	196	339.92
Middle ring suburb			221	305.71
Outer ring suburb			201	284.00
Total			618	309.5
Positive attitude towards the environment				
Inner ring suburb	10.486	0.005	196	343.43
Middle ring suburb			221	294.70
Outer ring suburb			201	292.69
Total			618	309.5
Negative attitude towards private vehicle				
Inner ring suburb	14.594	0.001	196	345.33
Middle ring suburb			221	307.06
Outer ring suburb			201	277.25
Total			618	309.5

TABLE 3.8

Pairwise Comparisons of Location (Public Transport)

Public transport	Sample 1–Sample 2	Test Statistic	Std. Error	Std. Test Statistic	Sig.	Adj. Sig.[a]
			Pairwise Comparisons of Location			
	Outer ring–middle ring	17.970	17.282	1.040	0.298	0.895
	Outer ring–inner ring	85.156	17.799	3.267	0.001	0.003
	Middle ring–inner ring	40.186	17.397	2.310	0.021	0.063
Smartphones and apps	Outer ring–middle ring	21.720	17.370	1.250	0.211	0.633
	Outer ring–inner ring	55.928	17.890	3.126	0.002	0.005
	Middle ring–inner ring	34.209	17.486	1.956	0.050	0.151
Environment	Outer ring–middle ring	2.017	17.312	0.117	0.907	1.000
	Outer ring–inner ring	50.739	17.830	2.846	0.004	0.013
	Middle ring–inner ring	48.722	17.427	2.796	0.005	0.026
Private vehicles	Outer ring–middle ring	29.810	17.343	1.719	0.086	0.257
	Outer ring–inner ring	68.085	17.862	3.812	0.000	0.000
	Middle ring–inner ring	38.275	17.458	2.192	0.028	0.085

Notes: Each row tests the null hypothesis that the Sample 1 and Sample 2 distributions are the same. Asymptotic significances (two-sided tests) are displayed. The significance level is .050.

[a] Significance values have been adjusted by the Bonferroni correction for multiple tests.

(mean rank = 277.25) and the inner ring (mean rank = 345.33) (p = <0.001), and the middle ring (mean rank = 307.06) and inner ring (mean rank = 345.33) (p = 0.028) but not between the middle and outer rings. In both cases, the mean rank of the inner ring is higher than the middle and outer rings. These scores reveal that positive attitudes towards PV are significantly higher in the outer and middle rings when compared to the inner ring.

Upon completion of the post hoc, it was revealed there are no statistically significant differences between any attitudinal factor and location within the middle and outer rings. Specifically, the completion of this test has demonstrated that respondents within the inner ring have more positive attitudes towards PT, smartphones, and the environment when compared to respondents in the middle and outer rings. Conversely, respondents in the middle and outer rings have more positive attitudes towards PV.

To further explore the underlying variables that could influence these differences between the inner ring and other areas the location variable was recoded into two groups—the inner ring and middle/outer ring. The two groups were then analysed using a chi-square test to identify any association with variables related to personal characteristics and transport habits. The purpose of this step was to understand if any of the underlying variables were likely to influence locational changes in attitudes towards PT, PV, the environment, and smartphones. The results of the chi-square test for association revealed statistically significant relationships and associations between location (inner and middle/outer) and the following variables: (a) dwelling type (Pearson 95.569, p = <0.001, Cramer's V = 0.395); (b) bus station coverage (Pearson 8.138, p = 0.043, Cramer's V = 0.115); (c) train station coverage (Pearson 46.310, p = <0.001, Cramer's V = 0.275); (d) weekly use of PT (Pearson = 41.428, p = <0.001, Cramer's V = 0.259); (e) typical transport mode (Pearson 34.180, p = <0.001, Cramer's V = 0.239); PV ownership (Pearson = 29.040, p = <0.001, Cramer's V = 0.217); and education (Pearson 25.991, p = <0.001, Cramer's V = 0.205).

With regard to dwelling type, 66.9% of all respondents in the inner ring live in apartments or semi-detached dwellings and only 28% in the middle/outer ring. Dwelling type is a good indicator of population density with high proliferation of apartments and semi-detached dwellings indicating a higher-density population, while more detached dwellings indicate a lower-density population. Furthermore, higher-density areas are typically easier to service with public and shared transport options due to lower travel distances and a built environment more conducive to active transport. This concept was further reflected following a cross-tabulation analysis where 12.8% of respondents in detached dwellings were considered frequent users of PT compared to 27.4% and 21.1% in semi-detached dwellings and apartments. Notwithstanding, density is unlikely the only key variable associated with land use and other factors such as land use mix, reduced distance to destination, and street and pedestrian connectivity within the inner city are also likely contributors to increased PT use (Ewing & Cervero, 2010).

Given the above, it was not unexpected that the inner ring areas had a better-estimated coverage of PT than the middle/outer rings. Specifically, inner ring respondents perceived a much higher coverage of train stations which provide more direct transportation than buses—which typically must contend with congestion and other delays. In fact, while the estimated coverage of bus stops is marginally higher in the inner ring—74% within 400 m compared to an average of 68.4% within the total case study area—it is when comparing train station coverage that the difference is more pronounced. 57.6% of all inner ring respondents estimated living within 800 m of a train station, compared to only 23.2% in the middle/outer ring.

While people generally walk further to a train when compared to a bus the average walking distance to PT is a significant contributor to the likelihood of using these services with the 400/800 m catchment a commonly cited indicator of accessibility (Daniels & Mulley, 2013; Saghapoura et al., 2016). This is further reflected in a cross-tabulation analysis which shows that 71% of those who use PT less than once a week live further than 800 m from a train station.

Potentially a result of greater coverage of services and land use distribution, the use of PT was significantly higher in the inner ring. In fact, a total of 26.5% of inner ring respondents were identified as frequent users of PT—more than 4 days per week—compared to only 12.6% in middle/outer ring areas. Similarly, in the middle/outer ring, a total of 60.7% of all respondents are identified as rarely using PT compared to 33.7% in inner city areas.

Carrying on from PT usage and looking at transport choices, the results emphasise that while PV is the primary mode in all areas (67.5%) this number is significantly higher in the middle/outer rings areas (77.1%) when compared to inner ring areas (54.5%). Furthermore, with regard to ownership while 16.8% of households in inner ring areas did not own a PV this number was reduced to only 7.3% of households in middle/outer ring areas. It is likely that PV use, and ownership, are positively influenced by the circumstances related to PT coverage and density.

These results reveal that better coverage and use of PT in inner ring areas is likely to contribute to the identified differences in attitudinal variables associated with PT which shows that inner ring respondents are more favourable than the middle/outer ring. This reflects previous studies including Forward (2019) and DeVos et al. (2021) who highlight the link between experiences, travel habits, and how improving positive experiences can contribute to changes in behaviour. Hence, while the implementation of smart mobility services to bridge the 'first/ last mile' gap is likely to be more successful in areas where PT services are already adequate, a key to its success in suburban and low-density areas may be increasing positive experiences with PT—though this may require unrealistic investment and resource allocation.

Finally, when looking at education results show that that 60.2% of all residents in inner city areas have a bachelor's or post-graduate degree compared to 48.6% in middle/outer areas. Previous research has shown that higher education levels reflect increased familiarity (Haboucha et al., 2017) and acceptance of technology (Rahimi et al., 2020), and this may be reflected in increased comfort and use of smartphones and associated technology. However, the use of smartphones—particularly in sprawling cities—is also likely influenced by the availability of a 4G/5G network and age distribution. Similarly, education levels may also be indicative of increased environmental awareness, although research by Baiardi and Morana (2021) found positive linkages with secondary education but a negative association with tertiary education. Potentially, increased education is a by-product of increased property prices and not necessarily an indicator of environmental awareness.

3.5 Findings and Discussion

This study focused on the overall research question of "how smart mobility can bridge the first/last mile gap". It provided new insights into smart mobility adoption based on a survey of participants in three Australian cities.

Research under the umbrella of 'smart mobility' has explored how strategies such as MaaS, SAS, and SAV improve accessibility and bridge the first/last mile gap between origin, destination, and PT. Bridging this gap remains critical to ensuring equitable transport systems—particularly when focusing on the spatial and temporal factors of transport disadvantage (Butler et al., 2020a). To explore the potential for these strategies to reduce disadvantages, this research has focused on user attitudes towards (a) technology; (b) PT; (c) sharing; (d) multimodality; (e) peer-to-peer transport; (f) smartphones and apps; (g) environment; and (h) PV, as potential determinants that are likely to shape user attitudes towards smart mobility.

The research shows that overcoming PV use, user aversion to multimodality, and reluctance to share rides with strangers present the most significant, city-wide barriers to smart mobility adoption. Furthermore, positive views towards PT, the environment, and smartphones—particularly in inner city areas with good PT coverage—present opportunities for promoting new approaches. The middle/outer rings have significantly more positive views towards PV which, combined with reduced accessibility to good quality PT and low population density, decreases the likelihood that existing behavioural patterns can be drastically altered. Based on these results the following insights into future planning and policymaking have been generated.

Most respondents have positive attitudes towards PV. They own more than one PV (89.6%), use PV as their main transport mode (67.5%), and rarely take PT (52.1%). Together with negative attitudes towards multimodality and shared transport—which are fundamental to smart mobility strategies such as MaaS—this presents a significant barrier for any future policy aimed at decreasing the use of PV and increasing public, shared, or active transport. Incremental, small-scale adjustments to mode choice are a more realistic expectation rather than mass transportation changes, especially given the required investment and resource allocation to bring good quality PT services into areas with little or no coverage, and difficulties with changing behaviours in areas where positive experiences with PT is limited.

Given this presumption, future planning should focus on reducing PV use in areas with adequate PT coverage. Based on this research those areas tend to be located within the inner-city or in locations where PT use is already significantly higher than the city-wide average. In this sense, planning in these areas should focus on providing efficient first/last mile catchments and feeder networks around existing PT hubs, including the continuation of compact city policy such as transit-oriented development where consideration of community interests and expectations is prioritised. Furthermore, given the high regard and use of PV, decreasing its use is unlikely to be a completely voluntary activity (Beirão & Cabral, 2007) and may require deterrents including congestion charges and car-free zones. Furthermore, where intermodal and shared options are encouraged, the operator should focus on user experience by optimising the efficiency of transfers, routes, frequencies, and comfort of passengers.

Promoting the use of smart mobility through strategies that focus on the use of smartphones to plan, book, and pay for trips and the environmental benefits may be beneficial. However, in lower-density suburban areas where coverage, use, and attitudes towards PT are significantly lower, improving commuters' experience with PT may be the most viable long-term strategy. For example, strategies could target commuters who make use of park-and-ride facilities and have the potential to replace PV with shared or active transport for the first/last leg of the trip chain. This would also free up land for future higher-density or mixed-use development around transport nodes, promoting more compact, transit-oriented development. Furthermore, the continued development of ITS, AV and passenger drones, and non-transport related trends such as WFH and improvements in virtual workspaces, should be explored to reduce the demand on transport infrastructure—particularly where providing PT is expensive and private travel is so entrenched that it is likely to remain the status quo.

Finally, each of the investigated cities covers multiple local government areas (or municipalities), each with its own role in the development of schemes and provisions for the future planning of land use and transportation systems. A coordinated approach to achieve regional integration of transport planning, through regional and state-wide strategies is critical to ensure the development of local planning schemes and instruments reflect the higher-level goals outlined in this chapter.

This study encountered some limitations: (a) The survey questions on PT coverage did not consider the location of respondents relative to tram stations which are common throughout Melbourne and parts of inner city Sydney; (b) The research did not analyse complex factors which may influence intention to use and acceptance of technology, including the theory of planned behaviour, theory of acceptance model, and unified theory of acceptance and use of technology (Golbabaei et al., 2020); (c) Due to the limited respondent numbers from each case city, the sample is not statistically representative of the population to compare and contrast the findings between the each city context; (d) Although the research relates to analysis of quantitative data there may still be bias associated with interpretation of results which could be addressed using model estimation frameworks as a supporting framework to remove bias (Paz et al., 2019); (e) Further analysis of qualitative data including analysis of opinions and views using AI and expressed through interviews and focus groups as opposed to quantitative variables would provide further insights into user attitudes and behavioural intention (Arteaga et al., 2020), and; (f) The use of the inner, middle, and outer rings based on location relative to the CBD did not consider presence of local and regional centres which provide may of the services and facilities, including employment and transport nodes common in the CBD. Future research could expand on the process of dividing the city into distinct areas by considering the location and connection with these dispersed centres.

Given the findings and insights generated by this chapter, future research could be expanded to: (a) Further analyse user attitudes towards PT, explore how these could shape future smart mobility planning, and identify any strategies to improve consumer experience with PT services; (b) Explore user expectations regarding smart mobility and discuss how this expectation may shape the future use of mobility services and technology; (c) Continue to analyse trials of smart mobility strategies such as MaaS and SAS to identify future opportunities and barriers to the ultimate adoption of services; (d) Develop multimodal strategies that optimise transfers between modes and seamless integration of services; and (e) Investigate ways to retrofit existing systems across a range of geographic areas and urban morphologies.

3.6 Conclusion

In conclusion, this chapter highlights the complexities and opportunities involved in bridging the first/last mile gap through smart mobility initiatives. Addressing challenges such as the public's strong preference for PV use, resistance to multimodal travel, and hesitancy towards ridesharing with strangers, is essential for advancing smart mobility adoption. Findings suggest that positive attitudes towards PT, environmental concerns, and the use of smartphones are particularly influential in encouraging smart mobility, especially within urban centres with well-developed PT infrastructure. Future strategies should focus on improving user experiences with multimodal and shared transit solutions, particularly in inner-city areas where PT accessibility is higher. Ultimately, this chapter underscores the need for incremental and user-centric approaches that integrate smart technology and sustainable practices to shift established behaviours towards more efficient and equitable urban mobility systems.

Acknowledgements

This chapter, with permission from the copyright holder, is a reproduced version of the following journal article: Butler, L., Yigitcanlar, T., Paz, A., & Areed, W. (2022). How can smart mobility bridge the first/last mile gap? Empirical evidence on public attitudes from Australia. *Journal of Transport Geography*, 104(1), 103452.

References

ABS (2021). Snapshot of Australia. Australian Bureau of Statistics. Retrieved on 18 January 2025 from www.abs.gov.au/statistics/people/people-and-communit ies/snapshot-australia/latest-release

Acheampong, R., & Cugurullo, F. (2019). Capturing the behavioural determinants behind the adoption of autonomous vehicles. *Transportation Research Part F*, 62, 349–375.

Ahern, Z., Paz, A., & Corry, P. (2022). Approximate multi-objective optimisation for integrated bus route design and service frequency setting. *Transportation Research Part B*, 155, 1–25.

Alonso-González, M., Hoogendoorn-Lanser, S., Oort, N., Cats, O., & Hoogendoorn, S. (2020). Drivers and barriers in adopting Mobility as a Service (MaaS). *Transportation Research Part A*, 132, 378–401.

Arteaga, C., Paz, A., & Park, J. (2020). Injury severity on traffic crashes. *Safety Science*, 132, 104988.

Baiardi, D., & Morana, C. (2021). Climate change awareness. *Energy Economics*, 96, 105163.

Beirão, G., & Sarsfield Cabral, J. (2007) Understanding attitudes towards public transport and private car. *Transport Policy*, 14, 478–489.

Bıyık, C., Abareshi, A., Paz, A., Ruiz, R.A., Battarra, R., Rogers, C., & Lizarraga, C. (2021). Smart mobility adoption. *Journal of Open Innovation*, 7, 146.

Boarnet, M., Giuliano, G., Hou, Y., & Shin, E. (2017). First/last mile transit access as an equity planning issue. *Transportation Research Part A*, 103, 296–310.

Bucchiarone, A., Battisti, S., Marconi, A., Maldacea, R., & Ponce, D. (2021). Autonomous shuttle-as-a-service (ASaaS). *IEEE Transactions on Intelligent Transportation Systems*, 22, 3790–3799.

Butler, L., Yigitcanlar, T., & Paz, A. (2020a). How can smart mobility innovations alleviate transportation disadvantage? *Applied Sciences*, 10, 6306.

Butler, L., Yigitcanlar, T., & Paz, A. (2020b). Smart urban mobility innovations. *IEEE Access*, 8, 196034–196049.

Butler, L., Yigitcanlar, T., & Paz, A. (2021). Barriers and risks of mobility-as-a-service (MaaS) adoption in cities. *Cities*, 109, 103036.

Butler, L., Yigitcanlar, T., & Paz, A. (2022). Factors influencing public awareness of autonomous vehicles. *Transportation Research Part F*, 82, 256–267.

Canitez, F. (2019). Pathways to sustainable urban mobility in developing megacities. *Technological Forecasting & Social Change*, 141, 319–329.

Coffee, N.T., Lange, J., & Baker, E. (2016). Visualising 30 years of population density change in Australia's major capital cities. *Australian Geographer*, 47, 511–525.

Contreras, S., & Paz, A. (2018). The effects of ride-hailing companies on the taxicab industry in Las Vegas, Nevada. *Transportation Research Part A*, 115, 63–70.

Daniels, R., & Mulley, C. (2013). Explaining walking distance to public transport. *Journal of Transport and Land Use*, 6, 5–20.

Dennis, S., Paz, A., & Yigitcanlar, T. (2021). Perceptions and attitudes towards the deployment of autonomous and connected vehicles: insights from Las Vegas, Nevada. *Journal of Urban Technology*, 28, 75–95.

De Vos, J., Singleton, P., & Garling, T. (2021). From attitude to satisfaction. *Transport Reviews*, 42, 204–221.

Dias, G., Arsenio, E., & Ribeiro, P. (2021). The role of shared e-scooter systems in urban sustainability and resilience during the Covid-19 mobility restrictions. *Sustainability*, 13, 7084

Dur, F., & Yigitcanlar, T. (2015). Assessing land-use and transport integration via a spatial composite indexing model. *International Journal of Environmental Science and Technology*, 12, 803–816.

Ewing, R., & Cervero, R. (2010) Travel and the built environment. *Journal of the American Planning Association*, 76, 265–294

Field, C., & Jon, I. (2021). E-scooters: A new smart mobility option? *Planning Theory & Practice*, 22, 368–396.

Forward, S. (2019). Views of public transport and how personal experiences can contribute to a more positive attitude and behavioural change. *Social Sciences*, 8, 47.

Francini, M., Chieffallo, L., Palermo, A., & Viapiana, M. (2021). Systematic literature review on smart mobility. *Journal of Planning Literature*, 36, 283–296

Gardner, B., & Abraham, C. (2007). What drives car use? *Transport Research Part F*, 10, 187–200.

Golbabaei, F., Yigitcanlar, T., Paz, A., & Bunker, J. (2020). Individual predictors of autonomous vehicle public acceptance and intention to use. *Journal of Open Innovation*, 6, 1–27.

Haboucha, C., Ishaq, R., & Shiftan, Y. (2017). User preferences regarding autonomous vehicles. *Transportation Research Part C*, 78, 37–49.

Heyns, W. (2021). Smart mobility—Keeping it real. *Civil Engineering*, 28, 8–11.

Hunecke, M., Richter, N., & Heppner, H. (2021). Autonomy loss, privacy invasion and data misuse as psychological barriers to peer-to-peer collaborative car use. *Transportation Research Interdisciplinary Perspectives*, 10, 100403.

Jain, T., Johnson, M., & Rose, G. (2020). Exploring the process of travel behaviour change and mobility trajectories associated with car share adoption. *Travel, Behaviour & Society*, 18, 117–131.

Kamruzzaman, M., Hine, J., & Yigitcanlar, T. (2015). Investigating the link between carbon dioxide emissions and transport-related social exclusion in rural Northern Ireland. *International Journal of Environmental Science and Technology*, 12, 3463–3478.

Kelley, S., Lane, B., Stanley, B., Kane, K., Nielsen, E., & Strachan, S. (2020). Smart transportation for all? A typology of recent US smart transportation projects in midsized cities. *Annals of the American Association of Geographers*, 110(2), 547–558.

Kim, Y., Kim, E., Jang, S., & Kim, D. (2021). A comparative analysis of the users of private cars and public transportation for intermodal options under Mobility-as-a-Service in Seoul. *Travel, Behaviour & Society*, 24, 68–80.

Lavieri, P.S. & Bhat, C.R. (2019). Modeling individuals' willingness to share trips with strangers in an autonomous vehicle future. *Transport Research Part A*, 124, 242–261.

Li, W., & Kamargianni, M. (2020). An integrated choice and latent variable model to explore the influence of attitudinal and perceptual factors on shared mobility choices and their value of time estimation. *Transportation Science*, 54, 62–83.

Loukopoulos, P., Jakobsson, C., Gärling, T, Schneider, C., & Fujii, S. (2005). Public attitudes towards policy measures for reducing private car use. *Environmental Science & Policy*, 8, 57–66.

Maheshwari, P., Khaddar, R., Kachroo, P., & Paz, A. (2015). Development of control models for the planning of sustainable transportation systems. *Transportation Research Part C*, 55, 474–485.

Mahrez, Z., Sabir, E., Badidi, E., Saad, W., & Sadik, M. (2021). Smart urban mobility. *IEEE Transactions on Intelligent Transportation Systems*. https://doi.org/10.1109/TITS.2021.3084907

Mohiuddin, H. (2021). Planning for the first and last mile. *Sustainability*, 13, 2222.

Neunhoeffer, F., & Teubner, T. (2018). Between enthusiasm and refusal. *Journal of Consumer Behaviour*, 17, 221–236.

Nikitas, A., Michalakopoulou, K., Njoya, E., & Karampatzakis, D. (2020). Artificial intelligence, transport and the smart city. *Sustainability*, 12, 2789.

Orro, A., Novales, M., Monteagudo, Á., Pérez-López, J., & Bugarín, M. (2020). Impact on city bus transit services of the Covid–19 lockdown and return to the new normal. *Sustainability*, 12, 7206

Paiva, S., Ahad, M., Zafar, S., Tripathi, G., Khalique, A., & Hussain, I. (2020). Privacy and security challenges in smart and sustainable mobility. *SN Applied Sciences*, 2, 1–10.

Paz, A., Arteaga, C., & Cobos, C. (2019). Specification of mixed logit models assisted by an optimization framework. *Journal of Choice Modelling*, 30, 50–60.

Paz, A., Maheshwari, P., Kachroo, P., & Ahmad, S. (2013). Estimation of Performance Indices for the Planning of Sustainable Transportation Systems. *Advances in Fuzzy Systems*, 2013, 601468.

Paz, A., & Peeta, S. (2009). Information-based traffic control strategies consistent with estimated driver behavior. *Transportation Research Part B*, 43, 73–96.

Ploeger, J., & Oldenziel, R. (2020). The sociotechnical roots of smart mobility. *Journal of Transport History*, 41, 134–159.

Rahimi, A., Azimi, G., & Jin, X. (2020). Examining human attitudes toward shared mobility options and autonomous vehicles. *Transportation Research Part F*, 72, 133–154.

Revington, N., & Townsend, C. (2016). Market rental housing affordability and rapid transport catchments. *Housing Policy Debate*, 26, 864–886.

Ristvej, J., Lacinák, M., & Ondrejka, R. (2020). On smart city and safe city concepts. *Mobile Networks and Applications*, 25, 836–845.

Saghapour, T., Moridpour, S., & Thompson, R. (2016). Public transport accessibility in metropolitan areas. *Journal of Transport Geography*, 54, 273–285.

Santos, G., & Nikolaev, N. (2021). Mobility as a service and public transport. *Sustainability*, 13, 3666.

Schulz, T., Böhm, M., Gewald, H., & Krcmar, H. (2021). Smart mobility: An analysis of potential customers' preference structures. *Electronic Markets*, 31, 105–124.

Scorrano, M., & Danielis, R. (2021). Active mobility in an Italian city. *Research in Transportation Economics*, 86, 101031.

Singh, Y. (2020). Is smart mobility also gender-smart? *Journal of Gender Studies*, 29, 832–846.

So, J., An, H., & Lee, C. (2020). Defining smart mobility service levels via text mining. *Sustainability*, 12, 9293.

Tirachini, A. (2019). Ride-hailing, travel behaviour and sustainable mobility. *Transportation*, 47, 2011–2047.

Vickerman, R. (2021). Will Covid-19 put the public back in public transport? A UK perspective. *Transport Policy*, 103, 95–102.

Vilathgamuwa, M., Mishra, Y., Yigitcanlar, T., Bhaskar, A., & Wilson, C. (2022). Mobile-energy-as-a-service (MEaaS). *Sustainability*, 14, 2796.

Yigitcanlar, T., & Cugurullo, F. (2020). The sustainability of artificial intelligence: An urbanistic viewpoint from the lens of smart and sustainable cities. *Sustainability*, 12, 8548.

Yigitcanlar, T., & Dur, F. (2013). Making space and place for knowledge communities: Lessons for Australian practice. *Australasian Journal of Regional Studies* 19, 36–63.

Yigitcanlar, T., Kankanamge, N., & Vella, K. (2021). How are smart city concepts and technologies perceived and utilized? A systematic geo-Twitter analysis of smart cities in Australia. *Journal of Urban Technology*, 28, 135–154.

Yigitcanlar, T., Wilson, M., & Kamruzzaman, M. (2019). Disruptive impacts of automated driving systems on the built environment and land use: An urban planner's perspective. *Journal of Open Innovation*, 5, 24.

Yuen, F., Chua, G., Wang, X., Ma, F., & Li, K. (2020). Understanding the public acceptance of autonomous vehicles using the theory of planned behaviour. *International Journal of Environmental Research and Public Health*, 17, 4419.

Zhang, M., Zhao, P., & Qiao, S. (2020). Smartness-induced transport inequality: Privacy concern, lacking knowledge of smartphone use and unequal access to transport information. *Transport Policy*, 99, 175–185.

4

Perceptions on Autonomous Demand-Responsive Transit Use

4.1 Introduction

Autonomous vehicles (AVs) have the potential to become a commonplace transport platform globally. However, the excessive or disorganised use of private AVs might increase traffic congestion and greenhouse gas emissions via several factors (Golbabaei et al., 2021). First, increased vehicle ownership and usage can result in the presence of more cars on the road, leading to overall higher vehicle miles travelled (VMT) and increased congestion. This increased traffic can lead to idling and stop-and-go driving, both of which contribute to higher emissions (Wadud et al., 2016). Second, AVs may encourage longer trips and more single-occupancy journeys, as people may be more willing to tolerate longer commutes if they can work or relax while the vehicle drives itself. Additionally, the convenience and comfort offered by AVs might reduce the appeal of public transportation, leading to a shift from shared mobility options to private vehicles (Harb et al., 2018). Finally, while AVs have the potential to improve fuel efficiency through better traffic flow and optimised driving patterns, the manufacturing and operational energy requirements, as well as the battery production and charging infrastructure, contribute to the life cycle emissions of these vehicles (Taiebat et al., 2018; Nunes et al., 2022; Silva et al., 2022).

The multilevel aspect of electric vehicles (EVs) regarding sustainability and their life cycle encompass various stages from production to end-of-life management. While EVs contribute to reduced greenhouse gas emissions during operation, concerns arise from their assembly and disposal processes. At the production level, the extraction of raw materials such as lithium, cobalt, and rare-earth metals for battery production has environmental and social implications. Additionally, the energy-intensive manufacturing process and associated emissions involved during vehicle assembly need to be considered. Furthermore, the end-of-life management of EV batteries poses challenges due to their recycling, reusability, and potential environmental

DOI: 10.1201/9781003605676-4

impacts if not properly handled. These multilevel aspects highlight the need for holistic approaches, including the sustainable sourcing of materials, efficient manufacturing processes, and effective recycling and disposal systems, to maximise the sustainability benefits of EVs while minimising their environmental footprint (Wu & Zhang, 2017; Vidhi & Shrivastava, 2018).

The excessive or disorganised use of private AVs additionally has the potential to stimulate urban sprawl (Fagnant & Kockelman, 2015; Milakis et al., 2017; Soteropoulos et al., 2019; Narayanan et al., 2020) by eliminating the stress of driving and enabling people to reside farther from their workplace, resulting in longer commuting distances and energy expenditure (Spurlock et al., 2019). Thus, without careful planning and regulation, the unchecked proliferation and haphazard use of private AVs can exacerbate greenhouse gas emissions. To minimise the detrimental effects, the widespread adoption of ridesharing using AVs should be publicly promoted to reduce traffic congestion by optimising routes and minimising empty trips towards creating a safer, more efficient, and sustainable transportation systems for the future (Golbabaei et al., 2020; Paddeu et al., 2020).

Autonomous demand-responsive transit (ADRT) is a recently introduced public transit mode and is predominantly available using autonomous shuttle buses (ASBs) (Ainsalu et al., 2018; Nordhoff et al., 2018; Rehrl & Zankl, 2018; Salonen, 2018; Golbabaei et al., 2020; Paddeu et al., 2020; Mouratidis & Cobeña Serrano, 2021). The implementation of ADRT has been stated as an applicable response to the climate change challenge (Nunes et al., 2022). ADRT has the potential to enhance mobility services and, as a result, enhance transit efficiency and reduce dependency on private vehicles (Millonig & Fröhlich, 2018; Nenseth, 2019). As a feeder mode of regular public transit, ADRT could provide first-/last mile services, supporting a transition to more sustainable mobility (Soteropoulos et al., 2019). The use of ASBs with transport capacities of up to 15 persons enables reasonably cost-effective, flexible on-demand 24/7 operation (Nordhoff et al., 2020). The use of ADRT in a more dynamic, mixed-traffic environment is evolving quickly (Beiker, 2019; Stocker & Shaheen, 2019; Iclodean et al., 2020).

Attitude may be explained as "a mental state of readiness, positively or negatively associated with a particular object. It is acquired through experience and is a precursor of behaviour related to the object" ((Pigeon et al., 2021), p. 251). Individuals' attitudes towards ADRT are crucial as they influence "the demand for the technology, governing policies and future investments in infrastructure" (Haboucha et al., 2017, p. 38). Nevertheless, if ADRT is to be deployed widely and embraced as an everyday travel mode, positive public attitudes are necessary.

In the past, many researchers have investigated public perceptions towards opportunities and challenges for AVs (Schoettle & Sivak, 2014; Kyriakidis et al., 2015; Bansal et al., 2016; König & Neumayr, 2017; Shabanpour et al., 2108; Gkartzonikas & Gkritza, 2019; Nastjuk et al., 2020; Mouratidis & Cobeña

Serrano, 2021), but few have focused on the Australian context (Regan et al., 2017; Pettigrew et al., 2018, 2019; Kaur & Rampersad, 2018; Cunningham et al., 2019; Pettigrew, Worral et al., 2019; Butler et al., 2021; Ledger et al., 2022). There is a lack of research about how Australians' socio-demographics affect their attitudes towards AVs, particularly ADRT. This study fills this research gap by fully classifying socio-demographic predictors of the publics' attitudes towards ADRT in the specific context of the South East Queensland (SEQ) region of Australia. The research method is founded on an online stated preference survey distributed across more than 250 postcodes across the region, complemented by a wide-ranging review of prior global studies, and descriptive and ordinal/binary logistic regression analysis using SPSS v.27. This study tries to address the following research question.

How are individuals' perceptions and attitudes towards ADRT influenced by gender, age, education, employment, income, household size, residential location, and having a driver's licence?

As a result, we shed more light on the social dynamics behind how potential adopters perceive different aspects of this innovative transit mode. The insights drawn from the current study may help alleviate concerns and encourage the future adoption of ASBs in this and other regions. Following this introduction, Section 4.2 provides an overview of relevant global studies. Section 4.3 explains the research method involving the questionnaire design, case study area, and the data collection process. Section 4.4 presents the descriptive statistics of the socio-demographic and attitudinal characteristics. This section then describes the analysis method along with the detailed results. Section 4.5 discusses the findings and implications for the transition to ADRT. Section 4.6 provides concluding remarks, study limitations, and suggestions for further research.

4.2 Literature Background

In this section, the study provides a concise review of the current literature on the association between attitudes towards ADRT, in particular ASBs, and the socio-demographic characteristics comprising gender, age, education, employment, income, household size, and residential location, adapting the reviews that were recently published in the AV context (Nordhoff et al., 2019; Golbabaei et al., 2020; Pigeon et al., 2021).

The literature on the subject of gender and attitude towards ADRT is mixed. According to Dong et al. (2017) and Winter et al. (2018), males have been demonstrated to be more open to using ASBs than females are, especially highly automated ASBs (Roche-Cerasi, (2019). Other research reported that males are generally more willing to use autonomous transport services

(Acheampong & Cugurullo, 2019), preferring ASBs over traditional vehicles (Alessandrini et al., 2014, 2016; Wien, 2019; Winter et al., 2019), and trust ASBs more than females do (Dekker, 2017). Similarly, males are found to be more confident to share a ride with strangers on ASBs than females are, but in terms of traffic safety or dealing with an emergency, there has been no major difference (Salonen, 2018). Females are less prone to believe that autonomous transit services are useful and have more concerns about them (Acheampong & Cugurullo, 2019). Furthermore, females may prefer to use ASBs themselves rather than allow their partners or children to use (Anania et al., 2018; Winter et al., 2019). Nevertheless, Madigan et al. (2017) and Nordhoff et al. (2017, 2018) found no significant difference between males' and females' intention to use ASBs, and neither did Pakusch and Bossauer (2017) in the autonomous transit context. No impact of gender was seen even when ASB service offerings were provided between transit hubs and parking lots or between the home and workplace (Roche-Cerasi, 2019).

Research findings on how age affects attitude and adoption are inconsistent. Some studies discovered no correlation between a person's age and willingness to use ASBs (Madigan et al., 2016, 2017; Moták et al., 2017; Kostorz et al., 2019) or other autonomous transit services (Pakusch & Bossauer, 2017), or even with the likelihood of preferring ASBs over other transport modes (Alessandrini et al., 2016; Wien, 2019). Similarly, Salonen (2018) reported an insignificant effect of age on concerns about safety on-board, in traffic or an emergency, as did Dekker (2017) regarding trust in ASBs. Even when ASBs offered mobility services between transit hubs and parking lots, or between the home and workplace, age was shown to not influence the willingness to use transit (Pigeon et al., 2021). ASBs seemed to be more popular among young individuals (Roche-Cerasi, 2019). According to Acheampong and Cugurullo (2019), there is a negative relationship between age and favourable attitudes towards technology, the perceived benefits of or intent to use autonomous transit services. Those of ages between 18 and 35 were more likely to use ASBs than those over 45 years old (Dong et al., 2017). Portouli et al. (2017) found that frequent customers of ASBs were younger than non-users, contrary to Nordhoff et al. (2018) who reported a higher acceptance of ASBs among older participants than among younger ones, though the former considered ASBs less efficient than their present transport mode.

Level of education was discovered to affect the intention to use ASBs (Roche-Cerasi, 2019), perceived usefulness, perceived ease of use, and willingness to use autonomous transit services (Acheampong & Cugurullo, 2109). ASBs were preferred by those with a higher education level over their traditional counterparts in some regions where ASBs were implemented in city centres (Alessandrini et al., 2014, 2016, 2017). In contrast, neither concerns about safety on-board, in traffic, nor an emergency seemed to be influenced by education levels (Salonen, 2018), nor did trust in ASBs (Dekker, 2017). The impact of education level between frequent users of ASBs and those who had never used such modes was insignificant (Portouli et al., 2017).

Employment was found not to affect preference for ASBs over their traditional counterparts (Alessandrini et al., 2016), or concerns regarding safety on-board, in traffic or in an emergency (Salonen, 2018). Nevertheless, Portouli et al. (2017) reported that students use ASBs more frequently than employees, unemployed persons, or retirees do. This might be attributed to the impact of ageing.

Household income was found not to affect preference for ASBs over their traditional counterparts (Alessandrini et al., 2016), or willingness of using ASBs (Kostorz et al., 2019). Similarly, Salonen (2018) reported an insignificant effect of income on concerns about safety on-board, in traffic or in an emergency, as did Dekker (2017) regarding trust in ASBs. Dong et al. (2017) argued that a person's greater income increases their intention to use ASBs, but only in the case of not considering the effect of AV knowledge.

Regarding residential location, some research stated that residents of densely populated regions have greater intentions to use ASBs at higher levels (Roche-Cerasi, I. (2019). Rural and urban populations in Germany had equal intentions to use ASBs (Kostorz et al., 2019). In contrast, US respondents were found to be more inclined to use ASBs than those in any of the other countries that were surveyed by Winter et al. (2018). Residents of areas with ASB services were shown to have more trust and intention to use ASBs than were residents of areas without ASB services in operation (Dekker, 2017). In a survey conducted on a German campus by Nordhoff et al. (2018), campus workers regarded ASBs as being less efficient than their existing transit mode compared to non-campus workers. Contrary to parents in the US, parents in India were more open to the idea of their children riding in ASBs. However, residents and tourists in La Rochelle (France) and Lausanne (Switzerland) were equally open to the idea of using ASBs (Madigan et al., 2016).

As noted, research findings differ concerning the influence of gender, age, education, employment, income, and residential location on attitudes towards ADRT, and in particular ASBs. Such discrepancies might be due to variations in research "methodology (qualitative interview or an online survey; involving a shuttle trial or not), nature and size of study samples, usage contexts (campus, city centre or rural environment) or vehicle considered (shuttle, buses)" (Pigeon et al., 2021, p. 268). Even though there are extensive detailed studies in this field, they have mostly focused on the US and European populations. The applicability of those findings to the Australian context, thus, is questionable. Empirical research is lacking for measuring public perceptions and attitudes towards ADRT in Australia, especially in the ASB context. Only a small number of individual characteristics have been explored, which restricts both the depth and breadth of our knowledge of the association between Australians' socio-demographics and their perceptions and attitudes towards ADRT. Cross-national differences may obscure individual differences in attitudes towards and adoption of ASBs according to Kyriakidis et al. (2015). Further, prior findings indicate that the public's perception towards AVs more generally varies between nations, namely between Australians (Schoettle & Sivak, 2014) and others, highlighting the need for further study in the regional context.

The current research builds upon prior studies, not only by measuring perceptions and attitudes towards ADRT among adult residents of SEQ but also by carrying out an in-depth exploration of how those perceptions and attitudes are associated with particular socio-demographic characteristics.

4.3 Research Design

A stated preference survey was designed and implemented to investigate challenges and opportunities in the adoption of ADRT services by adult residents of the case study region. SEQ is a metropolitan region centred on Brisbane that has a land area of 35,248 km^2 and a population of 3,817,573 million (2021). The per capita gross state product of Queensland is AUD 71,037 (USD 53,280) (Australian Bureau of Statistics, 2021). It has 12 adjoining local government areas (LGAs), where a LGA is a municipality administered by the third and lowest tier of government.

The survey respondent recruitment methodology involved only individuals living within the urban and peri-urban areas of SEQ (see Figure 4.1, highlighted in red and purple, respectively) including a total of 250 postcodes. Rural areas were excluded from the study because the implementation of this survey was not cost-efficient due to low density.

The questionnaire items were adapted and developed following a systematic literature review (Golbabaei et al., 2020) to verify the content's validity. Preliminary testing was conducted by surveying a group of higher-degree research students and the staff of the university because these people usually have broader knowledge regarding the application of surveys for reliable results. Preliminary and main survey participation was entirely voluntary. The final questionnaire was revised following the feedback provided by an expert supervisory panel review representing views of key informants in the field, specifically civil engineering and built environment academics specialising in transport systems and AVs. The questionnaire consisted of three sections: (a) questions about the respondent's socio-demographic characteristics, (b) questions relating to the respondent's existing travel habits—such factors being worth mentioning in understanding the attitudes towards ASBs, and (c) attitudinal indicators contained to assess perceptions that might affect the adoption of ASBs. The related indicators were measured by applying a 5-point Likert scale owing to its widespread usage in the

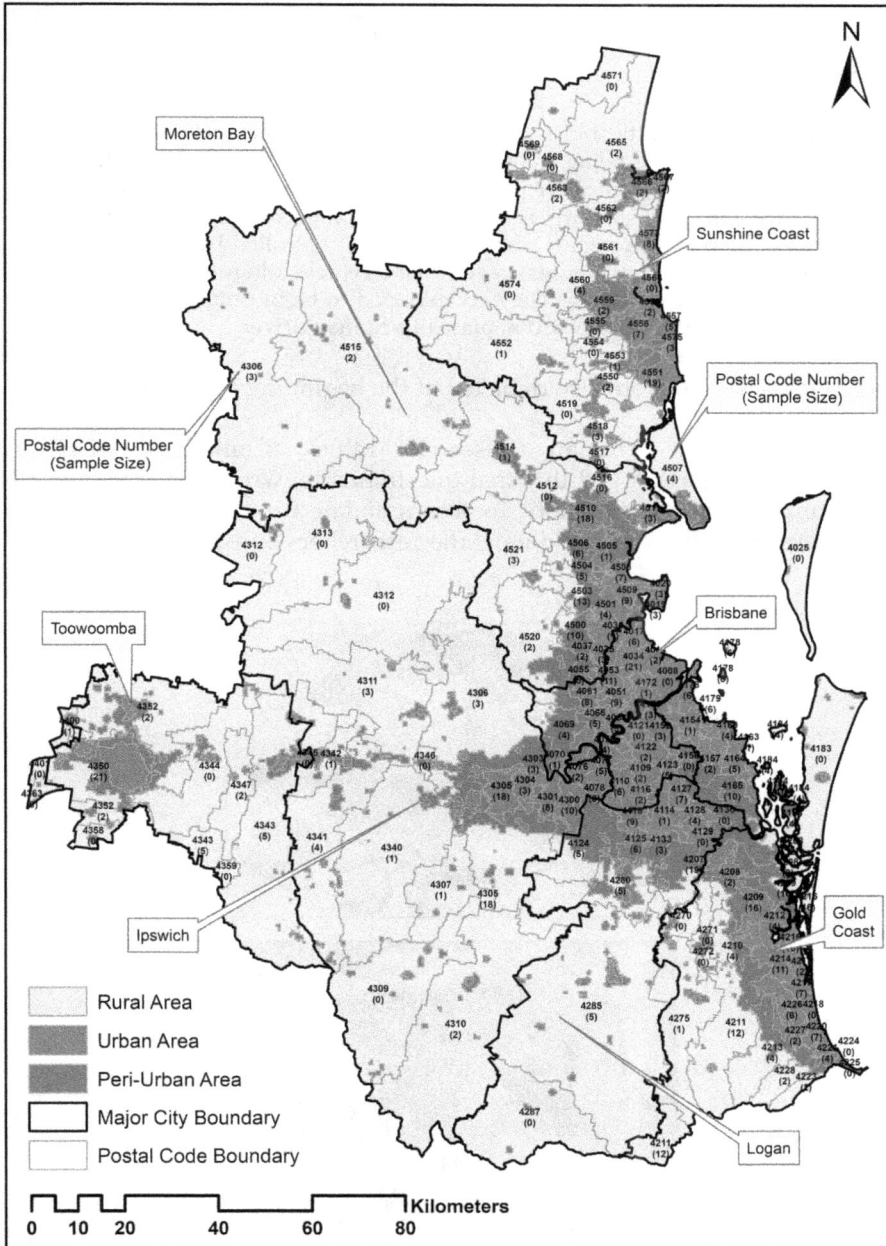

FIGURE 4.1
The map of the study area (Mortoja & Yigitcanlar, 2021).

literature and its ease of use in analysis. For a clearer interpretation of the typical usage of ASBs to deliver ADRT, an introductory paragraph was included stating that

> Autonomous shuttle buses are fully automated electrically powered vehicles which are being trialled in Australia including SEQ as a new travel mode. They can serve potentially similar markets to conventional shuttles and have similar passenger-carrying capacities. However, they are not driven by a person, instead, they are controlled by smart technology that safely optimises travel times, vehicle kilometres travelled, and energy consumption. They are expected to be in public use with a surveillance system on board in place of a human driver.

Two photos of ASB were also depicted at the beginning of the questionnaire (Figure 4.2).

The University Human Research Ethics Committee (UHREC RN: 2000000747) approved the final questionnaire, which was made available online for self-completion. To accommodate the limitations and risks posed by the COVID-19 pandemic, the researchers enlisted the services of

FIGURE 4.2
Introductory photos of ASBs presented to the survey participants regarding ADRT.

Qualtrics, a professional web-based survey platform provider, to employ a convenient random sampling method in reaching the target respondents and gathering data for the study. Each potential respondent received an email containing the survey link to ensure broad public access during May 2021. The email explicitly displayed a brief description of the academic purpose of the project and voluntary participation. The screening question ensured that only participants who were over 18 years old and residing in SEQ were asked to respond to the survey. Overall, 357 respondents finished the survey. Ultimately, after screening and cleaning the data, 300 responses with no missing values, invalid observations, or outliers were deemed to be valid for further analysis. Based upon Krejcie and Morgan (1970), a minimum sample size for a population above 1,000,000 (confidence = 95% and margin of error = 6%) is 300.

4.4 Analysis and Results

It is necessary to explore how the SEQ population is represented in our sample to fully understand both the background that yields the subsequent findings and the associated implications for the uptake of ADRT among Australians more broadly. The descriptive analysis was carried out using SPSS v.27 (George & Mallery, 2021).

4.4.1 Socio-Demographic Characteristics

The analysis results identified a total of seven socio-demographic predictor variables associated with personal characteristics including gender, age, education, employment, household income, residential location, and household size. Figure 4.3 illustrates the distribution of each personal characteristic. The multicollinearity assessment results indicated that predictor variable inflation factors (VIFs) were all acceptable at a level of < 2.50 (Daoud, 2017).

Out of the 300 survey participants, the age group between 18 and 35 years old constituted the largest proportion at 36%. Respondents aged 36–50 years old accounted for 17.7% of the participants, while those aged 51–65 and over 66 years old represented 19.3% and 27%, respectively. The number of female respondents was nearly twice that of males (65% compared to 35%). Regarding education, 27.4% of participants held tertiary degrees and almost the same portion of them completed high school (36.3%) or a vocational certificate (36.3%). The retired, homemaker, or not employed group accounted for 46% of respondents, while part-time or casual employees accounted for 24% and full-time or self-employed (30%) individuals accounted for the remainder. The median and mode annual income bracket was AUD 52,000–AUD 77,999.

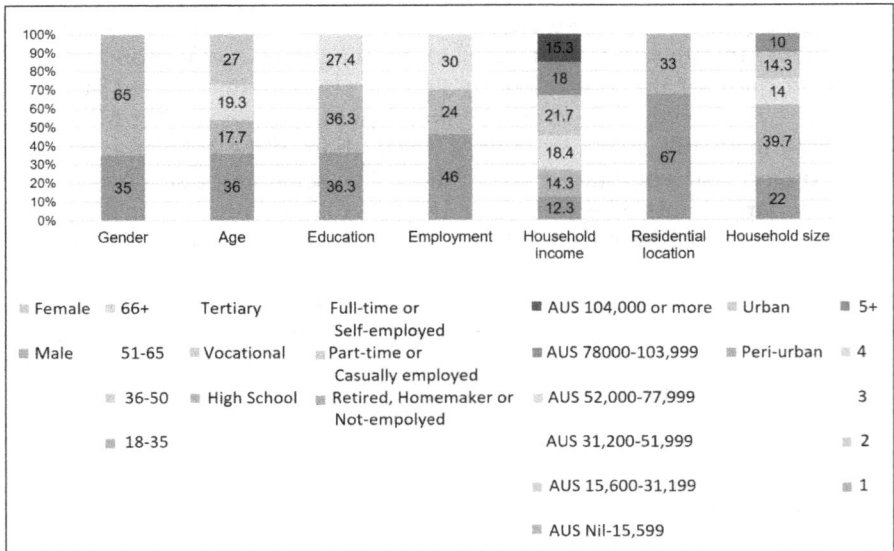

FIGURE 4.3
Demographic characteristics of SEQ respondents.

The survey also showed that two-thirds of respondents (67%) were living in peri-urban areas, and most of them were from two-person households (39.7%).

The distributions of existing travel characteristic response variables including (a) travel mode/frequency, (b) travel purpose/frequency, (c) driver's licence, (d) daily travel time, and (e) travel mode satisfaction are outlined in Figure 4.4. It can be seen that the most frequently used modes are walk, car, while the least frequently used modes are mobility scooter, motorcycle/moped, e-bike/e-scooter. The most popular transit modes are bus, train/tram, taxi, then ferry and conventional shuttle bus. Commercial vehicle and bicycle usages are similar. The majority of the survey participants hold a valid driver's licence. Almost the same portion of them had less than 30 min or 30 min–1 h of travel time. Only a small portion of them was neutral towards or dissatisfied with their current travel mode, while the rest were satisfied or very satisfied with it.

The summary of responses regarding exposure to AVs comprising AV knowledge and experience variables is shown in Figure 4.5. The AV knowledge variable was ordered in the following categorical range on the survey: not familiar, somewhat familiar, and very familiar. Only eight responses were recorded in the last category, so it was determined that the recoding of categories was appropriate. As can be seen, a percentage of the survey respondents were aware of AVs but very few of them had already used them.

HOW FREQUENTLY DO YOU USE EACH OF THE FOLLOWING TRIP MODES?

	Never	Fortnightly or less	Weekly or more	Daily	
Walk	4.3	29	21.3	45.3	
Mobility Scooter		93.7		2.7 2.7	1
Motorcycle/ Moped		91		4.7 4	0.3
e-Bike/ e-Scooter		88		7 4	1
Bicycle		66.3	19.3	13	1.3
Ferry		59.3	38	2	0.7
Train/ Tram	29		56.7	12.7	1.7
Bus	29.7		51.7	15.3	3.3
Conventional Shuttle		75.7	21	3	0.3
Taxi	35.3		57	7	0.7
Car	2	28.3	11.7	58	
Commercial Vehicle		58.7	25.3	6.3	9.7

0% 10% 20% 30% 40% 50% 60% 70% 80% 90% 100%

Never   Fortnightly or less   Weekly or more   Daily

HOW FREQUENTLY DO YOU MAKE ANY OF THE FOLLOWING TRIP TYPES?

	Never	Fortnightly or less	Weekly or more	Daily
Study	79.7		11	7 2.3
Medical	5.3	69	24.3	1.3
Personal	1.3	30.3	18.3	50
Work	48	20	6.7	25.3

0%   20%   40%   60%   80%   100%

Never   Fortnightly or less   Weekly or more   Daily

Driver's licence

12.3

Yes
No

87.7

Travel Time

> 2h

1h – 2h

< 30min

30mm – 1h

Travel mode satisfaction

3 2
10.7

50

34.3

Very Dissatisfied
Dissatisfied
Neutral/Not Applicable
Satisfied
Very Satisfied

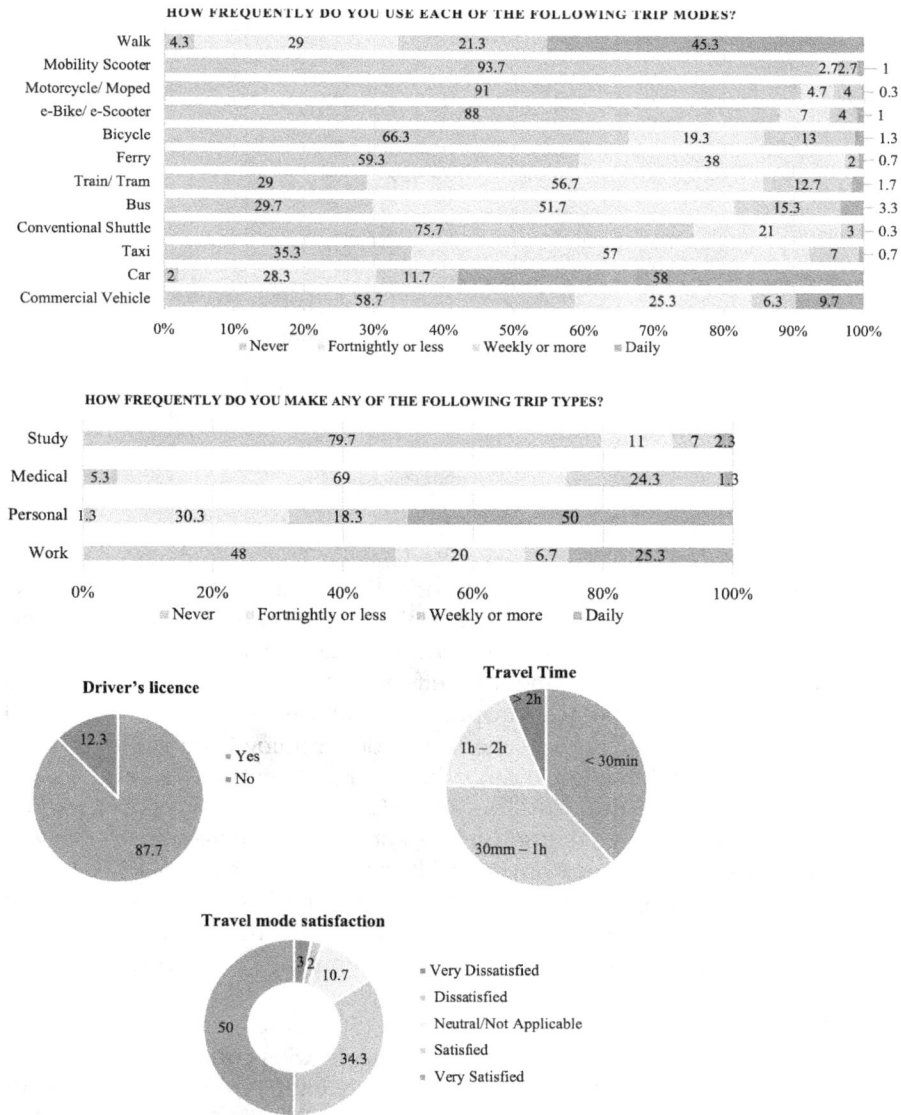

FIGURE 4.4
Summary of responses for travel characteristic variables.

4.4.2 Attitudinal Characteristics

Public attitude towards ASBs is a key factor that will shape the demand and market for them (Portouli et al., 2017). Since perception and attitudes "represent an individual's latent beliefs and values and unlike observable variables cannot be directly measured. These latent constructs, however, influence an individual's

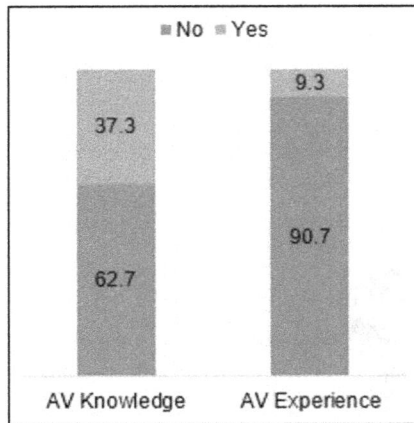

FIGURE 4.5
Summarised AV knowledge and experience response variables.

decision-making process" (Ben-Akiva et al., 1999; Thapa et al., 2019, p. 242). Psychometric indicators could be used to identify latent constructs (Thapa et al., 2019). Response variables in our study, which require respondents to rate certain statements on a scale, are psychometric indicators. In the literature response, variables are self-developed or modified effective statements. For each factor, the frequencies in each category were inspected and all were maintained for ordinal logistic regression. Reliability was checked by determining Cronbach's alpha (α) to assess the items' internal consistency. Each scale's Cronbach's α value should be greater than 0.7 (Pallant, 2020). The overall Cronbach's alpha was determined to equal 0.924 for perceived opportunities and 0.786 for perceived challenges, indicating strong consistency amongst all the response variables listed. The value of Cronbach's alpha for each item if deleted implies that the omission of none of the items could have substantively increased the reliability of this part of the survey (Briggs & Cheek, 1986); however, 'Higher fare' is less consistent than the others.

Public perception and attitudes towards ASBs were tested regarding the perceived opportunities and challenges of using ASBs compared to those of using conventional shuttles. The survey participants were presented with a list of opportunities to be expected by using ASBs. Their opinions on the agreement with the listed opportunities on the 5-point Likert scale ranging from 'strongly disagree' to 'strongly agree' are shown in Figure 4.6. The majority of the survey respondents gave responses ranging from neutral to agree, with each of the eight perceived opportunities listed. Of the opportunities that were agreed upon, the most appealing ones were 'Less congestion/emissions' (41%), and this was followed by "Easy to learn how to interact/travel" (39%), and 'Reduced fleet need' (38%). The least appealing ones were 'Safer' (18%) and 'More attractive' (23%).

The survey participants were presented with a list of challenges relating to the use of ASBs. Their opinions regarding the concerns about the listed

Response Variable / Caregory	Strongly Disagree	Disagree	Neutral	Agree	Strongly Agree	Cronbach's α if item deleted
More efficient	10.3	16	38	28	7.7	0.909
Reduced fleet need	7	13.3	35.7	38	6	0.916
Less congestion/emissions	6.7	13.3	30.3	41	8.7	0.918
Fewer driver errors	8	15.3	39	28	9.7	0.913
Easy to Learn How to Travel	6	9.3	35	39	10.7	0.919
Safer	15	26	35.7	18	5.3	0.91
More Attractive	12.3	20	39	23.7	5	0.914
More Positive Attitude	11	15.7	34.7	32	6.7	0.911

FIGURE 4.6
Perceived opportunities of autonomous shuttle buses (ASBs) compared to conventional shuttles (%).

Response Variable/ Caregory	Very Concerned	Concerned	Neutral	Not Concerned	Not Concerned at All	Cronbach's α if item deleted
Higher fare	18.7	44	24.3	11	2	0.867
Unreliable technology	24.7	51.3	14	9	1	0.673
Traffic accidents	38	39	11	11	1	0.649
Malfunction	43	40.7	9.7	5.7	1	0.694

FIGURE 4.7
Perceived concerns of autonomous shuttle buses (ASBs) (%).

challenges on the 5-point Likert scale ranging from 'very concerned' to 'not concerned' at all are shown in Figure 4.7. The majority of the survey respondents were concerned to very concerned with all listed challenges. Of the challenges, the most concerning was 'Unreliable technology' (51.3% were concerned and 24.7% were very concerned), followed by 'Malfunction' (43% were very concerned and 40% were concerned), 'Traffic accidents' (38% were very concerned and 39% were concerned), and 'Higher fare' (44% were concerned and 18.7% were very concerned).

The survey participants' intention to use ASBs if they become available is shown in Figure 4.8. More than two-thirds (38%) of the respondents stated that they would be happy to ride in ASBs that operate for special purposes, and about one-fifth (19%) did not consider using ASBs at all. Of the rest of the potential users of ASBs, 25% preferred to ride in ASBs that operate on all roads/streets, 10% preferred to ride on private streets, and 8% preferred to ride on local streets.

4.5 Findings and Discussion

Following the descriptive analysis of the survey, an ordinal logistic regression was employed using SPSS v.27 to develop a model to understand associations of the socio-demographic predictor variables for each response variable within each tabulated grouping of existing travel characteristics

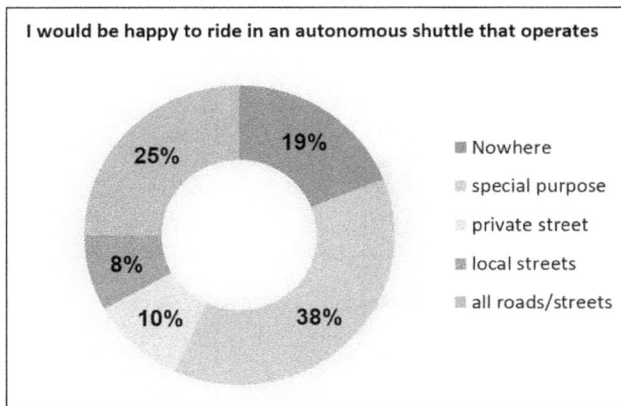

FIGURE 4.8
Autonomous shuttle bus (ASB) adoption choice.

and attitudinal variables. The link function that was used was Logit. For this exploratory study of each response variable, a backward elimination method was used to obtain the parsimonious model using only the predictor variables that pass the threshold significance of 0.05 (Bursac et al., 2008).

The next section presents the results of this modelling. For ordinal logistic regression and binary logistic regression, the omnibus test, like the likelihood-ratio chi-square test, is used to test whether or not the current model outperforms the null model as evidenced by $p \leq 0.05$. For the ordinal logistic regression, the parallel lines test is used to test the null hypothesis that the slope coefficients in the model are the same across response categories, and therefore that the one-equation model is valid, as evidenced by $p > 0.05$, suggesting the model fits well (Brant, 1990; Allison, 2012; Liu, 2015).

In our evaluation, we characterise relatively the predictor variables' odds ratios for decreasing odds: extremely strong > 0.2, 0.2 ≥ very strong > 0.4, 0.4 ≥ strong > 0.6, 0.6 ≥ moderate > 0.8, and 0.8 ≥ mild > 1.0. We use the inverses of these values for increasing odds (McHugh, 2009; Szumilas, 2010; Norton et al., 2018). Here, we present findings from our ordinal/binary logistic regression analysis.

4.5.1 Associations between AV Exposure and Socio-Demographic Predictor Variables

The results of the goodness of fit tests and statistics of the binary logistic regression models for these response variables are listed in Table 4.1.

Omnibus significance indicates that the AV knowledge model has a better fit than the intercept-only model. According to the odds ratio, evidence suggests that an increasing employment status across the scale from the

TABLE 4.1

Significant Binary Logistic Models of AV Exposure to Socio-Demographic Predictor Variables

Response Variable Model	Omnibus Sig.	Predictor Variable	Std. Error	Wald Sig.	OR	OR 95% Wald C.I.	
						Lower	Upper
AV knowledge	0.038	Employment	0.140	0.038	1.335	1.015	1.755
AV experience	0.002	Gender (male)	0.408	0.018	2.635	1.184	5.860
		Household income	0.142	0.015	1.411	1.069	1.861

retired, homemaker, or not employed level to the full-time or self-employed level is strongly associated with an increase in the likelihood of having knowledge about AV.

The AV experience model has a superior fit to the intercept-only model. Evidence suggests the following effects in the likelihood of having experienced AVs of any kind. Being male is very strongly associated with an increase. Increasing household income across the scale from the Nil to the AUD 15,599 level to the AUD 104,000 or more level is extremely strongly associated with an increase.

4.5.2 Associations between Attitudinal Characteristics and Socio-Demographic Predictor Variables

The results of the goodness of fit tests and statistics of the ordinal logistic regression model for each remaining response variable are listed in Table 4.2. All the response variable models have a superior fit to the threshold-only models, while the proportional odds assumption appears to have held.

Regarding perceived opportunities and challenges, evidence suggests the following, discussed according to the response variable.

Overall, the information highlights the advantages of ASBs over conventional shuttles, including their efficiency, reduced traffic congestion and emissions, the presence of fewer driver errors, ease of learning, safety, attractiveness, and positive attitudes towards ASBs. For each of the following response variables, no model was found to be significant via regression using the socio-demographic predictor variables from Table 4.1 of 'Reduced fleet Need', 'Unreliable technology', and 'Malfunction'.

Individuals with lower employment levels tend to have less familiarity with AVs. Therefore, it would be beneficial to explore ways to enhance their knowledge about AVs, especially if such an improvement can positively influence their willingness to adopt AVs when and where they are available. Additionally, females and individuals from lower-income households are less likely to have experienced riding in any type of AV. For these

TABLE 4.2

Significant OLM of Attitudinal Characteristics to Socio-Demographic Predictor Variables

Response Variable Model	Omnibus Sig.	Parallel Lines Sig.	Predictor Variable	Std. Error	Wald Sig.	OR	OR 95% Wald C.I. Lower	Upper
Perceived opportunities of ASBs								
More efficient	0.000	0.541	Age	0.087	0.001	0.742	0.626	0.881
			Household income	0.066	0.032	1.153	1.011	1.315
			Household income	0.068	0.001	1.263	1.106	1.442
Less congestion and emissions	0.001	0.493	Age	0.088	0.001	0.757	0.637	0.899
Fewer driver errors	0.003	0.153	Age	0.087	0.004	0.780	0.658	0.924
Easy to learn how to interact/ travel	0.000	0.783	Age	0.088	0.003	0.770	0.647	0.915
			Household income	0.068	0.001	1.263	1.106	1.442
Safer	0.007	0.052	Age	0.086	0.008	0.796	0.674	0.939
More attractive	0.000	0.131	Age	0.098	0.000	0.709	0.584	0.861
			Employment	0.138	0.036	1.337	1.014	1.762
			Drivers licence (yes)	0.331	0.009	0.423	0.221	0.809
			Age	0.098	0.000	0.690	0.570	0.834
			Household income	0.068	0.024	1.165	1.020	1.331
More positive attitude	0.000	0.124	Gender (male)	0.247	0.040	1.660	1.018	2.708
			Drivers licence (yes)	0.331	0.009	0.423	0.221	0.809
			Age	0.098	0.000	0.690	0.570	0.834
			Household income	0.068	0.024	1.165	1.020	1.331
Perceived challenges of ASBs								
Higher fare	0.008	0.529	Residential location (peri-urban)	0.228	0.044	1.584	1.012	2.479
			Education	0.136	0.031	0.746	0.571	0.973
Traffic accidents	0.003	0.472	Gender (male)	0.227	0.028	1.647	1.055	2.570
			Drivers licence (yes)	0.327	0.037	0.506	0.267	0.959

Note: OLM: ordinal logistic models; ASBs: autonomous shuttle buses; OR: odds ratio; C.I.: confidence interval.

socio-demographic groups, it may be worthwhile to investigate methods of increasing exposure to ASBs, such as through demonstrations, as increased exposure could potentially enhance their acceptance and the adoption of ASBs when and where they are deployed.

Male respondents exhibit a more favourable attitude towards ASBs, as do those who do not possess a driver's licence. Furthermore, younger respondents also demonstrate a more positive attitude towards ASBs. They perceive ASBs as more appealing, efficient, safe, and less congesting, with fewer emissions and driver errors compared to conventional shuttles. Younger respondents also believe that learning how to travel in an ASB is relatively easy. On the other hand, respondents with higher employment levels perceive ASBs as more attractive than conventional shuttles. Similarly, individuals from higher-income households hold a more positive attitude towards ASBs, perceiving them as more efficient and easier to learn how to use. These socio-demographic groups should be targeted to encourage the adoption of ASBs when and where they become available. For the socio-demographic groups that exhibit the opposite characteristics, it may be worthwhile to explore approaches to improve agreement regarding the benefits and opportunities associated with ASBs, as such improvements could enhance their willingness to adopt the use of these.

Respondents residing in peri-urban areas tend to be more concerned about fares when using ASBs compared to conventional shuttles. This concern is also observed among respondents with lower education levels. For both socio-demographic groups, it would be valuable to investigate whether or not fare structures based on spatial zones, time periods, and concession categories contribute to their concerns. Addressing these concerns related to fare structures, specifically in the context of ASB deployment, could help alleviate the worries and increase acceptance among these groups. Additionally, male respondents, those without a driver's licence, and individuals with higher employment levels are more concerned about traffic accidents when using ASBs compared to conventional shuttles. It is important to further explore how the automation of the driving task contributes to these perceived challenges and identify steps that can be taken to address these concerns specifically in the context of ASB deployment for these socio-demographic groups.

Addressing trust difficulties and worries about faulty technology can be carried out in several ways (Golbabaei et al., 2023):

- **Education and Awareness**: Policymakers should implement education campaigns that explain how ADRT works, its benefits, and the safety measures put in place. Transparency about technology can help alleviate fears.

- **Regulation and Standards**: Policymakers should establish stringent standards and regulations for ADRT systems. This would not only ensure safety but also promote public confidence in the technology.

- **Demonstrations and Trials**: Public demonstrations or pilot programmes can also help to increase public trust in ADRT. By seeing the technology in action and understanding its benefits first-hand, people might be more likely to trust and adopt it.
- **Addressing Equity Concerns**: A significant subset of the population that might be sceptical about ADRT could be those who worry about access and equity, particularly if they live in underserved areas or have limited mobility. Policymakers need to assure these communities that ADRT will be accessible and affordable to all, not just a privileged few.
- **Stakeholder Involvement**: Involving different stakeholders in the policymaking process can also build trust. This could include public forums or consultations where citizens can express their views and contribute to decision-making about ADRT.
- **Data Privacy and Security Measures**: Given the digital nature of ADRT, data privacy and cybersecurity are crucial. Policymakers should define clear guidelines to protect user data and ensure that robust cybersecurity measures are in place.

The ultimate goal for policymakers should be to foster a favourable public opinion towards ADRT while ensuring safety, accessibility, and trust in the technology. They should continuously gauge public sentiment and address concerns proactively to promote widespread acceptance and adoption.

4.6 Conclusion

The implementation of ADRT as a feeder to regular public transit holds the potential to enhance the effectiveness of public transportation. While autonomous trains and trams are already widely integrated into public transit systems worldwide (Fraszczyk & Mulley, 2017; Pakusch & Bossauer, 2017), the acceptability of pioneering ADRT services, such as ASBs, raises questions and concerns. To address these issues, our study focused on understanding the social dynamics behind how different groups perceive ADRT mobility, specifically ASBs, in the SEQ region, Australia. By exploring the perceptions and attitudes of individuals based on factors such as gender, age, education, employment, income, household size, residential location, and the possession of a driver's licence, we gained insights into the opportunities and challenges associated with these innovative transportation services in urban areas.

The findings from the present study provide valuable insights for alleviating concerns and increasing the adoption of automated driving and ride-sharing technologies (ADRT) in the SEQ region. These insights can serve as a useful guide for planners, suppliers, and policymakers, helping them cater to the demands and preferences of current and potential users, considering the variations in socio-demographic characteristics.

Our findings revealed the following key points: (a) The primary perceived opportunity of ASBs was the potential to reduce congestion and emissions, while the main anticipated challenge was related to concerns about the reliability of the technology; (b) fully employed respondents showed greater familiarity with AVs, while females and individuals from lower-income households had less experience riding in any form of AV; (c) male respondents, younger individuals, those with higher employment and incomes, and individuals without a driver's licence held a more favourable opinion of ASBs. Additionally, male respondents, those with higher employment and incomes, and those without a driver's licence expressed greater concern about traffic accidents when using ASBs. Less-educated respondents and individuals living in peri-urban areas were more concerned about fares.

We employed a methodological approach utilising binary and ordinal logistic regression modelling to understand the significance of socio-demographic variables in predicting changes in travel characteristics. This approach, supported by odds ratios, allowed us to analyse how variations in socio-demographic factors affected the likelihood of changes in travel characteristics. By identifying significant predictor variables and their odds ratios for each travel characteristic, our methodology provided valuable insights to inform policies and practices in order to address key issues (e.g., safety concerns) and target specific groups (particularly females) when planning public communication strategies to enhance receptiveness to ADRT.

To promote the adoption and future uptake of ADRT, policymakers should focus on fostering favourable attitudes (e.g., highlighting perceived opportunities) and addressing existing unfavourable attitudes (e.g., addressing perceived challenges). Our findings emphasise the importance of avoiding pilot operations that lead to negative experiences and fail to meet mobility demands. Providing reliable, effective, and convenient ADRT services is crucial for alleviating prospective users' concerns. Measures such as information screens and easy, obstacle-free access to vehicles can compensate for the absence of a driver, as suggested by Pigeon et al. (2021). In terms of deployment locations for ADRT, normal urban traffic conditions are currently perceived as less acceptable. Instead, deployment in secure contexts, such as dedicated routes, campuses, or areas with no existing public transport links such as peri-urban regions, is generally seen as desirable.

While conducting this research, certain simplifications were made, which may have resulted in limitations to the present study. Most survey respondents had no experience riding in an ASB, and thus, some of our conclusions are based on prospective users' perceptions (stated preference) rather than actual users' opinions (revealed preference), which may limit their generalisability. Future research could include individuals who have used these services once they become available, as demonstrated by Dennis et al. (2021) in their study on autonomous shuttles. Longitudinal studies exploring adoption attitudes over different time intervals could also provide valuable insights by recognising patterns over time and identifying significant outcomes (Haboucha et al., 2017).

It is worth noting that our study, like most previous quantitative surveys, primed respondents by listing specific potential opportunities and challenges associated with ASBs before assessing their opinions. This approach may lead individuals to perceive these issues as potential problems, even if they have minimal influence on their decision to use ASBs. An alternative approach to enhance ASB adoption could prioritise communication on aspects of deployment that users consider more important, to alleviate existing concerns, rather than addressing perceived problems of low importance. To facilitate this, data collection procedures should allow respondents to proactively raise issues, rather than directing their attention to aspects of ASB deployment they may not have considered otherwise (Haboucha et al., 2017; Pettigrew, Worrall et al., 2019).

While the target respondents of our study were the public, it is important to acknowledge that the benefits of ADRT might be particularly significant for the transport-disadvantaged population. Further studies can focus on specific socio-demographic groups in more detail, such as elderly individuals and people with disabilities, to better understand their demands and challenges (Nordhoff et al., 2019; Pigeon et al., 2021).

In future research, the methodology employed in this study can be replicated for the SEQ region. By comparing results between panel data, reasons for similarities and differences in travel characteristics over time can be investigated, particularly in response to geographical and socio-demographic shifts, as well as changes in policy and practice related to personal transport. This methodology is directly transferable to different regions, allowing for comparisons to identify similarities and differences in travel characteristics between different areas.

Further research will employ structural equation modelling with this dataset to gain deeper insights. Cross-referencing the results of this study will help determine the implications for each methodology and enable a comprehensive interpretation of findings. An extensive hypothesis testing approach is likely to benefit the analysis (Paz, 2019; Golbabaei et al., 2022; Beeramoole et al., 2023).

Acknowledgements

This chapter, with permission from the copyright holder, is a reproduced version of the following journal article: Golbabaei, M., Yigitcanlar, T., Paz, A.,

& Bunker, J., (2023). Perceived opportunities and challenges of autonomous demand-responsive transit use: What are the socio-demographic predictors? *Sustainability*, 15(15), 11839.

References

Acheampong, R.A., & Cugurullo, F. (2019). Capturing the behavioural determinants behind the adoption of autonomous vehicles: Conceptual frameworks and measurement models to predict public transport, sharing and ownership trends of self-driving cars. *Transportation Research Part F: Traffic Psychology and Behaviour*, 62, 349–375.

Ainsalu, J., Arffman, V., Bellone, M., Ellner, M., Haapamäki, T., Haavisto, N., & Åman, M. (2018). State of the art of automated buses. *Sustainability*, 10, 3118.

Alessandrini, A., Alfonsi, R., Site, P.D., & Stam, D. (2014). Users' preferences towards automated road public transport: Results from European surveys. *Transportation Research Procedia*, 3, 139–144.

Alessandrini, A., Delle Site, P., Stam, D., Gatta, V., Marcucci, E., & Zhang, Q. (2017). Using repeated-measurement stated preference data to investigate users' attitudes towards automated buses within major facilities. In Advances in Systems Science: Proceedings of the International Conference on Systems Science 2016 (ICSS 2016) 19 . Springer International Publishing: Berlin/Heidelberg, Germany pp. 189–199.

Alessandrini, A., Delle Site, P., Zhang, Q., Marcucci, E., & Gatta, V. (2016). Investigating users' attitudes towards conventional and automated buses in twelve European cities. *International Journal of Transport Economics*, 43, 413–436.

Allison, P.D. (2012). *Logistic Regression Using SAS: Theory and Application*; SAS Institute: Singapore.

Anania, E., Rice, S., Walters, N., Pierce, M., Winter, S., & Milner, M. (2018). The effects of positive and negative information on consumers' willingness to ride in a driverless vehicle. *Transport Policy*, 72, 218–224.

Australian Bureau of Statistics (2021). Available online: www.abs.gov.au/statistics/economy/national-accounts/australian-national-accounts-state-accounts/latest-release (accessed on 18 November 2022).

Bansal, P., Kockelman, K.M., & Singh, A. (2016). Assessing public opinions of and interest in new vehicle technologies: An Austin perspective. *Transportation Research Part C: Emerging Technologies*, 67, 1–14.

Beeramoole, P., Arteaga, C., Haque, M., Pinz, A., & Paz, A. (2023). Extensive hypothesis testing for estimation of mixed-Logit models. *Journal of Choice Modelling*, 47, 100409.

Beiker, S.A. (2019). Deployment of automated driving as an example for the San Francisco Bay area. In *Road Vehicle Automation 5*; Springer: Cham,, pp. 117–129.

Ben-Akiva, M., McFadden, D., Gärling, T., Gopinath, D., Walker, J., Bolduc, D., & Rao, V. (1999). Extended framework for modeling choice behavior. *Marketing Letters*, 10, 187–203.

Brant, R. (1990). Assessing proportionality in the proportional odds model for ordinal logistic regression. *Biometrics*, 46, 1171–1178.

Briggs, S.R., & Cheek, J.M. (1986). The role of factor analysis in the development and evaluation of personality scales. *Journal of Personality*, 54, 106–148.

Bursac, Z., Gauss, C.H., Williams, D.K., & Hosmer, D.W. (2008). Purposeful selection of variables in logistic regression. *Source Code for Biology and Medicine*, 3, 17.

Butler, L., Yigitcanlar, T., & Paz, A. (2021). Factors influencing public awareness of autonomous vehicles: Empirical evidence from Brisbane. *Transportation Research Part F: Traffic Psychology and Behaviour*, 82, 256–267.

Cunningham, M.L., Regan, M.A., Horberry, T., Weeratunga, K., & Dixit, V. (2019). Public opinion about automated vehicles in Australia: Results from a large-scale national survey. *Transportation Research Part A: Policy and Practice*, 129, 1–18.

Daoud, J.I. (2017). Multicollinearity and regression analysis. *Journal of Physics: Conference Series*, 949, 012009.

Dekker, M. (2017). Riding a Self-Driving Bus to Work: Investigating How Travellers Perceive ADS-DVs on the Last Mile. Master's Thesis, Delft University of Technology, Delft, the Netherlands.

Dennis, S., Paz, A., & Yigitcanlar, T. (2021). Perceptions and attitudes towards the deployment of autonomous and connected vehicles: Insights from Las Vegas, Nevada. *Journal of Urban Technology*, 28, 75–95.

Dong, X., DiScenna, M., & Guerra, E. (2017). Transit user perceptions of driverless buses. *Transportation*, 46, 35–50.

Fagnant, D.J., & Kockelman, K. (2015). Preparing a nation for autonomous vehicles: Opportunities, barriers and policy recommendations. *Transportation Research Part A: Policy and Practice*, 77, 167–181.

Fraszczyk, A, & Mulley, C. (2017). Public perception of and attitude to driverless train: A case study of Sydney, Australia. *Urban Rail Transit*, 3, 100–111.

George, D., & Mallery, P. (2021). *IBM SPSS Statistics 27 Step by Step: A Simple Guide and Reference*, 17th ed.; Routledge: New York, NY.

Gkartzonikas, C., & Gkritza, K. (2019). What have we learned? A review of stated preference and choice studies on autonomous vehicles. *Transportation Research Part C: Emerging Technologies*, 98, 323–337.

Golbabaei, F., Yigitcanlar, T., & Bunker, J. (2021). The role of shared autonomous vehicle systems in delivering smart urban mobility: A systematic review of the literature. *International Journal of Sustainable Transportation*, 15, 731–748.

Golbabaei, F., Paz, A., Yigitcanlar, T., & Bunker, J. (2023). Navigating autonomous demand responsive transport: Stakeholder perspectives on deployment and adoption challenges. *International Journal of Digital Earth*, 17(1), 2297848.

Golbabaei, F., Yigitcanlar, T., Paz, A., & Bunker, J. (2020). Individual predictors of autonomous vehicle public acceptance and intention to use: A systematic review of the literature. *Journal of Open Innovation*, 6, 106.

Golbabaei, F ., Yigitcanlar, T., Paz, A., & Bunker, J. (2022). Understanding autonomous shuttle adoption intention: Predictive power of pre-trial perceptions and attitudes. *Sensors*, 22, 9193.

Haboucha, C.J., Ishaq, R., & Shiftan, Y. (2017). User preferences regarding autonomous vehicles. *Transportation Research Part C: Emerging Technologies*, 78, 37–49.

Harb, M., Xiao, Y., Circella, G., Mokhtarian, P.L., & Walker, J.L. (2018). Projecting travelers into a world of self-driving vehicles: Estimating travel behavior implications via a naturalistic experiment. *Transportation*, 45, 1671–1685.

Iclodean, C., Cordos, N., & Varga, B.O. (2020). Autonomous shuttle bus for public transportation: A review. *Energies*, 13, 2917.

Kaur, K., & Rampersad, G. (2018). Trust in driverless cars: Investigating key factors influencing the adoption of driverless cars. *Journal of Engineering and Technology Management*, 48, 87–96.

König, M., & Neumayr, L. (2017). Users' resistance towards radical innovations: The case of the self-driving car. *Transportation Research Part F: Traffic Psychology and Behaviour*, 44, 42–52.

Kostorz, N., Hilgert, T., Kagerbauer, M., & Vortisch, P. (4–6 September 2019). What do people think about autonomous minibuses in Germany. In Proceedings of the Symposium der European Association for Research in Transportation (hEART), Budapest, Hungary.

Krejcie, R.V., & Morgan, D.W. (1970). Determining sample size for research activities. *Educational and Psychological Measurement*, 30, 607–610.

Kyriakidis, M., Happee, R., & de Winter, J.C.F. (2015). Public opinion on automated driving: Results of an international questionnaire among 5000 respondents. *Transportation Research Part F: Traffic Psychology and Behaviour*, 32, 127–140.

Ledger, S.A., Cunningham, M.L., & Regan, M.A (2022). Public opinion about automated and connected vehicles in Australia and New Zealand: Results from the 2nd ADVI public opinion survey. In ADVI Australia and New Zealand Driverless Vehicle Initiative Project; 2018. 28th ARRB International Conference – Next Generation Connectivity. Available online: https://trid.trb.org/view/1987 511 (accessed on 18 November 2022).

Liu, X. (2015). *Applied Ordinal Logistic Regression Using Stata: From Single-Level to Multilevel Modelling*; Sage Publications: Newbury Park, CA.

Madigan, R., Louw, T., Wilbrink, M., Schieben, A., & Merat, N. (2017). What influences the decision to use automated public transport? Using UTAUT to understand public acceptance of automated road transport systems. *Transportation Research Part F: Traffic Psychology and Behaviour*, 50, 55–64.

Madigan, R., Louw, T., Dziennus, M., Graindorge, T., Ortega, E., Graindorge, M., & Merat, N. (2016). Acceptance of automated road transport systems (ARTS): An adaptation of the UTAUT model. *Transportation Research Procedia*, 14, 2217–2226.

McHugh, M.L. (2009). The odds ratio: Calculation, usage, and interpretation. *Biochemia Medica*, 19, 120–126.

Milakis, D., van Arem, B., & van Wee, B. (2017). Policy and society related implications of automated driving: A review of literature and directions for future research. *Journal of Intelligent Transportation Systems*, 21, 324–348.

Millonig, A., & Fröhlich, P. (23–25 September 2018). Where autonomous buses might and might not bridge the gaps in the 4 A's of public transport passenger needs: A review. In Proceedings of the International Conference on Automotive User Interfaces and Interactive Vehicular Applications, Toronto, ON, Canada.

Mortoja, M.G., & Yigitcanlar, T. (2021). Public perceptions of peri-urbanism triggered climate change: Survey evidence from South East Queensland, Australia. *Sustainable Cities and Society*, 75, 103407.

Moták, L., Neuville, E., Chambres, P., Marmoiton, F., Monéger, F., Coutarel, F., & Izaute, M. (2017). Antecedent variables of intentions to use an autonomous shuttle: Moving beyond TAM and TPB? *European Review of Applied Psychology*, 67, 269–278.

Mouratidis, K., & Cobeña Serrano, V. (2021). Autonomous buses: Intentions to use, passenger experiences, and suggestions for improvement. *Transportation Research Part F: Traffic Psychology and Behaviour*, 76, 321–335.

Narayanan, S., Chaniotakis, E., & Antoniou, C. (2020). Shared autonomous vehicle services: A comprehensive review. *Transportation Research Part C: Emerging Technologies*, 111, 255–293.

Nastjuk, I., Herrenkind, B., Marrone, M., Brendel, A.B., & Kolbe, L.M. (2020). What drives the acceptance of autonomous driving? An investigation of acceptance factors from an end-user's perspective. *Technological Forecasting and Social Change*, 161, 120319.

Nenseth, V., Ciccone, A., & Kristensen, N.B. (2019). Societal consequences of automated vehicles–Norwegian scenarios; TØI report (1700/2019); Institute of Transport Economics: Oslo, Norway.

Nordhoff, S., Kyriakidis, M., van Arem, B., & Happee, R. (2019). A multi-level model on automated vehicle acceptance (MAVA): A review-based study. *Theoretical Issues in Ergonomics Science*, 20, 682–710.

Nordhoff, S., Stapel, J., van Arem, B., & Happee, R. (2020). Passenger opinions of the perceived safety and interaction with automated shuttles: A test ride study with 'hidden' safety steward. *Transportation Research Part A: Policy and Practice*, 138, 508–524.

Nordhoff, S., de Winter, J., Madigan, R., Merat, N., van Arem, B., & Happee, R. (2018). User acceptance of automated shuttles in Berlin-Schöneberg: A questionnaire study. *Transportation Research Part F: Traffic Psychology and Behaviour*, 58, 843–854.

Nordhoff, S., Van Arem, B., Merat, N., Madigan, R., Ruhrort, L., Knie, A., & Happee, R. (19–22 June 2017). User acceptance of driverless shuttles running in an open and mixed traffic environment. In Proceedings of the 12th ITS European Congress, Strasbourg, France.

Norton, E.C., Dowd, B.E., & Maciejewski, M.L. (2018). Odds ratios—Current best practice and use. *Journal of the American Medical Association*, 320, 84–85.

Nunes, A., Woodley, L., & Rossetti, P. (2022). Re-thinking procurement incentives for electric vehicles to achieve net-zero emissions. *Nature Sustainability*, 5, 527–532.

Paddeu, D., Parkhurst, G., & Shergold, I. (2020). Passenger comfort and trust on first-time use of a shared autonomous shuttle vehicle. *Transportation Research Part C: Emerging Technologies*, 115, 102604.

Pakusch, C., & Bossauer, P. (24–26 July 2017). User acceptance of fully autonomous public transport. In Proceedings of the 14th International Joint Conference on e-Business and Telecommunications (ICETE 2017), Madrid, Spain.

Pallant, J. (2020). *SPSS Survival Manual: A Step by Step Guide to Data Analysis Using IBM SPSS*; Routledge: New York, NY.

Paz, A., Arteaga, C., & Cobos, C. (2019). Specification of mixed logit models assisted by an optimization framework. *Journal of Choice Modelling*, 30, 50–60.

Pettigrew, S., Dana, L.M., & Norman, R. (2019). Clusters of potential autonomous vehicles users according to propensity to use individual versus shared vehicles. *Transport Policy*, 76, 13–20.

Pettigrew, S., Talati, Z., & Norman, R. (2018). The health benefits of autonomous vehicles: Public awareness and receptivity in Australia. *Australian and New Zealand Journal of Public Health*, 42, 480–483.

Pettigrew, S., Worrall, C., Talati, Z., Fritschi, L., & Norman, R. (2019). Dimensions of attitudes to autonomous vehicles. *Urban Planning and Transport Research*, 7, 19–33.

Pigeon, C., Alauzet, A., & Paire-Ficout, L. (2021). Factors of acceptability, acceptance and usage for non-rail autonomous public transport vehicles: A systematic literature review. *Transportation Research Part F: Traffic Psychology and Behaviour*, 81, 251–270.

Portouli, E., Karaseitanidis, G., Lytrivis, P., Amditis, A., Raptis, O., & Karaberi, C. (11–14 June 2017). Public attitudes towards autonomous mini buses operating in real conditions in a Hellenic city. In Proceedings of the 2017 IEEE Intelligent Vehicles Symposium (IV), Los Angeles, CA, USA, pp. 571–576.

Regan, M., Cunningham, M., Dixit, V., Horberry, T., Bender, A., Weeratunga, K., & Hassan, A. (2017). Preliminary findings from the first Australian national survey of public opinion about automated and driverless vehicles. *Transportation.* https://doi.org/10.13140/RG.2.2.11446.80967

Rehrl, K., & Zankl, C. (2018). Digibus©: Results from the first self-driving shuttle trial on a public road in Austria. *European Transport Research Review*, 10, 51.

Roche-Cerasi, I. (2019). Public acceptance of driverless shuttles in Norway. *Transportation Research Part F: Traffic Psychology and Behaviour*, 66, 162–183.

Salonen, A.O. (2018). Passenger's subjective traffic safety, in-vehicle security and emergency management in the driverless shuttle bus in Finland. *Transport Policy*, 61, 106–110.

Schoettle, B., & Sivak, M. (2014). *A Survey of Public Opinion about Autonomous and Self-Driving Vehicles in the US, the UK, and Australia*; University of Michigan, Transportation Research Institute: Ann Arbor, MI.

Shabanpour, R., Golshani, N., Shamshiripour, A., & Mohammadian, A.K. (2018). Eliciting preferences for adoption of fully automated vehicles using best-worst analysis. *Transportation Research Part C: Emerging Technologies*, 93, 463–478.

Silva, Ó., Cordera, R., González-González, E., & Nogués, S. (2022). Environmental impacts of autonomous vehicles: A review of the scientific literature. *Science of the Total Environment*, 830, 154615.

Soteropoulos, A., Berger, M., & Ciari, F. (2019). Impacts of automated vehicles on travel behaviour and land use: An international review of modelling studies. *Transport Reviews*, 39, 29–49.

Spurlock, C.A., Sears, J., Wong-Parodi, G., Walker, V., Jin, L., Taylor, M., & Todd, A. (2019). Describing the users: Understanding adoption of and interest in shared, electrified, and automated transportation in the San Francisco Bay Area. *Transportation Research Part D: Transport and Environment*, 71, 283–301.

Stocker, A., & Shaheen, S. (2019). Shared automated vehicle (SAV) pilots and automated vehicle policy in the US: Current and future developments. In *Road Vehicle Automation 5*; Springer: Cham, pp. 131–147.

Szumilas, M. (2010). Explaining odds ratios. *Journal of the American Academy of Child & Adolescent Psychiatry*, 19, 227–229.

Taiebat, M., Brown, A.L., Safford, H.R., Qu, S., & Xu, M. (2018). A review on energy, environmental, and sustainability implications of connected and automated vehicles. *Environmental Science & Technology*, 52, 11449–11465.

Thapa, D., Gabrhel, V., & Mishra, S. (2021). What are the factors determining user intentions to use AV while impaired? *Transportation Research Part F: Traffic Psychology and Behaviour*, 82, 238–255.

Vidhi, R., & Shrivastava, P. (2018). A review of electric vehicle lifecycle emissions and policy recommendations to increase EV penetration in India. *Energies*, 11, 483.

Wadud, Z., MacKenzie, D., & Leiby, P. (2016). Help or hindrance? The travel, energy and carbon impacts of highly automated vehicles. *Transportation Research Part A: Policy and Practice*, 86, 1–18.

Wien, J. (2019). An Assessment of the Willingness to Choose a Self-Driving Bus for an Urban Trip: A Public Transport User's Perspective. Master's Thesis, Delft University of Technology, Delft, the Netherlands.

Winter, K., Cats, O., Correia, G., & van Arem, B. (2018). Performance analysis and
    fleet requirements of automated demand-responsive transport systems as an
    urban public transport service. *International Journal of Transportation Science and
    Technology, 7*, 151–167.
Winter, K., Wien, J., Molin, E., Cats, O., Morsink, P., & van Arem, B. (5–7 June 2019).
    Taking the self-driving bus: A passenger choice experiment. In Proceedings of
    the 2019 6th International Conference on Models and Technologies for Intelligent
    Transportation Systems (MT-ITS), Cracow, Poland.
Wu, Y., & Zhang, L. (2017). Can the development of electric vehicles reduce the
    emission of air pollutants and greenhouse gases in developing countries?
    Transportation Research Part D: Transport and Environment, 51, 129–145.

5

Navigating Autonomous Demand-Responsive Transport

5.1 Introduction and Background

The emergence of electric and autonomous vehicles (AVs) is the biggest disruptor facing the Australian transport industry today. To turn this challenge into a greater opportunity, the industry could focus on embracing and integrating these technologies into their existing operations, developing new infrastructure and charging stations for electric vehicles, investing in the research and development of autonomous vehicles, and reducing their carbon footprint (Faisal et al., 2021). Additionally, embracing the sharing economy model could be an opportunity for the industry to revolutionise the way people move around cities, positioning themselves as leaders in the global movement towards more sustainable and efficient transportation systems (Golbabaei et al., 2023).

Government initiatives to provide more customer-focused services have elevated demand-responsive transit (DRT) from a specialty mode to a widely accepted mode of public transportation (Hensher, 2017; Butler et al., 2021). This mode provides door-to-door or stop-to-stop travel in response to passenger requests using vehicles that vary in size and operate on flexible routes and schedules. Despite its benefits, DRT has faced challenges as private and public providers have had to depend on subsidies to cover their costs, largely due to low demand (Currie & Fournier, 2020). The need for extra resources to schedule services on-demand and process requests, which are now made through technology, has added to costs during a time when low participation is a major concern. This, combined with the inflexible structure of many bus contracts, has led to DRT facing significant institutional challenges, resulting in some governments phasing out or prohibiting its provision (Perera et al., 2020).

Governments are committed to encouraging the adoption of eco-friendly transportation solutions that offer a more convenient and efficient experience for urban dwellers. With cities becoming more densely populated and

DOI: 10.1201/9781003605676-5

the overuse of personal vehicles contributing to traffic and environmental problems, the arrival of autonomous demand-responsive transport (ADRT) is a desired development. Leading-edge technology allows ADRT providers to deliver a seamless experience, and the business and academic communities are investing heavily in making autonomous driving a reality. ADRT also has the potential to significantly alleviate traffic congestion, especially in the form of autonomous shuttle buses (Palmer et al., 2004; Arbib & Seba, 2017; Golbabaei et al., 2020, 2021). If cost savings and the ease of not having to find parking spaces can convince people to switch, ADRTs have the chance to replace conventional cars. Some reports predict replacement rates between 5 and 30% (Bierstedt et al., 2014; Davidson & Spinoulas, 2016; Fernandez & Urbano, 2012). However, the continued use of public transportation, with policies aimed at reducing the use of personal vehicles, is assumed to play a big role in determining the success of this transition (Webb et al., 2019).

There have been over 30 automated vehicle trials in Australia to date, in every state and territory. Based on National Transport Commission (2020), the majority of trials have involved low-speed automated shuttle buses operating on set routes in various parts of Australia including coastal areas in the South East Queensland (SEQ) region, such as Karragarra Island, Cleveland, and Main Beach. SEQ, an Australian metropolitan area centred on the capital city of Brisbane (see Figure 5.1), has a land area of 35,248 km^2 and a population of 3,817,573 million (2021). The per capita gross state product of Queensland is AU\$71,037 (US\$53,280). Despite their advancement, public confidence in these leading-edge technologies is still in its early stages (Haboucha et al., 2015; Roche-Cerasi, 2019; Mouratidis & Cobeña Serrano, 2021). This presents a critical obstacle that must be overcome to ensure the success of the project, as the level of public trust plays a direct role in the shuttle's ridership and efficiency (Nordhoff et al., 2018; Chen et al., 2020; Papadima et al., 2020). The transportation industry can assess public support to make well-informed decisions on policy and management, such as creating pricing or subsidy models. The fully automated and renewables powered ADRT can then be tailored to meet public attitudes and preferences (Golbabaei et al., 2022; Wang et al., 2022).

While the role of government in shaping autonomous transport is widely recognised (Porter, 2018), a comprehensive understanding of how it impacts the field is lacking. Governments have several options to promote the next stage of automated transport. Additionally, there is a chance to evaluate Australia's preparedness for the widespread use of these vehicles by considering multiple factors, such as trials, regulations, infrastructure, and public perception. Previous studies on the topic have primarily focused on regulation (Shladover & Nowakowski, 2019; Mordue et al., 2020), ignoring other methods of governance. The literature also tends to concentrate on how to react to the technology rather than exploring proactive efforts commonly seen in other innovation areas (Salmenkaita & Salo, 2002). Governments are

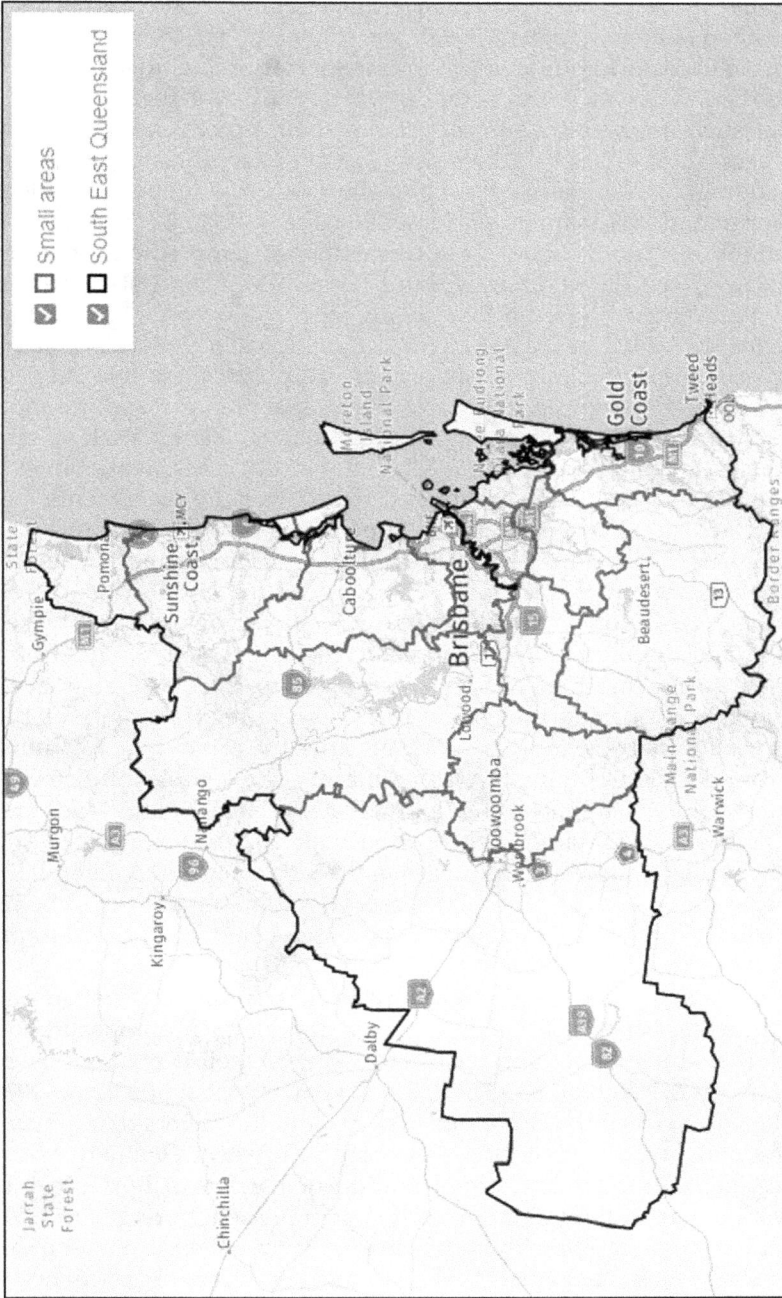

FIGURE 5.1
The South East Queensland local government area.

crucial in guiding technological advancements to prevent market failures and negative consequences, particularly concerning disruptive ADRT technology (Moon & Bretschneider, 1997; Stone et al., 2018).

Their role in fostering innovation in transportation includes both direct methods like R&D incentives and indirect ones like regulations. The timing of such interventions is pivotal for their effectiveness (Docherty, 2018; Freemark et al., 2019). Balancing the need to protect public safety and encourage autonomous transport innovation can be a challenge for governments (Shladover & Nowakowski, 2019). To promote the adoption of electric vehicles, governments can implement measures like emissions standards, charger safety regulations, and standardising vehicle requirements (Steinhilber et al., 2013). However, government approaches to ADRT innovation will differ globally, depending on the country's economic, technological, social, and political context (Li et al., 2019). ADRT technologies will require changes in planning, policy coordination, infrastructure investments, and revenue collection at all levels of government (Freemark et al., 2019; Mladenovic, 2019; Manivasakan et al., 2021). Sperling et al. (2018) recommend that national governments focus on vehicle design, while local governments focus on vehicle use.

The majority of literature on autonomous transport policy focuses on adapting current and future regulations to accommodate ADRT technology. There is a lot of interest in how safety, cybersecurity, privacy, licencing, and liability rules will need to be revised (Claybrook & Kildare, 2018; Freemark et al., 2019; Lee & Hess, 2020). Shladover and Nowakowski (2019) note that while road vehicle regulations have been stable, autonomous technology disrupts this by removing the human driver from decision-making. These regulatory challenges, influenced by ethical and value-based interpretations, hinder the technology's advancement (Claybrook & Kildare, 2018; Mordue et al., 2020). Furthermore, strict governmental oversight is slowing down the technology's commercialisation (Sperling et al., 2018). Hence, it is crucial to understand not only how ADRTs will affect regulations but also how these regulations will impact ADRT development.

Thus, to gain an in-depth exploratory understanding of enablers and hurdles in adopting ADRT in SEQ, expert interviews are appropriate to drive additional early adoption measures owing to their proficiency in this area (Grimsley & Meehan, 2007; Wu, 2011). As per Metaxiotis et al. (2019), qualitative approaches could provide insights into the factors that influence whether a new product or service is accepted or rejected. Qualitative methodologies enable us to discover opportunities and barriers to the deployment of innovation, particularly in the market launch phase (Quiring, 2006). This approach does not simply extract information from collected data but also treats the data as a living entity, leading to solid and trustworthy discoveries (Holloway & Todres, 2003; Sandelowski & Barroso, 2006).

Similar methods were employed in the latest adoption studies in the transport field to create a solid foundation for acceptance criteria (e.g., Amann, 2017). Mars et al. (2016) examined travel behaviour studies that employed qualitative analysis, while Simons et al. (2014) utilised grounded theory to uncover the forces that shape the travel mode choices of students and working young adults. Qualitative studies have also delved into the factors that influence the selection of specific modes of transportation, such as bicycles (Fishman et al., 2012; Sherwin et al., 2014), personal vehicles (Gardner and Abraham, 2007; Nguyen-Phuoc et al., 2018), and public transit (Beirão & Cabral, 2007). Additionally, in an earlier study, travel self-containment in Australian master planned estates was analysed (Yigitcanlar et al., 2007).

This exploratory study aims to investigate feasible ADRT service models and potential customer groups as well as deployment and adoption challenges and mitigations to establish this concept from transport stakeholders' perspectives. For our qualitative study, we designed a semi-structured questionnaire, to address three following research questions:

1. What do transport experts believe the feasible ADRT service models would be?

2. Who do transport experts believe the potential customer groups of ADRT would be?

3. What do transport experts believe the ADRT development and adoption challenges are?

4. How could transport decision-makers mitigate the ADRT deployment and adoption challenges?

Following this introduction, and review of relevant policy studies, Section 5.2 explains the adapted qualitative survey and analysis method. Section 5.3 details the findings regarding the ADRT service concept, potential customer groups, and potential use cases, followed by an explanation of the ADRT deployment and adoption challenges and the mitigation strategies. Section 5.4 discusses the main takeaways and exemplifies the policy recommendations and limitations and future research potentials. Section 5.5 presents the concluding remarks for the chapter.

5.2 Research Design

The following sections describe the data collection procedure of the interview sessions.

5.2.1 Questionnaire Design

There are three categories of interview approaches: (a) structured, (b) unstructured, and (c) semi-structured interviews. Structured interviews use a predetermined set of questions, while the unstructured type is conducted spontaneously, with no predetermined questions. Semi-structured forms are intertwined between these two categories and are the most used type in qualitative research (Robson, 2002).

The semi-structured interview incorporates predefined open-ended questions, which enable interviewees to convey their perspectives in their own words and give further detail about the main topics of the conversation (Robson, 2002; Smith et al., 2009). Associated sub-questions help to maintain the consistency of the structure of the study and data acquisition process, hence contributing to the accuracy of the information attained (Saunders et al., 2007). The semi-structured interview is considered to be the most appropriate method for exploratory research. According to Seidman (2013), semi-structured interviews enable us to uncover and acquire additional subjective findings. However, spontaneous additions were acceptable to capture more comprehensive and expressive responses where applicable (Liu et al., 2020).

The questionnaire guide is summarised in the following domains. For each domain, several questions were posed to cover the main aspects of that domain (Pancholi et al., 2019).

1. *ADRT user adoption perspectives*
   - Who do you believe would be the potential customer groups for autonomous DRT?
   - Do you believe that uptake of autonomous DRT would vary between potential customer groups, and if so, how?
2. *ADRT deployment challenges*
   - Do you believe that the needs of ADRT transit would vary between potential customer groups, and if so, how and why?
   - What characteristics of autonomous DRT do you believe are important for potential customers to accept this mode?
   - What characteristics do you believe are important to other road and transport system users to accept autonomous DRT in the fleet?
3. *ADRT planning recommendations*
   - What do you believe are the most significant barriers to the success of autonomous DRT, and how do you believe each barrier could be mitigated?
   - Do you believe that incentives would be important to encourage potential customer groups to use autonomous DRT, and if so, in what form/s?

5.2.2 Data Collection

The sample size in a qualitative study can vary, but for semi-structured interviews, a small sample can suffice (Qu & Dumay, 2011). Typically, most of the interviews per study fall between 11 and 20 (Marshall et al., 2013). Saturation, which is the point where all questions have been fully explored and no new ideas or themes emerge (Braun & Clarke, 2013), usually occurs after 12–24 interviews (Guest et al., 2006; Hennink et al., 2017). The data collection for the study was halted after 26 interviews (with 2 extra interviews) to ensure a diversity of viewpoints. According to Guest et al. (2006), most findings in qualitative research—over 70%—can be gained from the first six participants. By the time 12 interviews have been conducted, saturation is reached and 92% of the findings have been captured.

The first interview was the pilot interview, which uncovered a few small deficiencies that were later reviewed and modified (Hilgarter & Granig, 2020). All participants were interviewed individually to prevent the creation of any bias. At the beginning of each interview, the interviewer introduced a summary of the research and the purpose of the interview. The interviews were carried out in April/May 2022, with an average duration of 40 min. The interviews were audio-recorded and transcribed afterwards for further analysis. The general flow of the interview process is illustrated in Figure 5.2. Approaching appropriate interviewees is crucial for ensuring the credibility of the gathered information. According to Bolger and Wright (2011), multiple-perspective interviews help prevent one-sided assessments and give consistent insights and a broader spectrum of details.

As per Eisenhardt (1989), the current study employed a purposeful sampling technique to select the participants rather than random sampling (Kamruzzaman et al., 2015). According to Liu et al. (2020, p. 70),

> in-depth interviews with members of the scientific, and industry elites provide valuable insights that although could be critical to the exploration of a research topic, may not be obvious to the general public. This is because information on how elites perceive situations and make key decisions provides a unique perspective that often cannot be obtained through other data collection methods.

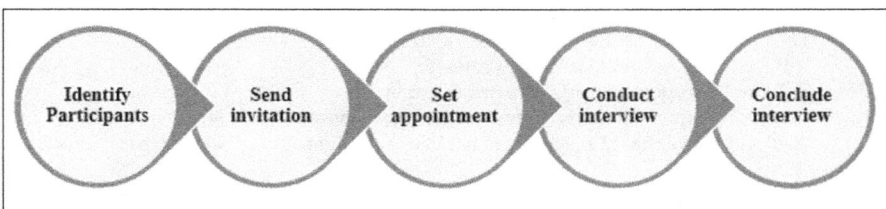

FIGURE 5.2
The general flow of the semi-structured interviews.

Hence, the target participants for this part of the study consisted of elites who occupy management and senior positions in various transport sectors including state transport authority, consulting, peak body, service providers, and academia who were engaged with Australia's industry environment. Table 5.1 outlines the key characteristics of the elite interview participants.

We recruited participants by sending email or LinkedIn messages with interview invitations to the target sample. We also used a degree of snowball sampling (Biernacki & Waldorf, 1981), whereby participants who were directly recruited referred colleagues who would be interested in participating in the research interview. No monetary participation bonuses were offered for recruiting purposes. In total 40 experts were selected and then officially invited to an online interview, of whom 26 replied and participated

TABLE 5.1

Expert Interviewee Details

No.	Affiliation	Sector
1	Department of Transport and Main Roads (TMR), Policy Director/Academic Partner	State transport authority
2	TMR, Executive Director (MaaS Program Management Office)	State transport authority
3	QUT, Deputy Director	Academia
4	ADVI-ARRB, Director	Consulting
5	ITS Australia, CEO	Peak body
6	ARRB, National Discipline Leader	Consulting
7	ITS Australia, FvaT Industry Reference Group	Peak body
8	TMR, Principal Business Analyst	State transport authority
9	Griffith University, Professor	Academia
10	NZTA, Technology Advisor	State transport authority
11	iMOVE, Education Manager	Academia
12	Department of Infrastructure, Director of Passenger Transport	State transport authority
13	Monash University, Director of Public Transport Research Group	Academia
14	APAC, Managing Director	Service provider
15	PATREC, Research Fellow	Academia
16	TMR, TransLink Division, Principal Network Planner	State transport authority
17	Technical principal transport planner	Consulting
18	Griffith University/Liftango	Academia
19	The University of Sydney, Senior Research Fellow	Academia
20	University of Western Australia, Professor	Academia
21	Deloitte, Transport and Infrastructure Partner	Consulting
22	RACQ, Transport Planning and Infrastructure Advisor	Service provider
23	TMR, Infrastructure Management and Delivery, Chief Engineer	State transport authority
24	University of South Australia, Professor	Academia
25	LaTrobe University, Professor	Academia
26	The George Institute, Professor	Academia

in the interview. Prior to their participation, the interviewees were told of the interview settings; they were advised and agreed that the sessions would be audio-recorded and transcribed, but the data would be anonymous (the requirements of the ethics approval from the university human research ethics committee; UHREC reference number: 2000000747) and used exclusively for the research purpose.

5.2.3 Data Analysis

We followed Braun and Clarke's (2006) impulsive thematic analysis technique, which aims to provide a dynamic foundation as opposed to the rigid structure used by the typical codebook approach. Figure 5.3 illustrates the six phases of our thematic analysis approach.

Thematic analysis was used in prior transportation research (Gössling et al., 2016; Hafner et al., 2017; Alyavina et al., 2020; Liu et al., 2020) and proved to be a

> sophisticated qualitative tool that allows for conducting research in a precise, consistent and exhaustive manner through recording, systematising, and disclosing the methods of analysis and the study results with enough detail to enable the reader to determine the credibility and validity of the process.
>
> *Nowell et al., 2017; Liu et al., 2020, p. 72*

Familiarizing yourself with your data	Transcribing data (if necessary), reading and re-reading the data, noting down initial ideas.
Generating initial codes	Coding interesting features of the data in a systematic fashion across the entire data set, collating data relevant to each code.
Searching for themes	Collating codes into potential themes, gathering all data relevant to each potential theme.
Reviewing themes	Checking if the themes work in relation to the coded extracts (Level 1) and the entire data set (Level 2), generating a thematic 'map' of the analysis
Defining and naming themes	Ongoing analysis to refine the specifics of each theme, and the overall story the analysis tells, generating clear definitions and names for each theme.
Producing the report	Selection of vivid, compelling extract examples, final analysis of selected extracts, relating back of the analysis to the research question and literature, producing a scholarly report of the analysis.

FIGURE 5.3

Phases of thematic analysis. (Braun & Clarke, 2006.)

According to Vaismoradi et al. (2013), qualitative descriptive methods like thematic analysis are better suited for research that calls for less interpretation, as opposed to the grounded theory which demands a more in-depth level of interpretation. Given the aim of this study, and the capability to uncover sufficient information with a minimal level of interpretation, thematic analysis was selected as the research method to identify, analyse, and present the overarching patterns (themes) found within the data (Braun & Clarke, 2006; Nikitas et al., 2018, 2019).

Thematic analysis in literature can be approached in two ways: the inductive and deductive methods. The deductive approach follows a pre-existing theory as a roadmap (Bengtsson, 2016; Lune & Berg, 2017), while the inductive approach allows analysing the contents with an open mind to uncover unanticipated topics and perspectives from interviews with expert participants that answer the research question. It also enables appreciation of the formal and informal dynamics involved in the policy-making process (Yigitcanlar, 2010). As suggested by Glaser and Strauss (1967) and Liu et al. (2020), the interviews were conducted, transcribed, and analysed through a manual process that involved multiple readings and annotating of the interview transcripts. Thematic analysis was chosen as the preferred method of analysis due to its versatility and the ability to uncover rich, complex accounts of the data. The freedom provided by the theoretical nature of thematic analysis allowed for an in-depth examination of the data, providing a detailed and nuanced understanding of the subject matter (Braun & Clarke, 2006).

As recommended by Pettigrew et al. (2019, p. 3), data coding is completed by a single coder due to "the little prior work in this area, the highly exploratory nature of the study, and the resulting emergent nature of the coding hierarchy". Erlingsson and Brysiewicz (2017) explain a code as "a label; a name that most exactly describes what this condensed meaning unit is about. Usually, one or two words long". In a similar vein to Yigitcanlar et al. (2023) and Dichabeng et al. (2021), the coding hierarchy covers deductive notions (factors extracted from the literature on AV adoption) and inductive notions (topics raised by experts). As a kind of member checking, emergent interpretations are discussed with the following interviewees to evaluate their utility (Wallendorf & Belk, 1989). The emerging themes were reviewed and reached consensus on the final interpretation.

Through the process of identifying themes, we note that some themes may have overlapping aspects, and a few statements may represent more than one theme. This, however, is not an issue as "the themes and the way these relate to each other do not have to be smoothed out or ignored but instead retain the tensions and inconsistencies within and across data" (Braun & Clarke, 2006; Liu et al., 2020, p. 73). It is important to emphasise that the themes were derived based on the importance of the interviewee's perspectives, not just the number of participants mentioning a theme (i.e., theme frequency). The frequency does not necessarily indicate the validity of the theme (Yeganeh et al., 2022).

5.3 Analysis and Results

This qualitative study revealed five main themes from the thematic analysis including

1. ADRT feasible service models
2. ADRT potential customer groups
3. ADRT deployment and adoption challenges
4. ADRT deployment and adoption policy recommendations

The findings of the study are presented in the following sections, which are organised according to the themes identified. Each theme is thoroughly described with its sub-themes and quoted examples from the participants to provide a clearer understanding of the concepts. The themes, while distinct, may have overlapping dimensions, which are reported as sub-themes in the scope of this study. To convey the thematic analysis effectively and objectively, selected relevant quotations are used to provide evidence for the themes presented. This method of presenting the findings through quotes is one of the most concrete ways to deliver the analysis (Nikitas et al., 2019; Liu et al., 2020).

5.3.1 Feasible Service Models

Across all stakeholder groups, there was agreement that there are several scenarios for offering ADRT service in SEQ, i.e., the first mile/last mile, access/egress mode from transport hubs, or localised on-demand transit on a more commercial note.

Generally, it is foreseen that the first/last mile integration is going to be more attractive to an on-demand service that feeds into a transit trunk line. So, the needs will be quite different and the service models that support their technology will also be quite different. This is probably about distinguishing different technology for every single customer growth segment. It will also indicate employment types, destination choice, and location of origin. These matters will depend upon the service provider and the technology provider to accommodate:

- The needs around trip facilities that need to be accommodated, i.e.,
  - If any disabled persons may require additional services.
  - If people are travelling in a group.
  - If there is luggage.
- The needs around the systems and technologies that need to be supporting this actual service, i.e.,

- Selection criteria that might make people feel more comfortable. Depending on the technology and the volume of subscribers, we can consider more tailored offerings for people to build their confidence.
- Understanding who else might be in the vehicle with the passenger.
- Being able to choose a vehicle with passengers of the same gender.
- The needs around trips, i.e.,
  - If they have multiple stops because they are setting down multiple passengers, reliability is important regarding when they are going to leave and when they are going to arrive.
  - There will be matters that might need to be tailored in the future because the timeliness of the trip, the cost of the trip, and the reliability and convenience of that will be the drivers of uptake.

Many experts agreed that the vehicle concept requires consideration of the trade-off between costs, feasibility, demand, and government subsidy. So the real factor in driving service design is the distribution of origins and destinations of passengers, as evidenced in the following quote:

> Let the private industry service the mass market for on-demand ADRT and then I'll let the right agencies deal with areas that don't stack up from a commercial player to operate or, as is the case, some agencies might have both and then they just subsidize or incentivize the road the transport provider to service those other lower volume routes with the money that they earn from the high-volume routes. These are all existing models for running public transport that I think cannot be adopted.
>
> *Consultant*

Of note is that dysfunctions in the network will drive the behaviour of people using the network that is offered. The challenge for adoption is to enable a system that is competitive with customers' private vehicles in terms of seamlessness and benefit.

5.3.2 Potential Customer Groups

The interviewees considered the potential customer groups from both the supply and demand sides.

5.3.2.1 The Supply Side

It is foreseen that the customer groups will be state government, local government, and public transit operators to expand their transit networks in low-density suburbs. There is a huge growth potential for these types of services.

In Queensland, public transport services are currently contract-based, and tenders are awarded by the state government (TransLink Division of Queensland Department of Transport and Main Roads). So, the state government will be very interested in how they can increase the efficiency of these services and run some trials in SEQ with providers. This may see current bus operators, fleet operators, and new providers entering the market.

5.3.2.2 The Demand Side

In terms of the demographic groups, as expressed by some experts, one notion exists that the customer group could be all citizens, as a full range of market opportunities exists for other groups including all age profiles (most people between the ages of 8 and 80 should be able to use these vehicles) and all genders, particularly to free up their time. For example, it would be ideal if a customer could board a Level 5 autonomous shuttle that is physically able to move in the network; they could perform other tasks quite comfortably, such as writing or typing or making phone calls. It would also fulfil the purposes of different customer groups including partygoers. Supply side could be influenced by variations that would depend on the range of choices available, and more importantly, system management.

Despite reforming the Disability Standards for Accessible Public Transport, not all transport systems have been retrofitted, meaning that equal access remains unavailable for everyone. It is still a necessity of the future value proposition of autonomous vehicles to be an all-inclusive transport mode and be able to provide mobility for everyone, whether they are elderly, have a disability, or live in a regional area.

Transport disadvantage and social exclusion are among the major problems many urban communities are facing today (Kamruzzaman et al., 2015; Yigitcanlar et al., 2019). In this case, as foreseen by the majority of experts, the early adopters will probably be transport-disadvantaged groups including

- Elderly people (senior citizens) and retirees, who do not drive and therefore have lost the ability to be independent travellers.
- Disabled people (with different types of disabilities) who are unable to travel on their own.
- Peri-urban/rural dwellers who don't have regular bus services or taxi services in fringe communities that are currently not well connected.
- Younger members of the community who are probably more technology enthusiasts; however, children are dependent on the decision of their parents, so it is needed to gain the trust of their families.
- People who do not have multiple cars in their household and who do not have car-based options.
- People from lower socioeconomic backgrounds, if the price is satisfactory and for those who are interested in a novel mobility solution.

In terms of the use case, based on the infrastructure requirements around AVs, most respondents viewed it as a precinct-specific solution for people.

They noted that as it is not a speedy form of transport, it would not be suited to long distances. It would be suitable for short-distance activities in low-speed environments with low volumes of traffic, including

- Nursing homes, retirement communities, and aged care facilities that are well suited to assisting people who need to move at slow speeds.
- Hospitals that are seeking to reduce missed appointments, as it is often quite difficult to either reach the venue and/or find a parking space, particularly where customers may be sick or injured.
- Airports where people want to limit their time on the ground side of the venue.
- University campuses, business parks, central business districts, or potentially the athlete villages.

Generally, from a safety standpoint, the operation of ADRT services is more feasible in locations where there is more control over the vehicle's interactions. These areas are the most appropriate locations to begin, and they are where most of this work has been done globally. However, it will make progress with time and more ADRT use cases would be developed in SEQ. As exemplified by one participant:

> In our review of the failure rates of DRT systems in the last three or four decades, the autonomous shuttle trials emphasize technology beyond practice. Since their capacity and speeds are not currently at all within any bounds of a reasonable quality of service, they can only really be deployed in very special situations where fixed-route services are too expensive. Higher capacity systems with lower marginal and operating costs compared to autonomous vehicles and capacity could work at the moment.
>
> *Academia*

Overall, there are different customer segments today, which will have different views of ADRT; however, in time and as customers become more familiar and comfortable with new technologies, it is feasible that uptake will increase.

5.3.3 Deployment and Adoption Challenges

During the interview, the experts were asked to provide their views regarding the challenges and barriers to ADRT development and adoption, particularly in SEQ. The major overarching challenges discussed by the interviewees emerged in four core themes comprising technical, financial, regulatory, and behavioural challenges (see Figure 5.4).

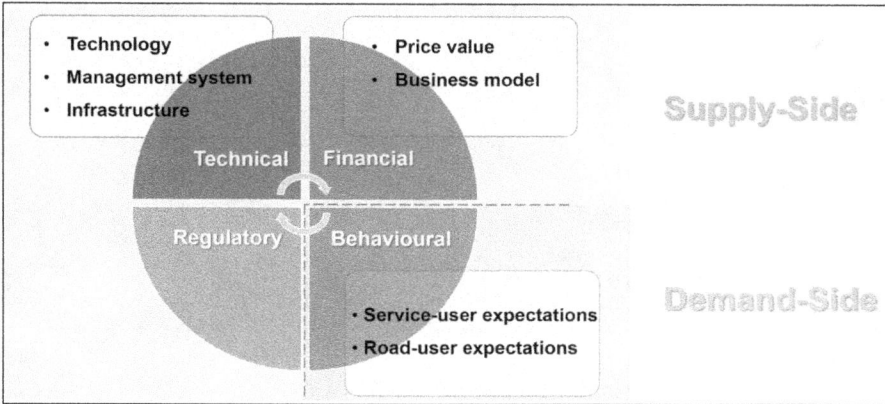

FIGURE 5.4
Overarching challenges in deployment/adoption of ADRT.

5.3.3.1 Technical Challenges

The technical challenges were categorised into three subcategories: technology, management system, and infrastructure, as depicted in Figure 5.5.

The technology: A recurring view was that technical development is not fully mature yet. Accordingly, there are a limited number of use cases and practical applications. The current technology is unable to accommodate all possible situations. For instance, one participant highlighted:

> The vehicles still struggle with rain, and snow, and have problems identifying potholes at times. There are all sorts of little things that are still being ironed out, but we are much closer than we've been at any time in history. The safety and security issue of vulnerable passengers is something that needs to be resolved again.
>
> *Academia*

Several industry-wide challenges have not been ameliorated, and more work is needed. The procurement of technology is a particular issue it presents. Australian DRT service trial in Shellharbour, NSW in the early 1990s failed because of technical issues stemming from poor planning and excessive reliance on unproven technology. This led to dissatisfaction among operators and a loss of enthusiasm, eventually resulting in the return to traditional service (Perera et al., 2020). The following are the main technology challenges mentioned by the interviewees:

- Unmature simulation environments/commercial models
- Shake-out of players, particularly after COVID
- Has taken longer than what was anticipated

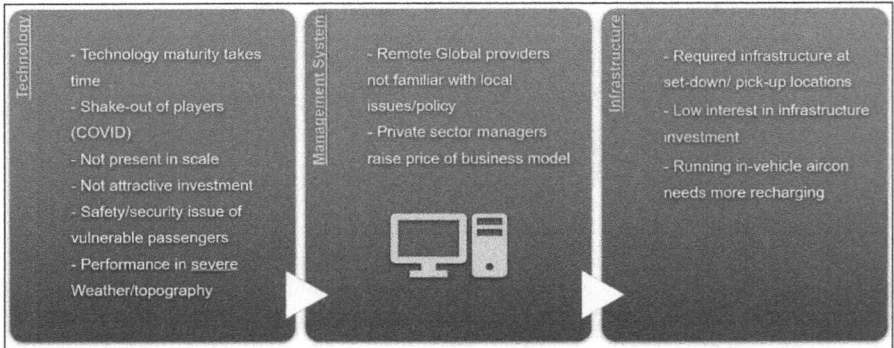

FIGURE 5.5
Technical challenges in deployment/adoption of ADRT.

- Not able to build it on the scale
- Not being an attractive market for foreign direct investment
- Safety and security issues of vulnerable passengers
- Struggling with severe weather and topography

The management system: During the discussions, the interviewees indicated that along with the technology, it is necessary to manage how to deploy ADRT systems because they need to interact between users, transport infrastructure, and other vehicles. For instance, one participant from academia highlighted:

> At the moment, most trials are pretty much like the entertainment park. They're just road based with very slow speeds, and everything is controlled and none of them is on demand. But this does not mean that an algorithm cannot slightly change.
>
> *Academia*

The Dial-a-Bus programme in Adelaide, South Australia was a flexible 'many-to-many' service without a set schedule and ultimately failed due to a lack of passenger demand and practicality. Although the elimination of spatial and temporal restrictions on passenger pickup and drop-off enhanced flexibility, it reduced the effectiveness of public transportation due to decreasing the capability of combining trips. In scenarios of low demand, a more rigid service, such as a pre-booked option or fixed-time operations, is more appropriate. High flexibility levels could be maintained just in case of ample travel demand (Perera et al., 2020). Furthermore, many system choices and protocols exist. The main management challenges extracted are as follows:

- Global providers are remote from the context where they are providing their apps
- Global providers do not understand local issues or are not concerned with the local policy
- Private sector managers of MaaS raise the price because of their business model

The infrastructure: The need for a clear picture of what infrastructure is required, whether on the roadside, mobile coverage, or connectivity, was strongly emphasised, with some participants worried that supporting infrastructure has not been fully addressed, particularly at set-down/pick-up locations. For example, one participant claimed that

> The autonomous shuttles cannot run everywhere by themselves and there should be a hub and spoke for high-volume public transport access into urban form areas. Because the autonomous shuttles of any type won't be too congested to meet the needs that your customers want.
>
> *State transport authority*

The following are the main infrastructure challenges mentioned by the interviewees:

- Supporting infrastructure has not been fully addressed, mainly at set-down/pick-up locations
- Nobody wants to invest in infrastructure as they do not know what the requirements are, which increases the cost of services
- Shuttle cannot deal with heated climate, so running the air con would need more recharging

5.3.3.2 Regulatory Challenges

The regulatory side is evolving as technology is evolving. A key challenge that has already been heavily discussed in the field and was mentioned in the interviews surrounds very few jurisdictions that have an agreed future regulatory framework. In some jurisdictions, some frameworks are not adapted to current technology. Accordingly, many respondents highlighted the importance of having consistent legislation and operational regulatory framework in one form or another. As indicated by one comment from a service provider,

> At the moment there are no two frameworks that are the same throughout the world, and Australia is more advanced than the rest of the world on this point as the National Transport Commission made the proposition

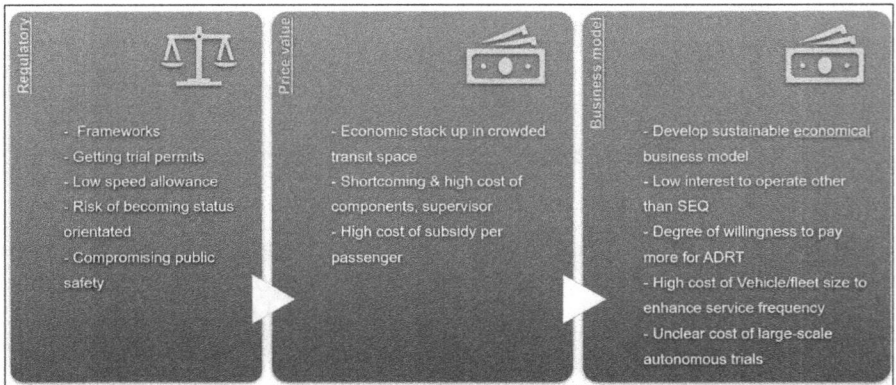

FIGURE 5.6
Technical challenges in deployment/adoption of ADRT.

to the federal government that was accepted last year, and that will kick in 2026. But there are some jurisdictions from which remains extremely complicated to get your vehicle certified amalgamated or to get a permit to operate on public roads.

Service provider

The main regulatory challenges as presented in Figure 5.6 are as follows:

- Frameworks not adapted to current technology
- Getting trial permits non-compatible due to current legislation
- Shuttles' low-speed allowance prevents operation in public
- Risk of ADRT becomes very income/status orientated
- Incentivise innovation without compromising public safety
- Caution in short-run makes technology deployment take longer

5.3.3.3 Financial Challenges

The financial challenges were categorised into two subcategories: price value, and business model evolution, as highlighted in Figure 5.6.

The price value: The uncertainty amongst experts regarding the feasibility of ADRT services concerning the cost-benefit of their deployment is evident in some reported opinions:

if you are using this DRT to fill the transport need that has not been met, then I think if it is somewhat fit for purpose, it's going to be held. If you are trying to make it a commercially successful service in an already crowded transit space, for example in Sydney, that's going to be much more challenging because the economics of it have to stack up, and I think the last few years have shown that the economics of anything autonomous is much harder to make the case for than people originally.

Academia

According to Perera et al. (2020), for ADRT services driven by public policy, having an operator who is investing in the success of the scheme is essential. The discontinuity of most prior DRT trials has often been attributed to a lack of operator enthusiasm. Conversely, commercially led DRT services may face opposition from local authorities. Hence, the success of ADRT programmes often depends on a productive partnership between multiple stakeholders. Overall, the following are the main challenges regarding the price value mentioned by the interviewees:

- Making commercially successful service in an already crowded transit space is challenging as economics must stack up
- Economics of anything autonomous is much harder to make case for initially
- Shortcomings and high cost of components, and supervisor
- Cost per passenger too high compared to traditional transit
- Cost of subsidy per passenger too hard to justify
- Removing driver cost saving but needs to have a human operator/steward

Business model evolution: The same issue translates into concerns that might occur for the evolution of a sustainable business model that can deliver profits to the service provider and benefits to the users and comply with government and community requirements (Perveen et al., 2017). It is challenging to find a profitable or economical business model, as another state department recognised,

Are people going to pay more for an autonomous vehicle if it's fundamentally no better than a human? That's the other big barrier, and maybe it won't make a profit, but maybe councils and governments will choose to subsidize because it gives more equity of access to them. It's a variance.

Peak body

The Dial-a-Bus trial in Milton Keynes, UK, ultimately failed due to fares that were set too low. After being scaled down to only operate during off-peak hours, the cost of running the programme was higher than initially budgeted. Finally, there was a lack of political commitment to maintain the programme, leading to its termination (Perera et al., 2020). The main challenges extracted regarding the business model evolution are the following:

- Develop a sustainable economic business model
- Lots of providers want to operate in SEQ because of commercial value
- Are people going to pay more for ADRT if not better than humans?
- Vehicle/fleet size, high costs of resources to service frequency
- Technology companies do not provide services to socioeconomically disadvantaged people who need them
- No large-scale trials to know the cost of having a completely autonomous system in the network and its feasibility

Hence, service providers must develop a viable business model that impacts the regulatory and operational framework and infrastructure readiness.

5.3.3.4 Behavioural Challenges

User adoption of ADRTs was stated in form of behavioural challenges during the interview. As claimed by one participant from academia, *"it is about knowing the markets, as different people have different requirements"* (Academia). All respondents considered how generic ADRT services are and whether they can cater to personalisation and expressed that tailoring the experience would be challenging. Accordingly, the success/failure of the services may be a consequence of the following (see Figure 5.7):

1. Service user expectations: How do ADRT services meet the customer expectations around the level of service quality, ease of use, safety, and privacy?

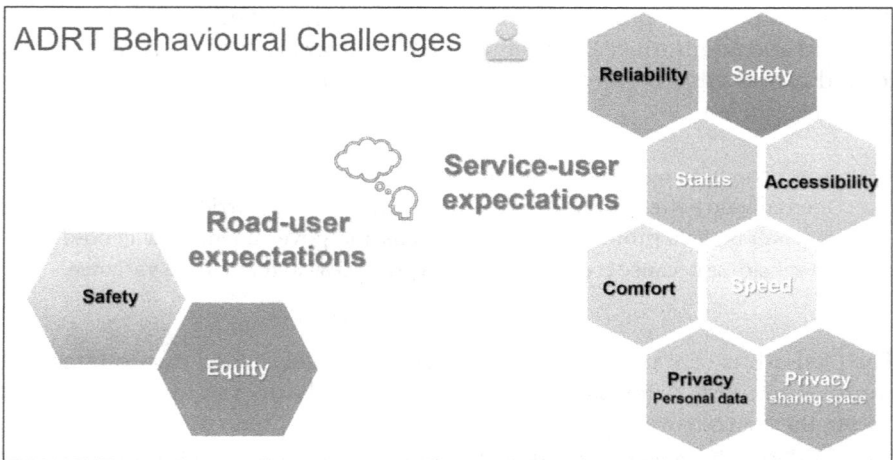

FIGURE 5.7
Behavioural challenges in deployment/adoption of ADRT.

2. Road user expectations: How do other road users interact with ADRT services and what are their expectations for safety and equity?

5.3.3.4.1 Service User Expectations

Service quality: All interviewees highlighted the need to provide good service quality. They stressed that passengers are very time-sensitive and time conscious. Ride quality, acceleration, deceleration, number of turns, and vehicle stops will be crucial to whether the technology is accepted. All participants referred to various aspects of service quality in one form or another as below (see Table 5.2):

- *Reliability:* Many respondents pointed out that it is important if potential users can rely on being able to book a trip and the timeframes involved, which affects how users would need to manage their time around travel. A recurring view was about the extra cost associated with extra travel time associated with deviations to collect multiple passengers during a revenue trip, which is the unplanned time from the users' perspective. Reliable pick-up and travel time estimations would provide a level of trust in the system.

- *Accessibility:* Coverage is important concerning the proximity of vehicle pick-up and set-down locations to potential users' origins and destinations. Near door-to-door convenience is desirable. Another point concerning accessibility that has been addressed in the literature and was mentioned in the interviews was how the user will get ADRT service. For example, they may book service via an app, and it may arrive at a certain time to enhance the ease of use in that entire value chain from waking up in the morning to deciding to go somewhere. The user will need to learn how to book service, reach the boarding location, wait for the required time, board the service, travel on the service, alight the service, and then reach their final destination. Many respondents discussed that accessibility is not just in terms of the convenience of location; emphasis is placed on the need for catering for the various mobility needs of our diverse community. According to an interviewee, in terms of accessibility needs one in five Queenslanders has a disability, so the business providing those shared services must meet regulations that ensure the protection of those users.

- *Speed:* Some respondents expressed that the biggest impediment to shuttles so far has been the speed with which they can travel within the traffic if they are on a dedicated lane. There will be some infrastructure in addition to running the service. However, this additional infrastructure cost could deliver some benefits in times for people to use it.

TABLE 5.2

The Frequency of Items Mentioned by Participants in Each Category (%)

Category	Subcategory	1	2	3	4	5	6	7	8	9	10	11	12	13	14	15	16	17	18	19	20	21	22	23	24	25	26	Total No.	(%)
Service quality	Reliability	•	•	•	•		•		•		•	•	•				•				•	•	•	•	•			14	54
	Accessibility		•					•	•	•								•				•	•	•	•	•	•	10	38
	Speed	•			•	•		•		•												•	•	•	•			8	31
	Comfort			•	•		•		•		•	•							•			•	•				•	8	31
	Status	•					•							•														4	15
Ease of use	Booking/travel		•							•	•								•					•				5	19
Perceived risks	Safety	•	•	•		•		•	•	•	•	•	•	•	•		•	•			•	•	•	•	•	•	•	23	85
	Privacy											•				•		•										3	11.5

- *Comfort:* Some experts underlined the importance of considering the demographics in providing a comfortable service.
- *Status:* Several respondents stated that there is a certain latency in the sense that people might not be so keen to take public transport, but they do not have any choice, particularly when the cost is concerned. But there is a latent demand if they had even the slightest chance to take a luxury bus or a luxury car that can carry five or six people. Maybe that is not expressed yet. So, there is another potential demographic group. The London Tube was discussed by analogy, where very wealthy individuals in high-status roles are using that public transport facility regularly because the surface network is congested.

Ease of use: Many respondents discussed ease of use, predominantly around the elderly. One interviewee, though, expressed a contrary view to the aforementioned ones (E2).

Safety: The uncertainty regarding the perceived risks amongst experts is evident in their reported opinions. All of them discussed safety and privacy. Across all stakeholder groups, there was agreement that it is crucial to ensure that issues are attended to for safety reasons and reliability of the technology that will impact the quality of the service, which would be the trade-offs between traffic flow improvements and route identification versus the actual smoothness of the ride.

Privacy: Privacy concerns regarding shared space as well as personal data were challenging. Several interviewees noted that some people preferred very specialised, individualised services. In the more individualised service, just a person or his/her family/immediate friends would be able to book and then they could use the service exclusively. Albeit that would be a challenge to government providers because they base their whole transport system and mass transit on the general public as a whole. Very few interviewees were worried about the privacy of the data. Figure 5.8 provides a visual representation of how the participants rated the service user expectations in each category.

5.3.3.4.2 Road User Expectations

Safety: Part of the safety issue translates into concerns that might occur from the sensitivity of other road users. During the interviews, the participants indicated that even though vehicles are safe, if they move in ways that are not familiar to humans, it may cause errors by human drivers of other vehicles.

Equity: People have a very strong sense of equity and fairness, and so if these vehicles are seen to be exploiting road rules or doing something that is not considered to be fair, that could cause resentment.

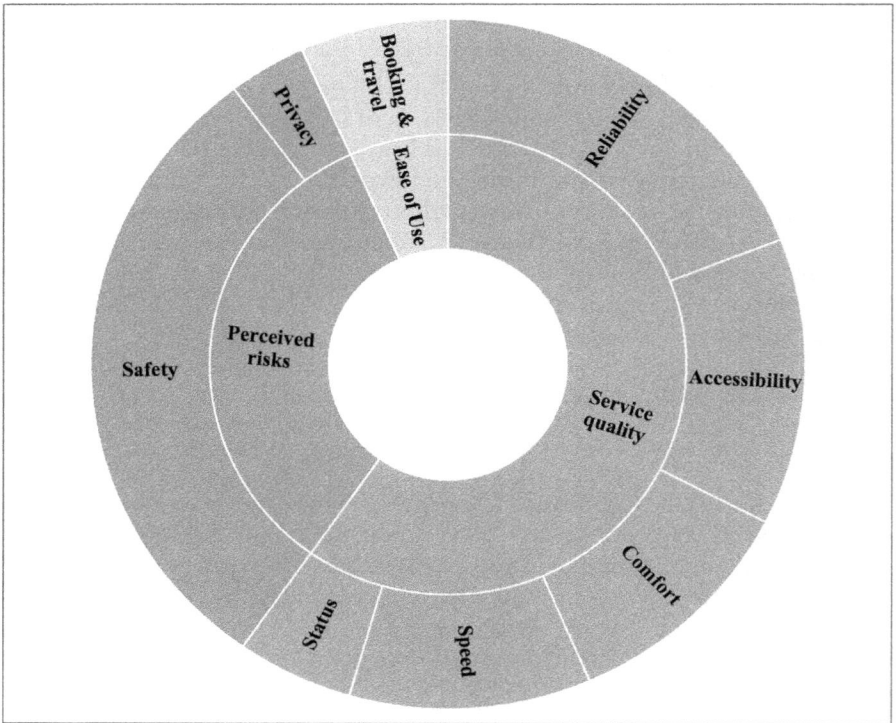

FIGURE 5.8
The frequency of main categories and subcategories.

For instance, one interviewee mentioned,

> Young people probably tend to make take more risks and they know that
> the vehicle is meant to yield to a human which those interventions or
> interactions can be if you are in a less controlled environment when you
> have got younger people, cyclists, and drivers. The automatic vehicle will
> always currently operationally yield which will have a detrimental impact
> on the customer experience in the vehicle that stops every few meters, so
> I think you'd need to have a designated operational design domain that
> limits deployment opportunities. I think when you try to allocate it a 24-
> hour bus lane or something like that, make sure that there's not a percep-
> tion that you are taking away from the vehicle space.
>
> *State transport authority*

People will have high expectations from the system, and if providers do not
address these matters properly, even if they are not users, it will likely drive
negative community reaction. As acknowledged by Perera et al. (2020), the
lack of marketing was a significant factor in the lacklustre patronage of the

Shellharbour, NSW trials. Previously, there was hesitation among customers to embrace ridesharing, but with the rise of sharing economies like Uber Pool, this attitude is rapidly changing. To ensure the success of the service, it is vital to implement strong marketing strategies and communicate with customers to manage their expectations.

5.3.3.5 Deployment/Adoption Mitigation Strategies

All interviewees mentioned the significance of the partnership between transport authorities, government, industry, and academia in one form or another to help increase societal acceptance of autonomous shuttles in preparation for ADRT. As can be seen in Figure 5.9, strategies provided by the experts can be classified into the following categories.

Mindset change: Several respondents highlighted the importance of getting people to shift away from an ownership mindset to a shared-use mindset. This is not dependent on technology; rather, on changing human behaviour, which is a much more difficult proposition. As indicated by one comment,

> if people understand that moving away from privately owned vehicles to shared, autonomous, and electric motor transportation to reduce accidents and emissions is important and urgent, then that drives adoption. Otherwise, it would become just an electronic gadget if they do not change their behaviour. A lot of people are saying we need special lanes for autonomous vehicles, but my argument is we need to just have fewer cars on the road, so the government needs to discourage car ownership and make all the cars extremely expensive just like they do in Singapore. If they make it expensive to own a car and public transport is good, then people will use public transport that needs to be shared, and I think that's going to be the biggest success factor
>
> *Academia*

Mindset change	• Shift away from **ownership** mindset to **shared-use** mindset • People **understand** moving away from privately owned **important and urgent**
Education **ADRT Exposure**	• Educate community through **campaigns** • Offering **tangible experience** through autonomous shuttle **trials**
Subsidies	• Offer discount to **encourage multimodal** public transit trips • Subsidize service in initial phases to strike **affordable fares** • Subsidize **trials** to avoid market **failure**
Pricing **Non-pricing** **Incentives**	• Incentive can be just **provided service** for those who didn't have any • Use of ADRT **covers commute/parking costs of regular patients** in health precinct • Upfront customer service/**passive surveillance** to offer **attractive service** & overcome concerns
Liability & **Insurance**	• Conducting **trials to test / ensure safety** & build **public confidence** • Build up **roadside equipment** & **visual signals/signs** around the vehicle • **Prevent entry** of several providers/operators in market with **different vehicles & products**

FIGURE 5.9
Proposed mitigation strategies for deployment/adoption of ADRTs.

Education and exposure: The majority of respondents acknowledged that the differences between the age groups regarding the uptake of ADRT reduce the levels of awareness and knowledge, which are essential in gaining individuals' trust. There should be awareness and educative campaigns as more vehicles enter the fleet. The community should be given a tangible experience via book/play with the system and obtain feedback from them. Regulators need to be prepared to understand how this technology is going to be regulated and introduced.

For example, one participant claimed that

> in the first instance, getting people to book and be able to play with the system is probably the critical aspect. So, you build that awareness in your pilot trial, which is what the government is currently doing. NSW did some good work on it. You've got to have a behavioural change aspect in your policy, and the engineering aspect of building these new things drives people's behaviours to change. So, like you've adopted the mobile phone over the decade, you are planting seeds for future opportunities with your work in these trials. That's the idea of leading and shaping expectations and communities. Appetite for these things will drive investment decisions and therefore if you've got general warmth towards the product, people will be much more responsive to taking that AV journey.
>
> *State transport authority*

Government subsidies: Transit systems throughout the world are mostly not financially viable proposals, so are subsidised by government as a community service. Many experts suggested that government could offer a discount to incentivise the use of traditional mass transit not to make existing investment redundant and encourage multimodal public transit trips rather than using a shuttle for the entire trip. They also mentioned that subsidy is needed at least in the initial phases to strike the right service fee for people. On the contrary, some experts expressed that any formal incentives/subsidies need to be justified just in case of some form of market failure.

One interviewee described that

> in a fairly liberal market democracy, the default assumption is the private sector is more efficient and the public sector needs only step in when there is a market failure. So, any formal incentives or subsidies need to be justified just in case of some form of market failure. Hence if ADRT services are commercially viable, then it's probably best left to the private sector to provide these services. Alternatively, the government might need to step in more selectively where they provide subsidies on particular routes where there is not as much demand to ensure a minimum level of service to all people as a government does more about equity or allocate

incentives for suppliers that offer ADRT services in future contracts/ tenders. Thus, the government right now should not be thinking about incentives and subsidies so much as just helping the process of development and deployment through integrating ADRT service within the current offering of public transport services and facilitating that shift away from human-driven buses to autonomous systems.

Academia

Consumer incentives: According to interviewees consumer incentives concern managing the expectations of the customer. Across all stakeholders, there was agreement that using incentives would not need to be too wide because the economics or usability of the service should sell itself and the government and providers should not need to add extra incentives. Notwithstanding, some forms of pricing and non-pricing incentivisation from a commercial perspective could be considered in the future. Some examples of non-pricing incentives were as follows:

- Offering attractive and faster service, not necessarily disrupting the cost.
- Incentives can be just provided service for those who did not have any.
- Use of ADRT covers commute/parking costs of regular patients with no carers in the health precinct.
- Provide upfront customer service/passive surveillance to make it an attractive offer to overcome concerns.

Liability and insurance: It was acknowledged during the interviews that we cannot overcome public acceptance challenges without building trust in the technology and properly integrating the system within the whole transport network. Diffusion into society will be much easier as people understand the benefits and potential risks of the technology through word of mouth. The strategies provided in this regard are as follows:

- Conducting trials to test/make sure technology works safely to build public confidence.
- Running autonomous delivery initially rather than carrying people will allow researchers to obtain a lot more data on the economics, reliability, and acceptance of other road users.
- Build up roadside equipment and visual signals/signs around the vehicle to ensure it is responding to danger.
- Overcome low insurance with the cooperation of Australian state governments to share insights to set up regulations.
- Prevent entry of several providers/operators in the market with different vehicles and offering different products.

5.4 Findings and Discussion

This study sought to develop an overall picture of opportunities and challenges of ADRT deployment and user adoption by employing qualitative techniques through the lens of actors involved in decision-making who were engaged with Australia's industry environment, comprising state transport authorities, consultants, peak bodies, service providers, and academia. The expert interviews generated significant new knowledge, opening up the possibility for further research in this area. The study has revealed the optimistic prospects of ADRT regarding market development and sustainability. These findings hold great importance for expanding research in the context of autonomous shuttle buses where user acceptance will play a pivotal part in the diffusion of these innovative technologies.

The comprehensive detailing of the implications of findings enables them to identify priority areas within the broad agenda which they can then investigate with the general public. Therefore, providers would be able to take measures to substantially increase the potential users' behavioural intention and ultimately actual usage. Interviewing the transport stakeholders and asking questions systematically regarding the adoption of ADRT services also engenders knowledge and heightens awareness in the sectors that have been interviewed. The major overarching challenges discussed by the interviewees emerged in four core themes comprising technical, financial, regulatory, and behavioural challenges. These are all critical areas that are linked together because they depend on each other and need to be focused on before the full-scale launch of ADRTs.

The main policy recommendations mainly contributing to ensuring the success of the ADRT scheme, especially autonomous shuttle buses, were extracted from the literature and the above findings as follows:

- Intervene throughout the innovation process by stimulating the use of shared transport modes among consumers or manufacturers and discourage private ownership. These transport system outcomes typically align with broader governmental goals, such as carbon reduction strategies highlighted by Sindi and Woodman (2021), which could provide tax increases on fuel, cars, licences and registration; or disincentivising private single-occupant car driving to dedicate road space to space-efficient modes or restricting one to four occupant vehicles to access/park in, e.g., airports or university campuses.

- Anticipate unfolding risk scenarios in regulatory approaches while providing a robust framework for innovators to operate, rather than lagging (Mordue et al., 2020). Local governments for instance hold key regulatory powers' (e.g., managing the right of way, articulating policies for use of land) and can plan for AVs (Freemark et al., 2019) if they have the goals and resources. Through deliberate and stringent regulation, governments can prompt technological responses, which influence the direction of innovation development (Beerepoot & Beerepoot, 2007). As Taylor et al. (2005) describe, the existence and anticipation of regulation, as well as the degree of stringency and certainty, are important drivers of innovation. The regulatory frameworks should not be too restrictive at this point to create a favourable environment for tech, changing the algorithm of road-based, slow-speed, controlled, none on-demand trials, and setting safety regulations, performance standards, and economic regulations around market entry/behaviours, and facilitating multi-operator, multi-tech trials in different locations with different use cases.

- Have a test and validation centre for benchmarking all kinds of autonomous technology, even through partnering with private companies as they may have private roads to deploy autonomous shuttles. Adjust the legal voids to enable the testing and operation of autonomous technology on public roads (Ferreira et al., 2020). The pace at which researchers and manufacturers can advance technology and suitability for practical integration of the ADRT service within current public transport environments relies on the efficiency of policymakers to keep up (Skeete, 2018). Further complicating these regulatory challenges are the often-misaligned interests across different levels of government, private sector, and public stakeholders at local, national, and international levels (Stone et al., 2018; Freemark et al., 2019). These pressures sit against a backdrop of multiple countries competing to become attractive markets for AV industry development by offering innovator-friendly regulatory environments (Lee & Hess, 2020 Schepis et al., 2023).

- Change the public transport subsidisation model to include ADRT where it fits best in the community. According to Currie and Fournier (2020), most ADRT failures are caused by excessive costs. It is essential to consider local factors when introducing a new ADRT service, as previous ADRT experiences in other locations may not be directly applicable (Papanikolaou et al., 2017). According to Davison et al. (2014), many unsuccessful projects result from insufficiently realistic cost estimates and a lack of understanding of the target market. Therefore, practitioners and policymakers need to pay more attention to the funding and commercial prospects of ADRT; e.g., providing subsidies on particular low-demand routes to ensure equity or allocating incentives for suppliers that offer ADRT services in future contracts or tenders, or offering a discount on ADRT fares for people who need point-to-point services on weekends or during work weeks to quit private vehicles to reach an entertainment venue, work, or home; or offering loyalty points to buy coffee along the way in the case of capturing ADRT.

- Build trust in technology and have a clear, definitive message about this technology or service offering through social media. Many consumer studies have highlighted low trust and safety perceptions relating to AVs, which, Cunningham et al. (2019) suggest, can be addressed by the government through controlled demonstrations or simulations. Moreover, it can be beneficial to identify early adopter consumer feedback loops that can drive uptake in the positive feedback loop or to identify breakdowns in feedback loops to incentivise reluctant technology adopters. Government is typically the sole actor available to sponsor critical basic research which achieves technological breakthroughs that later permeate the private sector (Salmenkaita & Salo, 2002). These investments complement corporate R&D, which tends to focus on applied research and commercial development, often in a closed manner that protects intellectual property. Governments can use policy instruments, such as taxation or grant incentives, to subsidise investments and reduce privacy risks while adding conditions that necessitate certain social benefits outside of capitalist motivations (Taylor et al., 2005; Pinske et al., 2014).

- Have a consolidated proactive policy view to support technology development as ADRT typically involves a greater focus on marketing and the development of strategic partnerships, compared to conventional bus services (Enoch et al., 2006, Perera et al., 2020). A strong market signal contributes to reputation and expectation effects that increase managerial willingness to invest in areas predicted to receive continuing government support. A coordinated network-wide provision of services such as investments in human capital or technical infrastructure is also necessary to accelerate the pace of innovation (Moon & Bretschneider, 1997). As suggested by Enoch et al. (2006), achieving a harmonious balance between providing customers with flexible services and managing technology costs is crucial for an operator. The most effective approach to achieving this balance is by adopting an incremental strategy. This approach allows the operator to gradually implement new services and technology, assess the impact, and make necessary adjustments. By taking this step-by-step approach, the operator can minimise upfront investment and ensure a smoother implementation process. The development and diffusion of AV rely on complimentary transport infrastructure developments such as communications and physical road networks, which are currently provided by the government (Li et al., 2019). They are also facilitating electrification and other business models (e.g., shared use) (Sperling et al., 2018). More broadly, companies benefit from government activities that are directed towards a more efficient and effective transfer of knowledge within emerging technological domains, as well as the promotion of common visions which can guide collaborative efforts (Salmenkaita & Salo, 2002). Additionally, it can be beneficial to provide ADRT services to meet demand for new housing developments that do not have many car parking spaces.

- The trials of automated vehicles have provided substantial valuable insights encompassing technology, safety, project management, road users, occupants, reporting, and infrastructure. However, it is worth noting that these trials are not solely focused on the technology itself. Therefore, governments must capitalise on these lessons to shape future trials. Industry, government, and universities can work together to seize the opportunities that have been uncovered, leading to a brighter future for the emerging trends that will shape the policy agenda (Liu et al., 2020). Yet very few local governments have started preparations for AVs and many city officials are sceptical about the benefits of AVs (Fraedrich et al., 2019; Freemark et al., 2019). Given the variations between the environments in which governments operate and the high degree of anticipated social and market disruption, governance approaches and the application of policy instruments will vary greatly.

Even though the qualitative method allowed for an in-depth discussion with a multitude of stakeholders from different sectors, the findings should be regarded as provisional. Hence, further work is required to generate broader data from which more nuanced recommendations could well be prepared. Moreover, future studies might go beyond the current work's primary focus on Australian stakeholders to include representatives from a broader variety of international organisations and sectors.

The snowball sampling technique used for the current study may have overlooked some sectors that could be considered for incorporation in future studies, particularly the perspectives of emerging new ride-sharing organisations, which are expected to deepen the study. The repercussions for developing countries are also of great interest since ADRT has the potential to develop and expand the current transport services and provide advantages to larger population groups.

5.5 Conclusion

The emergence of ADRT creates a responsibility for transport author-ities to give proper attention to public passenger transportation policy and to enhance the public transport efficiency. Based on National Transport Commission (2020), there are several reasons why governments and trialling organisations become involved in trials. Governments must be clear about the objectives they are trying to achieve through trials in their jurisdiction and evaluate them in light of those objectives. There is an opportunity to place Australia in a better position to be ready for the commercial deployment of automated vehicles through sharing learnings across jurisdictions.

Accordingly, we conducted an explorative qualitative study through interviews with transport experts in the field to gain a more profound insight into the challenges and opportunities of ADRT deployment and user adoption, especially autonomous shuttles in the SEQ metropolitan region. Thematic analysis was used to derive the main themes and provide comprehensive insights regarding the ADRT service concept, potential customer groups in terms of the supply and the demand side, and the potential use cases. Challenges regarding the development and adoption of ADRT challenges in terms of supply (technical, regulatory, and financial barriers) and demand (the behavioural challenges) were explained as depicted in Figure 5.10.

These are all critical areas that are linked together because they depend on each other and need to be focused on before the full-scale launch of ADRTs. Such novel findings, as well as further validation through wider studies, have the potential to enrich the knowledge of transportation specialists and add to the body of research on ADRT user adoption behaviour. In addition, the four main themes of challenges obtained in this study represent key areas of focus for future research regarding how the introduction of ADRT services can be best aligned with the needs of SEQ residents, especially

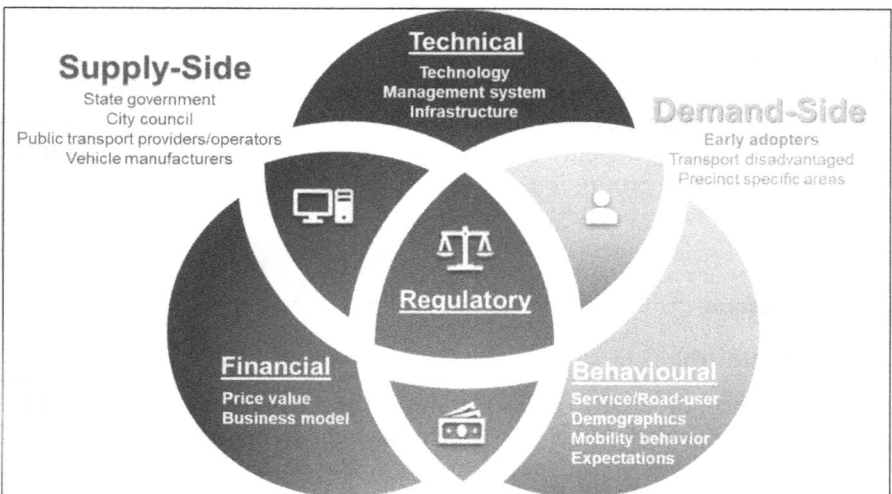

FIGURE 5.10
ADRT deployment/adoption challenge framework.

transport-disadvantaged populations. These findings open an avenue for developing a scale for measuring latent constructs underlying SEQ residents' transport mode choice behaviour and using them alongside present behavioural or psychological theories. It is important to emphasise that solving these issues will require a cooperative effort from sociologists, psychologists, engineers, planners, and transport and urban policymakers.

Acknowledgements

This chapter, with permission from the copyright holder, is a reproduced version of the following journal article: Golbabaei, M., Paz, A., Yigitcanlar, T., & Bunker, J. (2024). Navigating autonomous demand responsive transport: Stakeholder perspectives on deployment and adoption challenges. *International Journal of Digital Earth*, 17(1), 2297848.

References

Alyavina, E., Nikitas, A., & Njoya, E.T. (2020). Mobility as a service and sustainable travel behaviour: A thematic analysis study. *Transportation Research Part F: Traffic Psychology and Behaviour*, 73, 362–381.

Amann, V., (2017). Consumer Acceptance, Barriers and Success Factors of Peer-to-Peer Carsharing in Perspective of Connected Car Services and Autonomous Vehicles. WU Vienna University of Economics and Business.

Arbib, J., & Seba, T. (2017, May). Rethinking Transportation 2020–2030. *RethinkX*, 143, 144.

Beerepoot, M., & Beerepoot, N. (2007). Government regulation as an impetus for innovation: Evidence from energy performance regulation in the Dutch residential building sector. *Energy Policy*, 35(10), 4812–4825.

Beirão, G., & Cabral, J.S. (2007). Understanding attitudes towards public transport and private car: A qualitative study. *Transport Policy*, 14, 478–489.

Bengtsson, M. (2016). How to plan and perform a qualitative study using content analysis. *NursingPlus Open*, 2, 8–14.

Biernacki, P., & Waldorf, D. (1981). Snowball sampling: Problems and techniques of chain referral sampling. *Sociological Methods and Research*, 10(2), 141–163. https://doi.org/10.1177/ 004912418101000205

Bierstedt, J., Gooze, A., Gray, C., Peterman, J., Raykin, L., & Walters, J. (2014). Effects of next-generation vehicles on travel demand and highway capacity. *FP Think Working Group*, 8, 10–11.

Bolger, F., & Wright, G. (2011). Improving the Delphi process: Lessons from social psychological research. *Technological Forecasting and Social Change*, 78(9), 1500–1513.

Braun, V., & Clarke, V. (2006). Using thematic analysis in psychology. *Qualitative Research in Psychology*, 3(2), 77–101.

Braun, V., & Clarke, V. (2013). *Successful Qualitative Research: A Practical Guide for Beginners*; Sage: London.

Butler, L., Yigitcanlar, T., & Paz, A. (2021). Barriers and risks of Mobility-as-a-Service (MaaS) adoption in cities: A systematic review of the literature. *Cities*, 109, 103036.

Chen, J., Li, R., Gan, M., Fu, Z., & Yuan, F. (2020). Public acceptance of driverless buses in China: An empirical analysis based on an extended UTAUT model. *Discrete Dynamics in Nature and Society*, 2020, 1–13.

Claybrook, J., & Kildare, S. (2018). Autonomous vehicles: No driver… no regulation? *Science*, 361(6397), 36–37.

Cunningham, M.L., Regan, M.A., Ledger, S.A., and Bennett, J.M. (2019). To buy or not to buy? Predicting willingness to pay for automated vehicles based on public opinion. *Transportation Research Part F*, 65, 418–438.

Currie, G., & Fournier, N. (2020). Why most DRT/Micro-Transits fail – What the survivors tell us about progress. *Research in Transportation Economics*, 83, 100895.

Davidson, P., & Spinoulas, A. (2016). Driving alone versus riding together-How shared autonomous vehicles can change the way we drive. *Road & Transport Research: A Journal of Australian and New Zealand Research and Practice*, 25(3), 51–66.

Davison, L., Enoch, M., Ryley, T., Quddus, M., & Wang, C. (2014). A survey of demand responsive transport in Great Britain. *Transport Policy*, 31, 47–54.

Dichabeng, P., Merat, N., & Markkula, G. (2021). Factors that influence the acceptance of future shared automated vehicles – A focus group study with United Kingdom drivers. *Transportation Research Part F: Traffic Psychology and Behaviour*, 82, 121–140.

Docherty, I. (2018). New governance challenges in the era of 'Smart' Mobility. In Marsden, G. and Reardon, L. (Ed.) *Governance of the Smart Mobility Transition*; Emerald Publishing Limited: Bingley, pp. 19–32.

Eisenhardt, K. M. (1989). Building theories from case study research. *Academy of Management Review*, 14(4), 532–550.

Enoch, M., Potter, S., Parkhurst, G., & Smith, M. (2006). Why do demand responsive transport systems fail?. In: *Transportation Research Board 85th Annual Meeting*, 22–26 Jan 2006, Washington, DC. http://pubsindex.trb.org/view.aspx?id=775740

Erlingsson, C., & Brysiewicz, P. (2017). A hands-on guide to doing content analysis. *African Journal of Emergency Medicine*, 7(3), 93–99.

Faisal, A., Yigitcanlar, T., Kamruzzaman, M., & Paz, A. (2021). Mapping two decades of autonomous vehicle research: A systematic scientometric analysis. *Journal of Urban Technology*, 28(3–4), 45–74.

Fernandes, P., & Nunes, U. (2012). Platooning with IVC-enabled autonomous vehicles: Strategies to mitigate communication delays, improve safety and traffic flow. *IEEE Transactions on Intelligent Transportation Systems*, 13(1), 91–106.

Ferreira, A., von Schönfeld, K. C., Tan, W., & Papa, E. (2020). Maladaptive planning and the pro-innovation bias: Considering the case of automated vehicles. *Urban Science*, 4(3), 41.

Firoozi Yeganeh, S., Khademi, N., Farahani, H., & Besharat, M.A. (2022). A qualitative exploration of factors influencing women's intention to use shared taxis: A study on the characteristics of urban commuting behavior in Iran. *Transport Policy*, 129, 90–104.

Fishman, E., Washington, S., & Haworth, N. (2012) Barriers and facilitators to public bicycle scheme use: A qualitative approach. *Transportation Research Part F: Traffic Psychology and Behaviour*, 15, 686–698.

Fraedrich, E., Heinrichs, D., Bahamonde-Birke, F.J., & Cyganski, R. (2019). Autonomous driving, the built environment and policy implications. *Transportation Research Part A: Policy and Practice*, 122, 162–172.

Freemark, Y., Hudson, A., & Zhao, J., (2019). Are cities prepared for autonomous vehicles? Planning for technological change by US local governments. *Journal of the American Planning Association*, 85(2), 133–151.

Gardner, B., & Abraham, C. (2007). What drives car use? A grounded theory analysis of commuters' reasons for driving. *Transportation Research Part F: Traffic Psychology and Behaviour*, 10, 187–200.

Glaser, B., & Strauss, A. (1967). *The Discovery of Grounded Theory*; Aldine Publishing Company: Chicago.

Golbabaei, F., Yigitcanlar, T., & Bunker, J. (2021). The role of shared autonomous vehicle systems in delivering smart urban mobility: A systematic review of the literature. *International Journal of Sustainable Transportation*, 15(10), 731–748.

Golbabaei, F., Yigitcanlar, T., Paz, A., & Bunker, J. (2020). Individual predictors of autonomous vehicle public acceptance and intention to use: A systematic review of the literature. *Journal of Open Innovation*, 6(4), 1–27.

Golbabaei, F., Yigitcanlar, T., Paz, A., & Bunker, J. (2022). Understanding autonomous shuttle adoption intention: Predictive power of pre-trial perceptions and attitudes. *Sensors*, 22(23), 9193. https://doi.org/10.3390/s22239193

Golbabaei, F., Yigitcanlar, T., Paz, A., & Bunker, J. (2023). Perceived opportunities and challenges of autonomous demand-responsive transit use: What are the sociodemographic predictors? *Sustainability*, 15(15), 11839.

Gössling, S., Cohen, S.A., & Hares, A. (2016). Inside the black box: EU policy officers' perspectives on transport and climate change mitigation. *Journal of Transport Geography*, 57, 83–93.

Grimsley, M., & Meehan, A. (2007). e-Government information systems: Evaluation-led design for public value and client trust. *European Journal of Information Systems*, 16, 134–148.

Guest, G., Bunce, A., & Johnson, L. (2006) How many interviews are enough? An experiment with data saturation and variability. *Field Methods*, 18, 59–82.

Haboucha, C.J., Ishaq, R., & Shiftan, Y. (2015). User preferences regarding autonomous vehicles: Giving up your private car. IATBR 2015-WINDSOR.

Hafner, R.J., Walker, I., & Verplanken, B. (2017). Image, not environmentalism: A qualitative exploration of factors influencing vehicle purchasing decisions. *Transportation Research Part A: Policy and Practice*, 97, 89–105.

Hennink, M.M., Kaiser, B.N., & Marconi, V.C. (2017). Code saturation versus meaning saturation: How many interviews are enough? *Qualitative Health Research*, 27(4), 591–608.

Hensher, D.A. (2017). Future bus transport contracts under a mobility as a service (MaaS) regime in the digital age: Are they likely to change? *Transportation Research. Part A, Policy and Practice*, 98, 86–96.

Hilgarter, K., & Granig, P. (2020). Public perception of autonomous vehicles: A qualitative study based on interviews after riding an autonomous shuttle. *Transportation Research Part F: Traffic Psychology and Behaviour*, 72, 226–243.

Holloway, I., & Todres, L. (2003). The status of method: Flexibility, consistency and coherence. *Qualitative Research*, 3(3), 345–357.

Kamruzzaman, M.D., Hine, J., & Yigitcanlar, T. (2015). Investigating the link between carbon dioxide emissions and transport-related social exclusion in rural Northern Ireland. *International Journal of Environmental Science and Technology*, 12, 3463–3478.

Lee, D., & Hess, D.J. (2020). Regulations for on-road testing of connected and automated vehicles: Assessing the potential for global safety harmonization. *Transportation Research Part A*, 136, 85–98.

Li, S., Sui, P.C., Xiao, J., & Chahine, R., (2019). Policy formulation for highly automated vehicles: Emerging importance, research frontiers and insights. *Transportation Research Part A*, 124, 573–586.

Liu, N., Nikitas, A., & Parkinson, S. (2020). Exploring expert perceptions about the cyber security and privacy of connected and autonomous vehicles: A thematic analysis approach. *Transportation Research Part F: Traffic Psychology and Behaviour*, 75, 66–86.

Lune, H., & Berg, B.L. (2017). *Qualitative Research Methods for the Social Sciences.* London: Pearson.

Manivasakan, H., Kalra, R., O'Hern, S., Fang, Y., Xi, Y., & Zheng, N. (2021). Infrastructure requirement for autonomous vehicle integration for future urban and suburban roads – Current practice and a case study of Melbourne, Australia. *Transportation Research Part A*, 152, 36–53.

Mars, L., Arroyo, R., & Ruiz, T. (2016). Qualitative research in travel behavior studies. *Transportation Research Procedia*, 18, 434–445.

Marshall, B., Cardon, P., Poddar, A., & Fontenot, R. (2013). Does sample size matter in qualitative research? A review of qualitative interviews in is research. *Journal of Computer Information Systems*, 54(1), 11–22.

Metaxiotis, K., Carrillo, F.J., & Yigitcanlar, T. (2010). *Knowledge-Based Development for Cities and Societies: Integrated Multi-Level Approaches.* Hersey, PA: IGI Global.

Mladenovic, M.N. (2019). How should we drive self-driving vehicles? Anticipation and collective imagination in planning mobility futures. Chp. 6 In Finger, M. and Audouin, M. (Eds.) *The Governance of Smart Transportation Systems.* Zurich: Springer Nature, pp.103–124.

Moon, M.J. & Bretschneider, S. (1997). Can state government actions affect innovation and its diffusion?: An extended communication model and empirical test. *Technological Forecasting and Social Change*, 54(1), 57–77.

Mordue, G., Yeung, A., & Wu, F. (2020). The looming challenges of regulating high level autonomous vehicles. *Transportation Research Part A: Policy and Practice*, 132, 174–187.

Mouratidis, K., & Cobeña Serrano, V. (2021). Autonomous buses: Intentions to use, passenger experiences, and suggestions for improvement. *Transportation Research Part F, Traffic Psychology and Behaviour*, 76, 321–335.

National Transport Commission (2020). Lessons learned from automated vehicle trials in Australia. Research report, NTC, Melbourne.

Nguyen-Phuoc, D.Q., Currie, G., De Gruyter, C., & Young, W. (2018). How do public transport users adjust their travel behaviour if public transport ceases? A qualitative study. *Transportation Research Part F: Traffic Psychology and Behaviour*, 54, 1–14.

Nikitas, A., Avineri, E., & Parkhurst, G. (2018). Understanding the public acceptability of road pricing and the roles of older age, social norms, pro-social values and trust for urban policy-making: The case of Bristol. *Cities, 79*, 78–91.

Nikitas, A., Wang, J.Y., & Knamiller, C. (2019). Exploring parental perceptions about school travel and walking school buses: A thematic analysis approach. *Transportation Research Part A: Policy and Practice, 124*, 468–487.

Nordhoff, S., de Winter, J., Madigan, R., Merat, N., van Arem, B., & Happee, R. (2018). User acceptance of automated shuttles in Berlin-Schöneberg: A questionnaire study. *Transportation Research Part F: Traffic Psychology and Behaviour, 58*, 843–854.

Nowell, L.S., Norris, J.M., White, D. E., & Moules, N.J. (2017). Thematic analysis: Striving to meet the trustworthiness criteria. International Journal of travel behavior and perceptions of ridesharing in Denmark. *Transportation Research Part A: Policy and Practice, 78*, 113–123.

Palmer, K., Dessouky, M., & Abdelmaguid, T. (2004). Impacts of management practices and advanced technologies on demand responsive transit systems. *Transportation Research. Part A, Policy and Practice, 38*(7), 495–509.

Pancholi, S., Yigitcanlar, T., & Guaralda, M. (2019). Place making for innovation and knowledge-intensive activities: The Australian experience. *Technological Forecasting and Social Change, 146*, 616–625.

Papadima, G., Genitsaris, E., Karagiotas, I., Naniopoulos, A., & Nalmpantis, D. (2020). Investigation of acceptance of driverless buses in the city of Trikala and optimization of the service using Conjoint Analysis. *Utilities Policy, 62*, 100994.

Papanikolaou, A., Basbas, S., Mintsis, G., & Taxiltaris, C. (2017). A methodological framework for assessing the success of Demand Responsive Transport (DRT) services. *Transportation Research Procedia, 24*, 393–400.

Parsons, K., McCormac, A., Pattinson, M., Butavicius, M., & Jerram, C. (2014). A study of information security awareness in Australian government organisations. *Information Management & Computer Security, 22*(4), 334–345.

Perera, S., Ho, C., & Hensher, D. (2020). Resurgence of demand responsive transit services – Insights from BRIDJ trials in inner west of Sydney, Australia. *Research in Transportation Economics, 83*, 100904.

Perveen, S., Kamruzzaman, M., & Yigitcanlar, T. (2017). Developing policy scenarios for sustainable urban growth management: A Delphi approach. *Sustainability, 9*(10), 1787.

Pettigrew, S., Cronin, S. L., & Norman, R. (2019). Brief report: The unrealized potential of autonomous Vehicles for an aging population. *Journal of Aging & Social Policy, 31*(5), 486–496.

Pinkse, J., Bohnsack, R., & Kolk, A. (2014). The role of public and private protection in disruptive innovation: The automotive industry and the emergence of low-emission vehicles. *Journal of Product Innovation Management, 31*(1), 43–60.

Porter, L. (2018). The autonomous vehicle revolution: Implications for planning. *Planning Theory & Practice, 19*(5), 753–778.

Qu, S.Q., & Dumay, J. (2011). The qualitative research interview. *Qualitative Research in Accounting & Management, 8*(3), 238–264.

Robson, C. (2002). *Real World Research*, 2nd ed.; Blackwell Publishing: Malden.

Roche-Cerasi, I. (2019). Public acceptance of driverless shuttles in Norway. *Transportation Research. Part F, Traffic Psychology and Behaviour, 66*, 162–183.

Salmenkaita, J.P., & Salo, A. (2002). Rationales for government intervention in the commercialization of new technologies. *Technology Analysis & Strategic Management*, 14(2), 183–200.

Sandelowski, M., & Barroso, J. (2006). *Handbook for Synthesizing Qualitative Research*. Singapore: Springer Publishing Company.

Saunders, M., Lewis, P., & Thornhill, A. (2007). Research methods. *Business Students 4th Edition Pearson Education Limited, England*, 6(3), 1–268.

Schepis, D., Purchase, S., Ellis, N., Olaru, D., & Smith, B (2023). How Governments Influence Autonomous Vehicle (Av) Innovation. Available at SSRN 4013442.

Seidman, I. (2013). *Interviewing as Qualitative Research: A Guide for Researchers in Education and the Social Sciences*. New York: Teachers College Press.

Sherwin, H., Chatterjee, K., & Jain, J. (2014). An exploration of the importance of social influence in the decision to start bicycling in England. *Transportation Research Part A: Policy and Practice*, 68, 32–45.

Shladover, S.E., & Nowakowski, C. (2019). Regulatory challenges for road vehicle automation: Lessons from the California experience. *Transportation Research Part A: Policy and Practice*, 122, 125–133.

Simons, D., Clarys, P., De Bourdeaudhuij, I., de Geus, B., Vandelanotte, C., & Deforche, B. (2014). Why do young adults choose different transport modes? A focus group study. *Transport Policy*, 36, 151–159.

Sindi, S., & Woodman, R. (2021). Implementing commercial autonomous road haulage in freight operations: An industry perspective. *Transportation Research Part A*, 152, 235–253.

Skeete, J.P. (2018). Level 5 autonomy: The new face of disruption in road transport. *Technological Forecasting and Social Change*, 134, 22–34.

Smith, J.A., Flowers, P., & Larkin, M. (2009). *Interpretative Phenomenological Analysis: Theory, Method and Research*; Sage: London.

Sperling, D., van der Meer, E., & Pike, E. (2018). Vehicle automation: Our best shot at a transportation do-over? In Sperling, D. (Ed.) *Three Revolutions: Steering Automated, Shared, and Electric Vehicles to a Better Future*; Island Press: Washington, DC, pp. 77–108.

Steinhilber, S., Wells, P., & Thankappan, S. (2013). Socio-technical inertia: Understanding the barriers to electric vehicles. *Energy Policy*, 60, 531–539.

Stone, J., Legacy, C., & Curtis, C. (2018). The driverless city? *Planning Theory & Practice*, 19(5), 756–761.

Taylor, M.R., Rubin, E.S., & Hounshell, D.A. (2005). Regulation as the mother of innovation: the case of SO2 control. *Law & Policy*, 27(2), 348–378.

Vaismoradi, M., Turunen, H., & Bondas, T. (2013). Content analysis and thematic analysis: Implications for conducting a qualitative descriptive study. *Nursing & Health Sciences*, 15(3), 398–405.

Wallendorf, M., & Belk, R.W. (1989). Assessing trustworthiness in naturalistic consumer research. In Hirschman, E. (Ed.) *Interpretive Consumer Research*; Association for Consumer Research: Provo, UT (pp. 69–84).

Wang, N., Pei, Y., & Fu, H. (2022). Public acceptance of last-mile shuttle bus services with automation and electrification in cold-climate environments. *Sustainability*, 14(21), 14383.

Webb, J., Wilson, C., & Kularatne, T. (2019). Will people accept shared autonomous electric vehicles? A survey before and after receipt of the costs and benefits. *Economic Analysis and Policy*, 61, 118–135.

Wu, P.F. (2011). A mixed methods approach to technology acceptance research. *Journal of the Association for Information Systems*, 13, 172–187.

Yigitcanlar, T. (2010). *Rethinking Sustainable Development: Urban Management, Engineering, and Design*. Hersey, PA: IGI Global.

Yigitcanlar, T., Agdas, D., & Degirmenci, K. (2023). Artificial intelligence in local governments: Perceptions of city managers on prospects, constraints and choices. *AI & Society*, 38(3), 1135–1150.

Yigitcanlar, T., Dodson, J., Gleeson, B., & Sipe, N. (2007). Travel self-containment in master planned estates: Analysis of recent Australian trends. *Urban Policy and Research*, 25(1), 129–149.

Yigitcanlar, T., Mohamed, A., Kamruzzaman, M., & Piracha, A. (2019). Understanding transport-related social exclusion: A multidimensional approach. *Urban Policy and Research*, 37(1), 97–110.

6

Attitudes towards the Deployment of Autonomous Vehicles

6.1 Introduction and Background

The development, production, and deployment of partially and fully connected and autonomous vehicles (CAVs) are paving the way for significant changes in many ways (Faisal et al., 2019; Yigitcanlar et al., 2020). Nonetheless, some of the changes ahead are difficult to anticipate (Yigitcanlar et al., 2019; Faisal et al., 2020). From November 2017 to October 2018, City of Las Vegas deployed a CAV shuttle bus within the Fremont Entertainment District of downtown Las Vegas. This deployment was in partnership with Keolis North America, American Automobile Association (AAA), and the Regional Transportation Commission of Southern Nevada. At that time, this was one of the few completely driverless vehicles in the world operating on public roads in mixed traffic. The free-service shuttle bus allowed the public to experience the autonomous driving technology first-hand.

The deployment of this CAV in downtown Las Vegas is representative of how rapidly CAVs are developing and improving. Hence, it is important to study the feasibility of widespread use to plan accordingly. The use of these vehicles is largely subject to public perception, attitudes, and preferences, as they could dictate how, where, when, and if CAVs are widely adopted. There are many assumptions and models in the literature about CAVs but relatively little field data and real-world insights.

The CAV in downtown Las Vegas was connected and integrated into the traffic signals, allowing it to obey traffic rules without manual control. Hence, classifying this as a CAV is more appropriate than simply calling it an autonomous vehicle (AV). The term 'connected' can encompass many different types of connections, including vehicle to vehicle (V2V), to infrastructure (V2I), and many others. In this case, the shuttle would be considered a V2I-CAV (Christie, 2015).

DOI: 10.1201/9781003605676-6

A map of the shuttle route is presented in Figure 6.1—it included four stoplights and two stop signs. The route was pre-programmed and took approximately 6–8 min from start to finish. On this route, the shuttle interacted with mixed traffic and pedestrians. This route is in a high-tourist area and being a tourist was not considered exclusion criterion to participate in this study. The shuttle had varying hours of operation; however, it generally operated Tuesday through Saturday from 11 am to 7 pm. This route allowed shuttle-riders to experience the range of capabilities of the shuttle in dynamic situations (i.e., unpredictable behaviours from pedestrians or vehicles, interaction with traffic signals and stop signs). In addition, a CAV licenced attendant was on-board during all operations to ensure safety, discuss general shuttle information (technical and non-technical), and address rider questions and comments.

The shuttle was manufactured by Navya, EasyMile, Local Motors, May Mobility, and Coast Autonomous; it is owned and operated by Keolis. The pilot test was funded by Keolis and the AAA. The pilot test also involved the City of Las Vegas and the Regional Transportation Commission of Southern Nevada. The shuttle was advertised via social media and various stakeholder websites. Eight people could ride the shuttle at one time and over 32,000 passengers rode it during the operational period. The shuttle could travel at 25 miles per hour or more; however, the speed was limited to 10–15 miles per hour on the route shown in Figure 6.1.

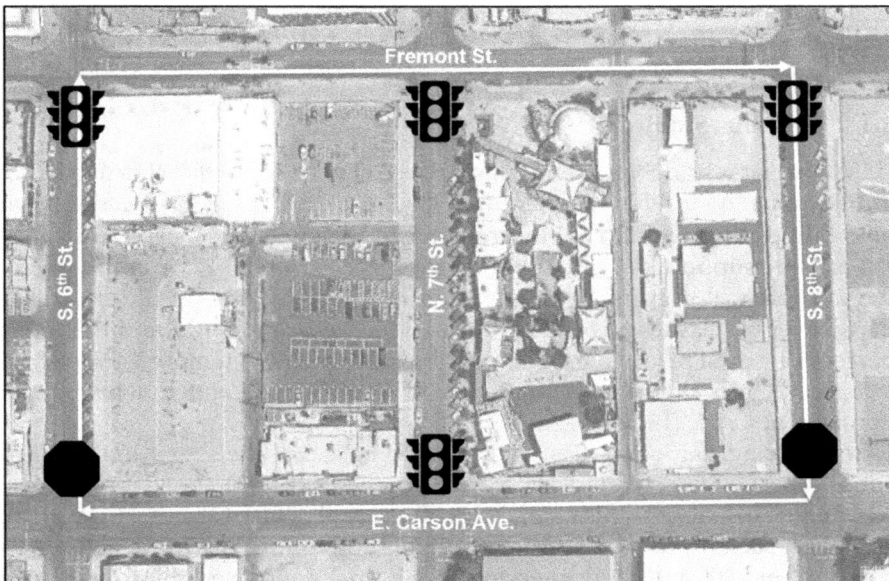

FIGURE 6.1
Map of the CAV route.

Attitudes and preferences towards the connected and autonomous shuttle bus can be used to investigate future adoption. If the public does not trust or is not comfortable with the technology, it is unlikely to be used. Similarly, knowledge about specific aspects of concern and/or segments of the population can be used to minimise barriers that could limit or delay the development and deployment of CAVs. This study seeks to gain knowledge about perceptions and attitudes regarding CAVs considering people with and without experience with the technology.

To accomplish these objectives, the general public was surveyed from various community centres within the City of Las Vegas and via university communication channels. Similarly, people who had ridden the CAV in downtown Las Vegas were surveyed. Considering that the shuttle was an early representation of the type of vehicle that is expected to become more widespread within the next several decades, it provided an opportunity for capturing opinions and attitudes from people who have some experience with the technology.

6.1.1 Study Objectives

This chapter contributes to the existing body of knowledge regarding CAVs and their effects by comparing the experiences and attitudes of people who have and have not ridden on a CAV. The CAV in this study was available to the public, operated on public roads, and was a great instrument to illustrate the potential of the technology. The primary objectives of this study were to evaluate public perceptions towards CAVs. Specifically, how they differed between those who had and had not ridden a CAV.

6.1.2 Previous Studies

CAVs are largely debated and studied because of their difficult-to-predict implications. A study by Fagnant and Kockelman (2015) considered several possibilities through an extensive literature review and conceptual exploration. They supported the motivation for this current study by emphasising the importance of users' perceptions and attitudes. Liu et al. (2019) stated that perception issues have often been known to drive policy and that poor perception could result in near-impossible adoption. Even more, how people feel about CAVs can be more predictive of future adoption than what people think about CAVs (Liu et al., 2019).

Considering demographics, men tend to be more likely to adopt, pay to adopt, and use CAVs (Payre et al., 2014; Schoettle & Sivak, 2014; König & Neumayr, 2017; Kyriakidis et al., 2015; Hulse et al., 2018; Liljamo et al., 2018). Sensation-seeking drivers were more likely to be interested in using CAVs (Payre et al., 2014). Females tended to be more concerned than their male counterparts and less likely to believe that the anticipated benefits from the widespread use of CAVs would occur (Schoettle & Sivak, 2014).

Younger respondents were another demographic that tended to lean more favourably towards CAVs. They were more likely than older participants to think that the anticipated benefits of CAVs would be met (Schoettle & Sivak, 2014; König & Neumayr, 2017; Hulse et al., 2018; Shabanpour et al., 2018). Similarly, older people tend to prefer non-automated vehicles (Haboucha et al., 2017).

With higher levels of education, participants were more likely to indicate signs of acceptance for CAVs (Haboucha et al., 2017; Liljamo et al., 2018). This includes participants showing more interest in having autonomous technology in their own vehicles and they anticipated many benefits (Schoettle & Sivak, 2014).

Some other factors that appeared to contribute to participants' willingness-to-accept CAVs included knowledge (Schoettle & Sivak, 2014), access to higher levels of autonomous capabilities, lack of a personal vehicle (Schoettle & Sivak, 2014; Kyriakidis et al., 2015; König & Neumayr, 2017), and living in urban settings in comparison to rural locales (König & Neumayr, 2017). Pro-CAV sentiments and interest in technology were indicators that a person would likely choose to use a CAV compared to a non-automated vehicle (Haboucha et al., 2017). Participants with previous crash-experience (Shabanpour et al., 2018) and people with knowledge of CAVs (König & Neumayr, 2017) tend to have more positive views of their safety capabilities. People with disabilities are less sensitive to higher CAV purchase prices (Shabanpour et al., 2018). Higher income appears to be related to someone's willingness-to-adopt (Shabanpour et al., 2018) and the amount they are willing to pay (Kyriakidis et al., 2015). When a person had direct experience with a CAV, they reported higher levels of perceived use and ease of use of CAVs (Xu et al., 2018). In addition, people who felt safer while riding a CAV tended to show more willingness to use CAVs (Xu et al., 2018).

Although the overall results in most cases appeared to show that most people felt positively about CAVs, there were also many important and common concerns identified. When a definition of Level 4 automation was provided, 36% of people interviewed in the US were very concerned riding in a vehicle with this level of automation, which was the most severe negative category (Schoettle & Sivak, 2014). Although 36% is not the majority, it was the most frequent choice from four given options (Schoettle & Sivak, 2014). A study by Nielsen and Haustein (2018) revealed that only 25% felt that CAVs would improve safety. Schoettle and Sivak (2014) identified that only 16.4% of participants in their study indicated negative feelings towards CAVs, while Hulse et al. (2018) recorded 7% of participants being conditionally negative, and 3% being negative.

Although there are certainly some issues to be addressed in terms of widespread adoption, public acceptance, and comfort with CAVs, approximately two-thirds of participants in one study would be willing to use CAVs (Payre et al., 2014). Liu et al. (2019) reported that experience with CAVs resulted in positive feelings. Schoettle and Sivak (2014) disclosed that 56.3% of participants in the US have positive feelings towards CAVs. This finding is very comparable to Liljamo et al. (2018) who found that over 60% of participants felt positive about the development of CAVs. They also reported only about 7% of participants expressing very negative feelings towards CAVs (Liljamo et al., 2018). Zhang et al. (2019) reported that participants tended to feel that CAVs were moderately trustworthy, had a positive attitude in regard to them, and conveyed an intention to use CAVs. They also found that initial trust towards CAVs largely predicted if people would accept CAVs (Zhang et al., 2019). Identifying the proportions of people who are accepting and rejecting CAVs is important as it has been estimated that fully automated vehicles will take up a significant portion of the market space by the year 2030 (Kyriakidis et al., 2015; Nielsen & Haustein, 2018).

Public acceptance and use are based largely around trust. A study with similar aims and design to the one discussed in this manuscript was completed by Eden et al. (2017). Preliminary discussions showed that all participants who had stated safety concerns before riding a driverless shuttle no longer expressed those concerns after their ride. Passengers also expressed a general feeling of safety due to the low speed of the shuttle. However, they expressed greatly reduced convenience due to associated long travel times. To our knowledge, at the time that this was written, this study was the only surveying participants who had direct experience in riding on a CAV.

6.2 Research Design

6.2.1 Stated Preferences

CAVs are emerging across the world under experimental conditions. Hence, the general public have little-to-no personal experience with them. In addition, although vehicles with semi-automated features such as adaptive cruise control and lane correction are increasingly more common, fully AVs are not yet commercially available. For this reason, a revealed preference approach is not feasible to study these emerging technologies. Rather, a stratified stated preference (SP) approach is proposed to gain insights about people's attitudes and preferences towards CAVs. The SP approach is a well-known technique used to gain similar insights typically associated with emerging technologies (Peeta et al., 2008; Nordland et al., 2013).

The SP approach potentially suffers from a hypothetical bias that results from participants responding differently to the hypothesised circumstance than they would to the real-world case. With emerging technologies there is a potential that participants are learning about the technology through the survey questionnaire and therefore their responses are based on limited information. However, this study included participants who have ridden on the CAV to evaluate SPs with and without the experience. This allows for comparison between how SPs change based on experience with CAVs. Comparing responses between those who did and did not ride the shuttle helps account for some bias by allowing interpretation of the change rather than just the opinion.

6.2.2 Survey Design and Implementation

Two survey questionnaires were used to collect data. Both surveys include a brief word-bank to clearly define any terms used within the surveys that may be unclear or ambiguous. Included were terms such as AV transportation, mode, ride-hailing (Contreras & Paz, 2018), and business trips versus personal trips. The surveys were collected by the authors and volunteers.

The first survey, referred to as 'the general survey', was intended for the general public. This survey included 30 questions and was conducted at community centres within the City of Las Vegas to capture opinions representing the entire community. There were 236 survey responses collected for this study. An identical online version of the survey was shared via outlets provided by the University of Nevada, Las Vegas. This survey consisted of four types of questions: (a) Demographic information; (b) Current transportation modes and practices; (c) Participant's criteria for the use of CAVs as well as the circumstances and situations in which they would choose autonomous transportation in some form; and (d) Perceptions and opinions towards CAVs.

The second survey, referred to as 'the shuttle-rider survey', was intended only for participants who had ridden the shuttle in downtown Las Vegas. These surveys were conducted in-person and on-site at the shuttle location, and 153 surveys were collected. Participants completed the survey either during or immediately after their ride. Throughout this, the participants of this survey are referred to as the 'shuttle-riders'. This version of the survey included 17 questions, all of which fell into the following categories: (a) Demographic information; (b) Perceptions and opinions towards CAVs; and (c) If/how the shuttle impacted their perceptions or opinions.

Given the different sample populations for each survey, the following considerations are important. First, it is expected that a portion of the shuttle-riders were likely not Las Vegas locals. This could be an important consideration in comparing the results of the two surveys. However, the socio-demographics of the two sample groups were not that different. Las Vegas is a very cosmopolitan city with many residents relocating constantly from all over the world. Second, shuttle-riders elected to ride the CAV; hence, they may already have more positive sentiments towards this type of technology resulting in some bias within the results. If people felt a certain level of negativity about CAVs, they likely would have never selected into riding on the shuttle. Nonetheless, the experience with the shuttle is not necessarily perceived as positive for some of them. For example, the shuttle is very small, and people were all very tight together. Similarly, the operation speed is very low and even then, there is uncertainty about the vehicle capabilities to navigate autonomously. For the general survey, people may have elected to participate in the survey because they had strong feelings (positive or negative) about CAVs. This could also result in some self-selection bias.

In addition, Puhr et al. (2017) were able to show that the Firth's penalised logistic regression (PLR), used in this study and discussed in the following sections, has predicted probabilities that can be biased towards 0.5. This makes the coefficient estimates more conservative, pulling high and low coefficients closer to the 'middle' or more neutral values.

6.2.3 Survey Sample Characteristics

The demographics associated with each survey are presented in Table 6.1.

6.2.4 Methodology

6.2.4.1 Discrete Choice Models

Participants were asked to choose between several alternatives. Descriptive statistics are used to analyse the data and responses. In addition, discrete

TABLE 6.1

Demographics of General (n=236) and Shuttle-Rider (n=153) Survey Respondents

Demographics		Variable Names	Survey 1	Survey 2
			(Percentage of Participants)	
Gender	Male	–	42	44
	Female	FEMALE	57	56
	Other	–	1	0
Age	Less than 25	AGE25L.	30	37
	25–34	AGE2534.	30	29
	35–44	AGE3544.	19	18
	45–54	AGE4554.	11	6
	55–64	AGE5564.	7	5
	65 or higher	AGE65M.	4	5
Income	Under $30,000	INC30L.	24	31
	$30,000–$59,999	INC3059.	29	22
	$60,000–$99,999	INC6099.	20	21
	$100,000–$150,000	INC100150.	16	13
	Over $150,000	INC150M.	8	9
Education	High school or less	EDUCHS	8	10
	Some college	EDUCSC	38	45
	College graduate	EDUCCG	27	33
	Post-graduate	EDUCPG	27	10

choice models are estimated to study correlations and the simultaneous effects of multiple characteristics and attributes. A common assumption is to expect that people are naturally inclined to choose the alternative that maximises their satisfaction. Utility functions seek to capture the satisfaction with each alternative and are typically given by Equation 6.1:

$$u_{i,j} = v_{i,j} + \varepsilon_{i,j} = \beta_j x_{i,j} + \varepsilon_{i,j} \tag{6.1}$$

where

i = individual 1, 2, ... , N

j = alternative choice 1, 2, ... , J

$u_{i,j}$ = utility for participant i choosing alternative j

$v_{i,j}$ = observed utility

$\varepsilon_{i,j}$ = unknown error

$x_{i,j}$ = vector of observable characteristics of participant i or attributes of alternative j

$\beta_{i,j}$ = vector of regression coefficient for $x_{i,j}$

Maximum log-likelihood estimation is used to estimate β and obtain the best model specification. For this study, responses were binary-coded. Hence, binary choice models were created to analyse the acceptance of CAVs. Assuming that ε are normally distributed, and without loss of generality, the probability of a participant i choosing alternative j is (Equation 6.2)

$$P_i(j) = \Phi\left(\frac{\beta_j x_{i,j} - \beta_{j+1} x_{i,j+1}}{\sigma}\right) \tag{6.2}$$

where $\Phi(\cdot)$ is the standardised cumulative normal distribution.

6.2.4.2 Data Considerations

Some of the categorical independent variables have very few samples for some levels of their domain. For example, for the general survey there were 236 survey responses collected, out of which nine participants were over 65 years old, four participants drove a work vehicle, and six felt that CAVs would have only downfalls. Similar observations were made for the shuttle-rider survey. Due to these small samples for select responses, estimated models using standard approaches, such as Logit and Probit models, resulted in perfect separation errors.

Perfect separation can occur with small datasets or when a specific exposure is 'rare' within the data (Albert & Anderson, 1984; Puhr et al., 2017). Rarity in the statistical sense is not strictly defined; however, where possible and appropriate for this study, an exposure was considered rare if it obtained less than 10% of the given sample space. Some regression models are shown to address perfect separation or rarity in a dataset, although more basic solutions, such as increasing sample size, combining similar categories of variables or omitting a category (Heinze & Schemper, 2002; Allison, 2004) are first attempted.

For this study, increasing sample size was not a viable option, the 'other' category was the only one that could be omitted, but several categories were combined. For example, in the general survey the 65 and older age group was combined with the 55–64 years old category. The result was an age category of 55 years or older appearing within the estimated model (Table 6.2). Similarly, the shuttle-rider survey also included the 45–54 age group (Table 6.2). This resulted in the highest age group being 45 years or older for the shuttle-rider survey. The same approach was used for 'current mode of transportation'. All mode variables, aside from 'drive self' were combined. These variable combinations are detailed in Table 6.2. Finally, the general survey included two questions asking how much exposure participants had to CAVs. Participants were asked what their level of exposure was on a scale of one to five, with five being the highest level of exposure, and if they had ever ridden a CAV. Nevertheless, only a small sample (9%) had an experience level of four or five so of the five exposure levels, three combination categories were created. They are defined in Table 6.2.

Unfortunately, this combination of categories still sometimes resulted in perfect separation errors. Hence, models that are capable of addressing rarity in a dataset were tested. Exact Logistic Regression and PLR were tested using the same model variables. Exact Logistic Regression is highly sensitive to a large number of independent variables within the model (Corcoran et al., 2001). After testing several specifications, Exact Logistic Regression was deemed inappropriate for this study. PLR is one of the more subjectively complicated ways of addressing the bias resulting from perfect separation (Firth, 1993). However, PLR is only slightly more complex than maximum likelihood (ML) estimation (Heinze, 2006) and it was capable of estimating models with this dataset. Therefore, PLR was the selected modelling method for this study.

6.2.4.3 Penalised Logistic Regression

Although PLR was originally developed to create reliable models for rare events, it has also been shown to reduce bias in models experiencing separation (Heinze & Schemper, 2002). In outlining the penalisation that takes place in PLR, a description of the binary Logit model is a helpful starting

TABLE 6.2

Reference Table of Variable Names and Definitions

Binary Variables		
AGE55M.	*age 55 or older, oldest age group for general survey*	
AGE45M.	*age 45 or older, oldest age group for shuttle-rider survey*	
WALKBIKE	*current mode of transportation walking or biking*	
DRIVESELF	*current mode of transportation driving themselves*	
BUSRAIL	*current mode of transportation riding a bus or rail*	
RHWVO	*current mode of transportation ride-hail, work vehicle, or other*	
EXP_LVL1.	*the participant ranked their exposure as level 1 (lowest)*	
SOME_EXP	*the participant had some exposure to AV (level 2 or 3)*	
HI_EXP	*the participant had some exposure to AV (level 4 or 5) (highest)*	
AV_BEN_V	*CAVs will have vast benefits*	
AV_BEN_MB	*CAVs will have many benefits, some downfalls*	
BENS	*CAVs will have benefits (a combination of AV_BEN_V and AV_BEN_MB)*	
AV_BEN_BD	*CAVs will have just as many benefits as downfalls*	
AV_BEN_MD	*CAVs will have many downfalls, some benefits*	
AV_BEN_OD	*CAVs will have only downfalls*	
DWNFL	*CAVs will have downfalls (a combination of AV_BEN_MD and AV_BEN_OD)*	
AV_BEN_NS	*not sure what level of benefits CAVs will have*	
MA_Y	*would like to CAVs more widely available*	
MA_N	*would not like to see CAVs more widely available*	
MA_X	*no opinion on if they would like to see CAVs more widely available*	
TECH_ALOT	*the shuttle enhanced the participants understanding of CAV:*	*technology a lot*
TECH_SMWT		*technology somewhat*
TECH_LTL		*technology a little*
TECH_NO		*technology not at all*
CPBL_ALOT		*capabilities a lot*
CPBL_SMWT		*capabilities somewhat*
CPBL_LTL		*capabilities a little*
CPBL_NO		*capabilities not at all*
SAFE_ALOT		*safety a lot*
SAFE_SMWT		*safety somewhat*
SAFE_LTL		*safety a little*
SAFE_NO		*safety not at all*
BNFT_ALOT		*benefits a lot*
BNFT_SMWT		*benefits somewhat*
BNFT_LTL		*benefits a little*
BNFT_NO		*benefits not at all*
FEEL_H	*participants felt happy about CAVs becoming more widespread*	
FEEL_U	*participants felt unhappy about CAVs becoming more widespread*	
IMPV_AGREE	*the participant agreed that CAVs will improve their life*	
IMPV_DIS	*the participant disagreed that CAVs would improve their life*	

point. The binary Logit model is represented by the following (Equation 6.3) (Wooldridge, 2016):

$$P\left(Y_i = 1 \mid x_i\right) = \rangle \left(\beta_0 + \beta_1 x_{1,i} + \beta_2 x_{2,i} + \ldots + \beta_N x_{N,i}\right) = \rangle \left(\beta_0 + x^2\right) \qquad (6.3)$$

where

i = index for a given decision-maker

Y_i = choice of decision-maker *i*; it can take values 0 or 1

$P(\cdot)$ = the probability of decision-maker *i* choosing alternative *j*

$X_{n,i}$ = characteristic or attribute *n* for decision-maker *i* ($n = 1, 2, \ldots, N$ = total number of independent variables)

β_0 = model constant or intercept

β_n = coefficient for X_n

x = vector of observed variables to be considered in the model

$\boldsymbol{\beta}$ = vector of estimated or estimable parameters

$\Lambda(z) = \exp\left(x_i\boldsymbol{\beta}\right) / \left[1 + \exp\left(x_i\boldsymbol{\beta}\right)\right]$ = the logistic CDF function that takes values strictly between 0 and 1, $0 < \Lambda(z) < 1$ for all real numbers z.

The coefficients β for logistic regression are estimated using ML. To facilitate the calculation of the ML and estimation of the coefficients, the likelihood function of the logistic regression model is (Equation 6.4)

$$\mathcal{L}(\boldsymbol{\beta}) = \prod_{i}^{N} P\left(x_i\right)^{y_i} \left(1 - P\left(x_i\right)^{1-y_i}\right) \qquad (6.4)$$

A convenient starting point to derive the coefficient estimates using Firth's PLR method is the likelihood function for PLR (Equation 6.5) (Firth, 1993):

$$\mathcal{L}_{PLR}\left(\boldsymbol{\beta}\right) = \mathcal{L}(\boldsymbol{\beta}) \times \left|i(\boldsymbol{\beta})\right|^{0.5} \qquad (6.5)$$

where

$\mathcal{L}_{PLR}\left(\boldsymbol{\beta}\right)$ = the likelihood function for PLR

$\mathcal{L}(\boldsymbol{\beta})$ = the logistic regression likelihood function

$i(\boldsymbol{\beta})$ = the Fisher information matrix

$\left|i(\boldsymbol{\beta})\right|^{0.5}$ = Jeffrey's invariant prior, in this case, the penalisation factor

Jeffrey's invariant prior is the penalisation applied to the binary logistic regression likelihood function to formulate the PLR likelihood function. As in the logistic regression model, the resulting coefficients β from PLR can be interpreted as the change in the log-odds of the output, given a one-unit change in the dependent variable.

As previously mentioned, both surveys sought various types of information from participants. For this study, the selected variables to include in the model are intended to relate the use, acceptability, and openness towards CAVs (dependent variables) to demographics, socioeconomic indicators, CAV exposure (general survey), mode of transportation, and acceptability of CAVs (independent variables). Models are described in detail in the following section.

6.3 Analysis and Results

Descriptive statistics and PLR model results are presented in this section. The following presentation of results includes only descriptions of model coefficients statistically significant at the 10% level. Table 6.2 is a reference table for variable definitions of each variable included in the models for this study.

Each black horizontal line separates the results of one model from the next. The coefficient estimate, standard error, and chi-squared statistic are reported. The likelihood ratio test (LRT) statistic is reported as well as the respective p-value. These results are reported for all independent variables that were statistically significant for each model, even if the LRT statistic was not significant. Finally, for ease-of-interpretation, the odds ratio and calculated probabilities are also reported.

6.3.1 Perceptions/Opinions

There are two perception-related questions shared by both surveys. The findings related to these questions are critical, given that Liu et al. (2019) reported that perceived benefits predict the adoption of CAVs more strongly than perceived risks do. That is, if participants perceive CAVs as beneficial, they may be more likely to adopt CAVs even under some perceived risk.

The first question was framed as follows: "In general, I think autonomous transportation will have". Then, participants were asked to select one of the following responses: 'vast benefits', 'many benefits and some downfalls', "just as many downfalls as benefits", 'many downfalls and some benefits', 'only downfalls', or 'I'm not sure'. Two combined-variable responses (BENS and

DWNFL in Table 6.2) referred to as 'benefits' and 'downfalls' were created for this question. A summary of the responses to this question is provided in Figure 6.2 and model outputs are presented in Table 6.3. Note that the totals in each figure in this section do not equal 100% due to participants failing to respond, responding as neutral, or choosing an opt-out response.

From Figure 6.2, one of the most obvious differences between the two surveys is that for the general survey, for nearly all classifications of respondent, the general survey has a larger portion of participants indicating they feel CAVs will have downfalls. In total, 12% of general survey participants anticipated downfalls while only 4% of shuttle-riders anticipated downfalls. Another notable outcome is that the percentage of participants who anticipated benefits increased for all income categories, especially those who make less than $30,000 per year. In addition, while there are no classifications in the general survey where zero participants selected 'downfalls', the shuttle-rider survey has several categories where this is true. This includes all income categories of $60,000 per year and above, ages 35–44 and 55 or older, and college graduates and post-graduates. Even more, the maximum percentage of participants selecting 'downfalls' in the shuttle-rider survey is 3% and it only occurs with those who make less than $30,000 per year or those with some college education.

The second question asked, "Would you like to see autonomous transportation more widely available in the future?" Participants are asked to respond with 'yes', 'no', or 'no opinion'. These outputs are provided in Table 6.4.

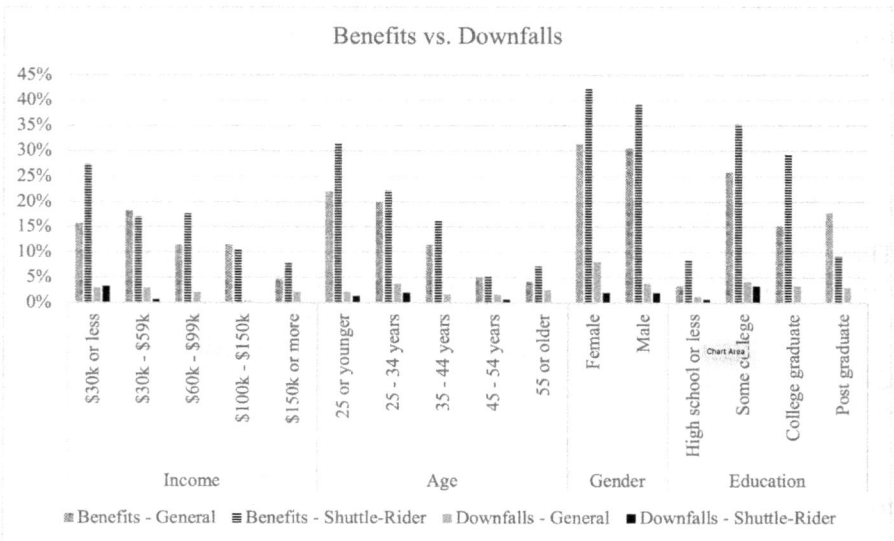

FIGURE 6.2
Percentage of respondents' perception of benefits by socioeconomic factor for each survey.

TABLE 6.3

Model Outputs for If Participants Foresee Benefits from CAVs

Survey	Dependent Variable	Independent Variable	Coefficient	Std. Error	Chi-Squared	LRT	Odds Ratio	Probability
General	AV_BEN_V	FEMALE	-0.91*	0.40	5.54	53.36 ‡	0.40	0.29
		MA_Y	2.20*	1.30	4.63		9.06	0.90
		MORE_EXP	1.17*	0.58	4.47		3.22	0.76
General	AV_BEN_MB	FEMALE	0.80**	0.33	7.00	69.27 ‡	2.22	0.69
		INC150M.	2.45***	0.79	11.51		11.62	0.92
		WALKBIKE	0.75~	0.41	3.73		2.12	0.68
		RHWVO	-1.22~	0.70	3.49		0.29	0.23
		MA_Y	2.63***	0.66	22.67		13.90	0.93
		MORE_EXP	-1.06*	0.58	3.87		0.35	0.26
General	AV_BEN_BD	AGE2534.	-1.10~	0.63	3.17	36.79 †	0.33	0.25
		INC150M.	-1.57~	0.91	3.40		0.21	0.17
		MA_Y	-2.40***	0.57	19.81		0.09	0.08
		MA_N	-1.20*	0.62	4.20		0.30	0.23
General	AV_BEN_MD	AGE2534.	1.89~	0.97	3.12	54.04 ‡	6.63	0.87
		EDUCCG	-3.16*	1.28	6.07		0.04	0.04
		EDUCPG	-2.26~	1.21	2.92		0.10	0.09
		MA_N	3.61***	1.21	12.51		36.96	0.97
General	AV_BEN_OD	AGE3544.	3.34~	1.60	3.14	24.43	28.12	0.97
General	AV_BEN_NS	AGE2534.	2.64*	1.30	3.95	29.19	14.04	0.93
		AGE3544.	2.37~	1.35	2.95		10.72	0.91
		AGE4554.	3.22*	1.44	5.02		24.96	0.96
		AGE55M.	3.72**	1.44	7.22		41.09	0.98
		MA_Y	-2.22**	0.71	9.79		0.11	0.10
		MA_N	-1.78*	0.87	4.45		0.17	0.14
General	AVBENS	AGE4554.	-1.75*	0.80	4.87	128.18 ‡	0.17	0.15
		AGE55M.	-2.51**	0.83	9.51		0.08	0.08

(*continued*)

TABLE 6.3 (Continued)

Model Outputs for If Participants Foresee Benefits from CAVs

Survey	Dependent Variable	Independent Variable	Coefficient	Std. Error	Chi-Squared	LRT	Odds Ratio	Probability
General	DWNFL	INC150M.	2.01*	0.92	5.25		7.46	0.88
		RHWVO	−1.47~	0.82	2.85		0.23	0.19
		MA_Y	3.83***	0.75	41.68	74.97‡	46.09	0.98
		AGE2534.	3.73**	1.34	7.29		41.54	0.98
		AGE4554.	2.64~	1.41	3.25		14.03	0.93
		AGE55M.	2.79~	1.52	2.99		16.31	0.94
		MA_N	4.45***	1.33	18.63	60.10‡	85.59	0.99
Shuttle-Rider	BENS	FEMALE	−2.09*	0.92	5.51		0.12	0.11
		INC3059.	−4.12**	1.48	10.38		0.02	0.02
		INC6099.	−3.53*	1.50	5.96		0.03	0.03
		INC150M.	−4.63*	2.06	4.02		0.01	0.01
		RHWVO	−2.03*	1.02	4.43		0.13	0.12
		FEEL_H	2.24*	0.95	5.35		9.37	0.90
		FEEL_U	−6.46**	2.52	8.59		0.00	0.00
		MA_Y	3.07**	1.19	8.99		21.55	0.96
Shuttle-Rider	AV_BEN_V	FEMALE	−0.99*	0.45	5.12	28.43	0.37	0.27
Shuttle-Rider	AV_BEN_MB	INC3059.	−0.90~	0.55	3.00	22.91	0.41	0.29
		FEEL_U	−4.95*	3.04	4.73		0.01	0.01
		MA_Y	1.43~	0.76	4.18		4.18	0.81
Shuttle-Rider	AV_BEN_BD	FEMALE	1.27~	0.75	2.82	31.38‡	3.55	0.78
		INC3059.	3.01**	1.19	7.07		20.35	0.95
		INC6099.	2.40*	1.21	3.97		11.05	0.92
		INC100150.	2.66*	1.25	4.70		14.24	0.93
		INC150M.	3.92*	1.67	4.84		50.52	0.98
		FEEL_H	−1.73~	0.80	4.55		0.18	0.15
		MA_Y	−1.50~	0.89	3.47		0.22	0.18

Coefficient: ~ $p < 0.10$, * $p < 0.05$, ** $p < 0.01$, *** $p < 0.001$

LRT: $p < 0.10$, † $p < 0.05$, ‡ $p < 0.001$

TABLE 6.4

Model Outputs for If Participants Would Like CAVs More Available

Survey	Dependent Variable	Independent Variable	Coefficient	Std. Error	Chi-Squared	LRT	Odds Ratio	Probability
General	MA_Y	FEMALE	–1.13 *	0.46	5.74	134.18 ‡	0.32	0.24
		INC150M.	–3.27 **	1.03	10.82		0.04	0.04
		AVBENS	3.94 ***	0.84	27.79		51.38	0.98
		AVDWNFL	–2.08 *	1.05	4.60		0.13	0.11
		MORE_EXP	2.62 ~	1.62	3.07		13.73	0.93
General	MA_N	AGE2534.	–2.54 *	1.05	6.04	100.45 ‡	0.08	0.07
		AGE3544.	–1.54 ~	0.86	3.12		0.21	0.18
		INC150M.	2.23 ~	1.16	3.20		9.32	0.90
		AVBENS	–2.81 *	1.07	5.92		0.06	0.06
		AVDWNFL	3.65 ***	1.11	14.95		38.47	0.97
General	MA_X	AGE2534.	1.68 *	0.82	3.88	60.22 ‡	5.35	0.84
		EDUCCG	–1.96 *	0.96	4.10		0.14	0.12
		AVBENS	–3.56 ***	0.90	18.22		0.03	0.03
		AVDWNFL	–3.27 **	1.13	10.38		0.04	0.04
Shuttle-Rider	MA_Y	AGE45M.	–2.21 ~	1.27	2.89	64.03 ‡	0.11	0.10
		INC6099.	2.75 ~	1.35	3.83		15.70	0.94
		INC150M.	4.47 ~	2.33	3.73		87.39	0.99
		BENS	4.51 *	2.50	4.38		91.26	0.99
		IMPV_AGREE	2.34 **	0.89	7.20		10.37	0.91
		IMPV_DIS	–3.15 *	1.83	4.49		0.04	0.04
Shuttle-Rider	MA_X	AGE45M.	2.41 ~	1.24	3.53	49.42 ‡	11.08	0.92
		INC6099.	–2.55 ~	1.26	3.60		0.08	0.07
		INC150M.	–4.46 *	2.24	3.99		0.01	0.01
		BENS	–5.07 *	2.53	5.58		0.01	0.01
		IMPV_AGREE	–2.04 *	0.89	4.79		0.13	0.12

Coefficient: ~ p < 0.10, * p < 0.05, ** p < 0.01, *** p < 0.001
LRT: p < 0.10, † p < 0.05, ‡ p < 0.001

A summary of the responses to this question for each survey is presented in Figure 6.3.

Similar to Figure 6.2, for all categories, the general survey has a higher portion of respondents not wanting to see CAVs more available than those participating in the shuttle-rider survey. Overall, 19% of the general survey participants did not want to see CAVs more available and only 4% of shuttle-riders selected this response. Interestingly, from both surveys, those with some college education responded that they would not like to see CAVs more available at a higher frequency than any other education level. For gender, females respond negatively at a higher portion than men

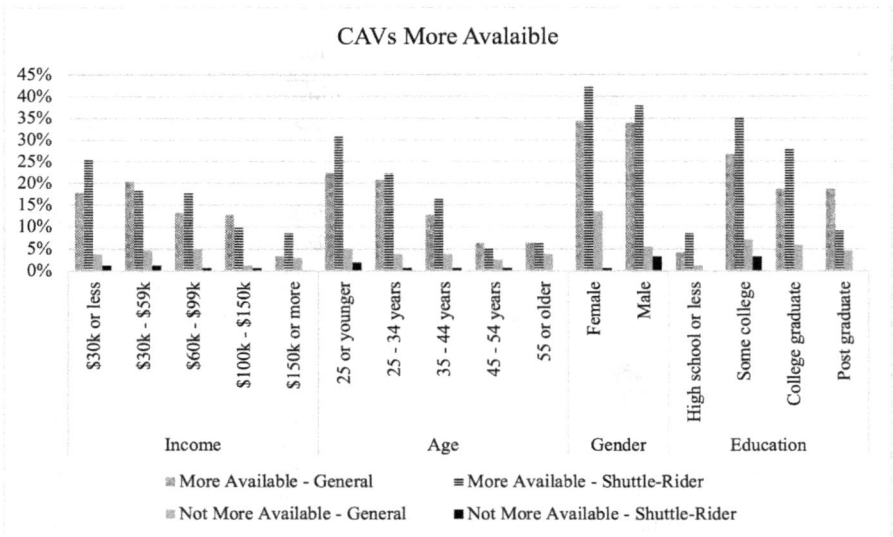

FIGURE 6.3
Percentage of participants indicating they would like CAVs more available, according to socioeconomic factor for each survey.

in the general survey, but the opposite is true for the shuttle-rider survey. In fact, in the general survey, 14% of participants selected 'not more available', while the shuttle-rider survey had only 1% select this response. The general survey had no socioeconomic categories with zero 'not more available' responses; however, zero shuttle-riders selected this response in the $150,000 income, 55 or older, high school or less, college graduate, or postgraduate categories.

In addition, shuttle-riders were asked how much the shuttle enhanced their understanding of self-driving technology, capabilities, safety, or benefits (variable names summarised in Table 6.2). Model outputs for this question are presented in Table 6.5. The results of this question are summarised in Figure 6.4.

From this series of questions, over half of participants responded that the shuttle bus enhanced their experience of CAV technology (58% 'a lot', 6% 'a little'), capabilities (59% 'a lot', 6% 'a little'), and safety (54% 'a lot', 12% 'a little'). For the benefits category, 32% of participants felt that the shuttle enhanced their understanding of CAV benefits 'a lot', 10% said 'a little', and 21% said 'somewhat'. Another major finding is that remarkably few participants felt that the CAV did not enhance their understanding in any category (2% for all categories). In addition, for nearly all socioeconomic categories, a higher portion of the group responded that the CAV enhanced their understanding 'a lot'. Interestingly, this is untrue for only three specific

TABLE 6.5

Model Outputs for How Shuttle-Riders Feel the Shuttle Enhanced Their Understanding of Autonomous Technology, Capabilities, Safety, and Benefits

Survey	Dependent Variable	Independent Variable	Coefficient	Std. Error	Chi-Squared	LRT	Odds Ratio	Probability
Shuttle-Rider	TECH_ALOT	AGE2534.	0.96~	0.59	3.04	27.58	2.61	0.72
		AGE3544.	1.1 ~	0.69	3.19		3.14	0.76
		INC6099.	−1.44*	0.66	5.50		0.24	0.19
		INC100150.	−1.42*	0.71	4.58		0.24	0.19
Shuttle-Rider	TECH_SMWT	AGE2534.	−1.19*	0.63	4.07	25.02	0.30	0.23
		INC6099.	1.33*	0.69	4.18		3.77	0.79
		INC10050.	1.61*	0.74	5.34		4.99	0.83
		INC150M.	1.74~	1.02	3.20		5.69	0.85
Shuttle-Rider	CPBL_ALOT	AGE3544.	1.22~	0.73	3.34	28.46	3.39	0.77
		IMPV_AGREE	1.08*	0.56	4.37		2.93	0.75
Shuttle-Rider	CPBL_SMWT	AGE3544.	−1.16~	0.76	2.73	25.12	0.31	0.24
		INC100150.	1.35*	0.73	3.89		3.87	0.79
Shuttle-Rider	CPBL_LTL	BENS	−6.89*	3.17	5.38	17.34	0.00	0.00
		AV_BEN_BD	−5.24*	2.86	3.92		0.01	0.01
Shuttle-Rider	SAFE_ALOT	AGE2534.	0.88~	0.57	2.84	24.53	2.42	0.71
		AGE3544.	1.73**	0.71	7.43		5.63	0.85
		IMPV_AGREE	0.95~	0.55	3.49		2.59	0.72
Shuttle-Rider	SAFE_SMWT	MA_Y	1.49~	0.95	3.05	18.81	4.43	0.82
Shuttle-Rider	SAFE_LTL	AGE2534.	−1.91*	0.89	5.17	29.44	0.15	0.13
		AGE3544.	−2.60*	1.24	5.19		0.07	0.07
		IMPV_AGREE	−1.26~	0.70	3.28		0.28	0.22
Shuttle-Rider	BNFT_ALOT	AGE3544.	1.10~	0.68	3.09	26.24	3.01	0.75
Shuttle-Rider	BNFT_SMWT	INC100150.	1.42*	0.76	3.90	29.92	4.13	0.81
		DRIVESELF	3.16*	1.82	4.28		23.47	0.96
		FEEL_U	−6.86*	3.79	4.90		0.00	0.00
		MA_Y	2.33*	1.20	4.73		10.30	0.91
		MA_N	4.64*	2.63	5.31		104.05	0.99
Shuttle-Rider	BNFT_LTL	INC150M.	2.21~	1.33	2.72	35.13 ‡	9.14	0.90

Coefficient: ~ p < 0.10, * p < 0.05, ** p < 0.01, *** p < 0.001
LRT: p < 0.10, † p < 0.05, ‡ p < 0.001

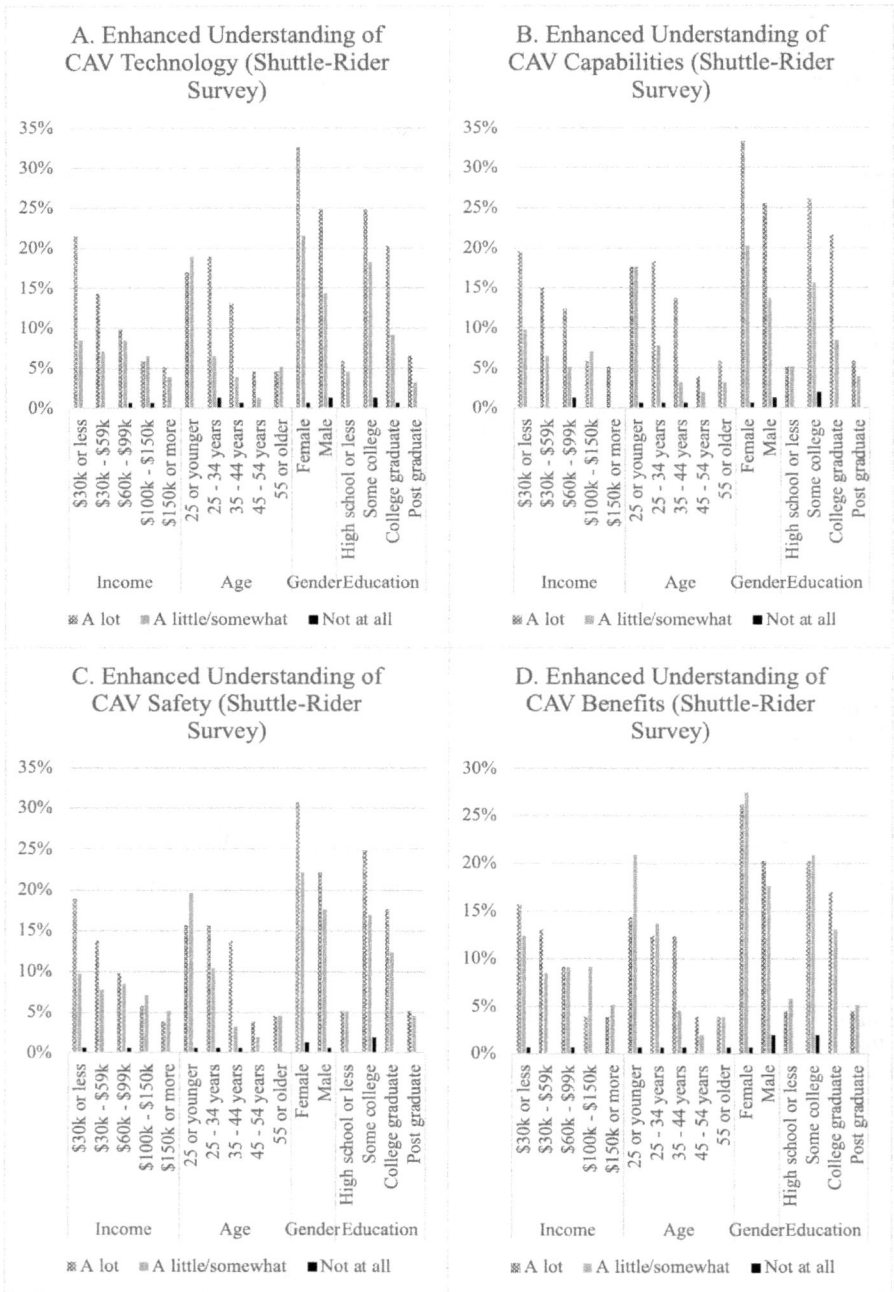

FIGURE 6.4
Summary of how much participants felt their experience of riding the CAV impacted their understanding of the technology (A), capabilities (B), safety (C), and benefits (D).

group and category combinations. These are the $100,000–$150,000 income group for the technology, capabilities, and benefits categories. The other is the $150,000 or more income group for the safety category. In each of these scenarios, the groups responded that the CAV 'somewhat' enhanced their knowledge in these areas. One reason that this result might make sense is that higher-income people likely drive more expensive and/or newer vehicles. More expensive and newer vehicles are more likely to have at least some semi-autonomous capabilities (such as adaptive cruise control or lane correction), so these participants already have some experience with their vehicle reacting without their intervention.

6.3.2 Model Outputs

Consistent with results from previous studies (Payre et al., 2014; Schoettle & Sivak, 2014; Kyriakidis et al., 2015; König & Neumayr, 2017; Hulse et al., 2018; Liljamo et al., 2018), females had lower odds of expressing positive sentiments towards CAVs than males. This is true for both the general survey and the shuttle-rider survey. For this section, 'positive sentiments' includes anticipating CAVs to have benefits or wanting to see them more available. Predictably, females also had a higher probability than their male counterparts to express negative feelings towards CAVs and their eventual outcomes. General survey females were less likely than males to feel that CAVs would have vast benefits; however, they were over two times more likely to feel that they would have many benefits and some downfalls (Table 6.3). Perhaps this is suggestive that female participants are maintaining more uncertainty than males, but still expect to see positive outcomes.

Shuttle-riders between the ages of 25 and 44 had high probabilities of indicating that riding on the shuttle enhanced their understanding of the self-driving technology. They also were less likely to feel positively about CAVs than those younger than 25. Shuttle-riders between 35 and 44 years old had respective probabilities of 77% and 75% for feeling the shuttle enhanced their understanding of the capabilities and benefits of the technology 'a lot'. Those from the general survey between 25 and 34 and 45 or older had a high probability (over 93%) of anticipating downfalls from CAVs becoming more widespread. Older people who had participated in the shuttle-rider survey had a very high probability (92%) of indicting that they had no opinion on whether they would like to see CAVs more available. The older participants who rode the shuttle indicate more uncertainty than negativity towards CAVs.

The general survey participants who identified as being within the highest income bracket ($150,000 a year or more) were over nine times more likely than those with an income of $30,000 or less to not want to see CAVs more available. This is interesting given that they had an 88% probability of believing CAVs would be beneficial. This is contrary to the findings of previous studies (Shabanpour et al., 2018).

Conversely, in the shuttle-rider survey the opposite is true. Lower-income shuttle-riders tended to express more positive sentiments than higher-income participants. This is a relationship that may result from a lack of accessibility to CAVs (or semi-automated technology) for lower-income individuals. Higher-income shuttle-riders also indicated that riding on the shuttle enhanced their understanding of self-driving vehicle benefits just a little bit. From Table 6.3, all shuttle-riders with an annual income of $30,000 or more had a high probability (over 91%) of indicating that they felt CAVs would have some benefits and some downfalls. Shuttle-riders who had an income of $60,000–$99,999 and those who make over $150,000 per year had high probabilities (94% and 99%, respectively) of wanting to see CAVs more available. Shuttle-riders with incomes $60,000 or higher had at least a 79% probability of saying that the shuttle somewhat enhanced their understanding of self-driving technology. Those who made between $100,000 and $150,000 per year also had a 79% probability of saying that the shuttle somewhat enhanced their understanding of the capabilities of CAVs.

From the general survey, college graduates and post-graduates had lower odds than those with a high school education or less of selecting that CAVs would have many downfalls.

Shuttle-riders who drive themselves had a 96% probability of feeling that riding the shuttle somewhat enhanced their understanding of the benefits associated with CAVs. Interestingly, participants who used ride-hail, a work vehicle, or other mode of transportation had a relatively low probability (19% for the general survey and 12% for shuttle-riders) of selecting that CAVs would have benefits. From the general survey, those who walk or bike had a 68% probability of selecting that CAVs would have many benefits and some downfalls.

As would be expected, when a participant felt that CAVs would have benefits, they also had a high probability (98% for the general survey and 99% for shuttle-riders) of wanting to see them more available. When general survey participants felt CAVs would have downfalls, they were highly likely (97%) to not want CAVs to be more available. Predictably, when general survey participants felt CAVs would have downfalls, they had a low probability (11%) of wanting to see CAVs more available. From the general public we can see that higher levels of exposure or having ridden on a CAV predict higher probabilities of positive sentiments towards CAVs. When a shuttle-rider felt positively about CAVs, they had a high probability of feeling that the shuttle somewhat enhanced their understanding of the safety and benefits of CAVs. Interestingly, shuttle-riders who did not want to see CAVs more widely available also had a 99% probability of saying that the shuttle somewhat enhanced their understanding of the benefits associated with CAVs.

6.4 Findings and Discussion

This study held the distinct objectives of contributing to the conglomerate of knowledge on public perceptions and opinions about CAVs. Two surveys were conducted to capture these perceptions and opinions including one for the 'general public' who were those without experience using CAVs and the one for people who experienced a shuttle CAV, 'shuttle-riders', deployed in Las Vegas, Nevada.

In terms of public opinions, many of the relationships seen in this study are comparable to what previous studies had identified. For example, in the two surveys completed as part of this study, females were more resistant to CAVs than males (Payre et al., 2014; Schoettle & Sivak, 2014; Kyriakidis et al., 2015; König & Neumayr, 2017; Hulse et al., 2018; Liljamo et al., 2018), more highly educated people were more open to CAVs (Haboucha et al., 2017; Liljamo et al., 2018), and younger people tend to be more accepting than older people (Schoettle & Sivak, 2014; Haboucha et al., 2017; König & Neumayr, 2017; Shabanpour et al., 2018; Hulse et al., 2018).

Nevertheless, in terms of income, this study differs from previous studies. From the general survey, it appears that higher-income participants feel more positively about CAVs, while in the shuttle-rider survey, lower-income people feel more positively. Shabanpour et al. (2018) found that people with a higher income tended to be more willing to adopt CAVs than their counterparts. Although these do not allow direct comparison, people are unlikely to accept a CAV if they have negative sentiments towards them. This relationship is worth future research to better understand if this is indicative of lower-income people making a larger 'leap' to positive sentiments. It could also be indicative of shuttle-riders with higher incomes not following the general trend of feeling more positively about CAVs after having experienced them. Perhaps higher-income people having better access to semi-autonomous technology in their vehicles and therefore the experience of a fully AV does not change their previous perceptions and opinions as dramatically as it does for other income levels.

The current mode of transportation was also included as an independent variable which, to the knowledge of these authors, has not previously been considered. Although the mode of transportation was rarely a statistically significant variable, there were a few statistically significant relationships. Mainly, participants from both surveys who use ride-hail, work vehicles, or other had low probabilities of anticipating CAVs to have benefits. This could be the result of people who use a work vehicle feeling that these changes are unlikely (they will not receive a new vehicle) or unfavourable (their job may be replaced by CAVs). For ride-hail users, this could be a result of their transportation being highly comparable to CAVs already (i.e., their action and attention are not required). This finding is deserving of future research to better understand why this relationship exists and if or how it can be counteracted.

The shuttle-rider survey provided interesting insights, given the access to a fully AV carrying members of the public who were interviewed. From this survey, there was a distinctive positive trend. Remarkably few participants who had ridden the shuttle expressed any negative sentiments towards CAVs, suggesting that experiencing the CAV is largely related to positive feelings about this technology. This is also supported by largely positive sentiments expressed by people who participated in the general survey but indicated having experience with CAVs. This finding is important for several reasons. First, entities interested in deploying CAVs should consider a trial period with CAVs prior to larger deployment—just as City of Las Vegas has done. Second, and perhaps more importantly, this could suggest that negative and uncertain feelings about CAVs are alleviated once people can experience the capabilities of this type of technology. Further, a trial period gives wary or curious members of the public an opportunity to trust CAVs.

Lastly, it appears that riding in a CAV helped enhance the understanding of many aspects of CAVs and could have been the reason the shuttle-riders felt more positive overall. This is a very useful finding because cities and states contemplating allowing CAVs to flourish in their communities can ensure that deployment of these vehicles is comfortable and favourable to their constituents.

Although this study resulted in important and highly useful information, there are some important considerations to keep in mind. For one, survey data can be lacking due to people's inaccuracy or uncertainty in reporting. Additionally, the scope of questions resulted in perfect separation errors. Although PLR was used to account for this, it is still a limitation of this study that a penalised model needed to be used to account for some uncertainty. The location of residence (i.e., Las Vegas resident or tourist) of participants was not collected. This means that people from all over the country or world likely participated and could have differing comfort levels with CAVs. Finally, there may be some resulting bias due to shuttle-riding participants electing to ride the shuttle. That is, participants could have felt positively about CAVs and that is why they rode the shuttle. However, there is no guarantee that the experience was positive and strengthened or weakened their original views. Future studies should attempt to provide further insights into these considerations.

This study confirms findings from previous CAV studies and contributes to the existing literature in several ways. First, this study identifies ways in which opinions or perceptions of CAVs can vary depending on personal characteristics, behaviours, or exposure levels to CAVs. This includes that people who use a work vehicle or ride-hail anticipate fewer benefits from CAVs. In addition, in the general survey, high-income people feel more positively about CAVs than lower-income people, but low-income people express more positive sentiments than high-income people when they have experienced the CAV.

This chapter also provides more detailed information than previous studies about how opinions on certain CAV aspects change after riding a CAV (technology, capabilities, safety, and benefits). This information is valuable to relevant industry parties both in terms of CAV development and marketing. Policymakers will also certainly find this information valuable, particularly for drafting CAV-related legislation. Next, the comparison between two surveys conducted in the same metropolitan area has value in gaining more direct comparisons of CAV perceptions. This is especially true given that the authors and participants had access to a free, publicly accessible, fully AV. Finally, perhaps the most valuable contribution of this is the policy implications. The CAV utilised for this study was part of a trial period for the public to gain direct experience. Given the largely positive results associated with the shuttle-rider survey, this may imply that the trial period eased stress and improved comfort levels for people experiencing autonomous technology.

6.5 Conclusion

In conclusion, this chapter underscores the critical role of public perceptions and attitudes in shaping the future of CAV deployment. Through a comparative analysis of general public attitudes and experiences of shuttle-riders in Las Vegas, this study reveals that direct experience with CAVs positively influences acceptance and reduces concerns about safety, technology, and utility. The findings highlight demographic variations, with income, education, and age affecting views on CAV adoption, and emphasise the impact of firsthand exposure on shifting perceptions. This analysis suggests that a public trial phase could be essential for fostering trust and addressing hesitancy around CAVs, offering a pathway for cities to strategically integrate autonomous technology into public spaces while accommodating community concerns.

Acknowledgements

This chapter, with permission from the copyright holder, is a reproduced version of the following journal article: Dennis, S., Paz, A., & Yigitcanlar, T. (2021). Perceptions and attitudes towards the deployment of autonomous and connected vehicles: Insights from Las Vegas, Nevada. *Journal of Urban Technology*, 28(3–4), 75–95.

References

Albert, A., & Anderson, J.A. (1984). On the existence of maximum likelihood estimates in logistic regression models. *Biometrika*, 71(1), 1–10.

Allison, P. (2004). Convergence problems in logistic regression. In Shewhart, W.A. & Wilks, S.S. (Eds.) *Numerical Issues in Statistical Computing for the Social Scientist*; Wiley & Sons, Inc.: Hoboken, NJ, 238–252.

Christie, B. (2015). "Status Summary of Connected Vehicle Related ITS Standards," presentation at the U.S. Department of Transportation Connected Vehicle Reference Implementation Architecture (CVRIA) Workshop. www.standards.its.dot.gov/Content/documents/workshop_CVRIA_pres_06-10-15.pdf

Contreras, S., & Paz, A. (2018). The effects of ride-hailing companies on the taxicab industry in Las Vegas, Nevada. *Transportation Research – Part A: Policy and Practice*, 115, 63–70.

Corcoran, C., Mehta, C., Patel, N., & Senchaudhuri, P. (2001). Computational tools for exact conditional logistic regression. *Statistics in Medicine*, 20, 2723–2739.

Eden, G., Nanchen, B., Ramseyer, R., & Evequoz, F. (September 2017). "Expectation and Experience: Passenger Acceptance of Autonomous Public Transportation Vehicles," paper presented at 16th IFIP Conference on Human-Computer Interaction (INTERACT) (Bombay, India, September 2017). pp. 360–363.

Fagnant, D.J., & Kockelman, K. (2015). Preparing a nation for autonomous vehicles: Opportunities, barriers and policy recommendations. *Transportation Research Part A*, 77, 167–181.

Faisal, A., Yigitcanlar, T., Kamruzzaman, M., & Currie, G. (2019). Understanding autonomous vehicles: A systematic literature review on capability, impact, planning and policy. *Journal of Transport and Land Use*, 12(1), 45–72.

Faisal, A., Yigitcanlar, T., Kamruzzaman, M., & Paz, A. (2020). Mapping two decades of autonomous vehicle research: A systematic scientometric analysis. *Journal of Urban Technology*. https://doi.org/10.1080/10630732.2020.1780868

Firth, D. (1993). Bias reduction of maximum likelihood estimates. *Oxford Journals*, 80(1), 27–38.

Haboucha, C.J., Ishaq, R., & Shiftan, Y. (2017). User preferences regarding autonomous vehicles. *Transportation Research Part C*, 78, 37–49.

Heinze, G. (2006). A comparative investigation of methods for logistic regression with separated or nearly separated data. *Statistics in Medicine*, 25, 4216–4226.

Heinze, G., & Schemper, M. (2002). A solution to the problem of separation in logistic regression. *Statistics in Medicine*, 21(16), 2409–2419.

Hulse, L.M., Xie, H., & Galea, E.R. (2018). Perceptions of autonomous vehicles: Relationships with road users, risk, gender and age. *Safety Science*, 102, 1–13.

König, M., & Neumayr, L. (2017). Users' resistance towards radical innovations: The case of the self-driving car. *Transportation Research Part F: Traffic Psychology and Behaviour*, 44, 42–52.

Kyriakidis, M., Happee, R., & de Winter, J.C.F. (2015). Public opinion on automated driving: Results of an international questionnaire among 5000 respondents. *Transportation Research Part F: Traffic Psychology and Behaviour*, 32, 127–140.

Liljamo, T., Liimatainen, H., & Pöllänen, M. (2018). Attitudes and concerns on automated vehicles. *Transportation Research Part F*, 59, 24–44.

Liu, P., Xu, Z., & Zhao, X. (2019). Road tests of self-driving vehicles: Affective and cognitive pathways in acceptance formation. *Transportation Research Part A*, 124, 354–369.

Nielsen, T.A.S., & Haustein, S. (2018). On sceptics and enthusiasts: What are the expectations towards self-driving cars? *Transport Policy*, 66, 49–55.

Nordland, A., Paz, A., & Khan, A. (2013). Public perceptions and preferences towards a VMT Fee System in Nevada. *Transportation Research Record*, 2345, 39–47.

Payre, W., Cestac, J., & Delhomme, P. (2014). Intention to use a fully automated car: Attitudes and *a priori* acceptability. *Transportation Research Part F: Traffic Psychology and Behaviour*, 27, 252–263.

Peeta, S., Paz, A., & DeLaurentis, D. (2008). Stated preference analysis of a new microjet on-demand air service. *Transportation Research – Part A*, 42(4), 629–645.

Puhr, R., Heinze, G., Nold, M., & Geroldinger, A. (2017). Firth's logistic regression with rare events: Accurate effect estimates and predictions? *Statistics in Medicine*, 36(14), 2302–231.

Schoettle, B., & Sivak, M. (2014). "A Survey of Public Opinion about Autonomous and Self-Driving Vehicles in the US, the UK, and Australia," *University of Michigan Transportation Research Institute*. UMTRI-014-21.

Shabanpour, R., Golshani, N., Shamshiripour, A., & Mohammadian, A.K. (2018). Eliciting preferences for adoption of fully automated vehicles using best-worst analysis. *Transportation Research Part C: Emerging Technologies*, 93, 463–478.

Wooldridge, J.M. (2016). *Introductory Econometrics*, 6th ed.; Cengage Learning: Boston.

Xu, Z., Zhang, K., Min, H., Wang, Z., Zhao, X., & Liu, P. (2018). What drives people to accept automated vehicles? Findings from a field experiment. *Transportation Research Part C*, 95, 320–334.

Yigitcanlar, T., Desouza, K., Butler, L., & Roozkhosh, F. (2020). Contributions and risks of Artificial Intelligence (AI) in building smarter cities: Insights from a systematic review of the literature. *Energies*, 13(6), 1473.

Yigitcanlar, T., Wilson, M., & Kamruzzaman, M. (2019). Disruptive impacts of automated driving systems on the built environment and land use: An urban planners' perspective. *Journal of Open Innovation: Technology, Market, and Complexity*, 5(1), 24.

Zhang, T., Tao, D., Qu, X., Zhang, X., Lin, R., & Zhang, W. (2019). The roles of initial trust and perceived risk in public's acceptance of automated vehicles. *Transportation Research Part C*, 98, 207–220.

7

Understanding Autonomous Vehicle Adoption

7.1 Introduction and Background

The driverless car has been identified as one of the six main smart mobility innovations (Butler et al., 2020a) and it has potential to improve accessibility, ensure first/last mile connectivity to public transport (Butler et al., 2022; Irannezhad & Mahadevan, 2022), and alleviate transportation disadvantage (Butler et al., 2020b). As predicted by the Australia and New Zealand Driverless Vehicle Initiative (ADVI), Level 4 (highly automated) vehicles will be ready for Australian roads between 2026 and 2030, and Level 5 (fully automated) vehicles could comprise 50%–75% of the Australian fleet mix by 2035–2045 (ADVI, 2021). According to the Australian National Transport Commission (NTC, 2022), the Infrastructure and Transport Ministers' Meeting (ITMM) agreed on the key elements and remaining elements of the in-service framework for automated vehicles in June 2020 and February 2022, respectively.

At the May 2021 ITMM, infrastructure and transport ministers agreed to a "regulatory implementation roadmap with the goal of end-to-end regulation in place to support the safe commercial deployment and operation of automated vehicles at all levels of automation in force in 2026" (NTC, 2022, p. 1). In July 2019, the NTC (2019) predicted that the release of Level 3 or above automated driving system (ADS) models would be commercially available to overseas markets between 2019 and 2022. While technology maturity and regulations supporting the real-time operation of driverless cars are due by 2026 (NTC, 2022), customers' attitudes towards driverless car adoption must be understood because the extent of the potential benefits of driverless cars is strongly linked to how widely they will be accepted and adopted (Golbabaei et al., 2020; Faisal et al., 2021). Considering the prospects, an examination of the hidden causes and contributing factors towards driverless car acceptance and adoption is necessary.

DOI: 10.1201/9781003605676-7

Though no studies were found on driverless car adoption and acceptance modelling in an Australian context, a sprawling and heavily private motor vehicle-concentrated urban context (Yigitcanlar et al., 2007; Perveen et al., 2017), few related studies were found on factors influencing public awareness of driverless cars (Butler et al., 2021), perceptions and attitudes towards the deployment of driverless cars (Dennis et al., 2021), and predictors of driverless car public acceptance (Golbabaei et al., 2020; Nastjuk et al., 2020). Due to study limitations, three major metropolitan areas—Brisbane, Melbourne, and Sydney—were selected for this study. Although all three respondent cohorts were evaluated against the same characteristics, such as employment in the household, driverless car trip purpose, and driver's licence duration, the adoption timeline due to a specific characteristic was expected to vary between regions. Thus, using a combined dataset could omit important differences in contributing factors between Brisbane, Melbourne, and Sydney preferences.

This study adopted a random parameter ordered probit approach to provide a deep understanding of the influence of contributing factors on driverless car adoption. This model considers the ordinal nature of driverless car adoption data, and it is statistically superior to the fixed parameters ordered probit model (FOPM) as it accounts for possible unobserved factors (Revelt & Train, 1998; Bhat, 2001; Chen et al., 2019). The findings of this study highlight potential differences and similarities in driverless car adoption factors between Brisbane, Melbourne, and Sydney. It can, therefore, be used to guide customer groups in relation to car features, awareness, and knowledge-based programmes and policies to promote the acceptance of driverless cars among Australian consumers.

7.2 Research Design

Sayer (1992) summarised that questionnaires are one of the preferred data construction methods used in extensive research, and data interpretation is typically done through statistical analyses. Questionnaires are usually part of a wider, quantitatively driven social survey strategy, where representative samples of people can be questioned to produce numeric measures of behaviour, attitude, attributes, and so on (Cloke et al., 2004). Questionnaire surveys have been described as an "indispensable tool when primary data are required about people, their behaviour, attitudes and opinions and their awareness of specific issues" (Parfitt, 2013, p. 78).

7.2.1 Data Description

In addition, online questionnaire surveys have the capacity to reduce and remove potential bias, which is inherent in all data collection as respondents are asked identical questions in the same order (Cloke et al., 2004). The absence of an interviewer provides the opportunity to interpret a question without emphasis on any specific part. The online survey comes with some inherent disadvantages as well; some of these are the inability to connect with people from remote areas, high chances of survey fraud, sampling issues, response bias, survey fatigue, an increase in errors, and many unanswered questions (Jared, 2022).

The questionnaire used in this study was designed and distributed through the online survey platform, Key Survey (Table 7A.1). The world's largest first-party data platform company Dynata was hired to send and prompt the survey questionnaire link to the Australian target customer groups to flog the survey through Key Survey platform and collect data in a reasonable amount of time. This is how the intended number of responses was gathered between July and August 2021. The Key Survey platform offered number of advantages, including convenience (since it could be accessed through the host university); security (compliant with industry best practices); and accessibility, as it can be used on mobile and computer devices. Through its respondent panel, Dynata reached out to a diverse group of Australian customers, representing a range of professions, ages, income levels, academic backgrounds, family sizes, and geographic locations, among other parameters.

The structure of the whole survey questionnaire and a short description of each section is outlined in Table 7.1. The survey questionnaire sections applicable to this study i.e., driverless car adoption modelling, included 27 questions divided into Sections A, B, C, D, E, and M (Table 7.1). These created a multifaceted approach for addressing driverless car adoption. The rest of the questions and sections are not applicable to this chapter.

In total, each respondent had to answer 50–51 closed/Likert scale questions, which were broken down as follows:

- Q2–35: 28 unique multiple-choice questions.
- Q2–35: 6 Likert scale/matrix/rating questions.
- Q36–41: 1 multiple-choice question.
- Q42–68: 1–2 multiple-choice question(s).
- Q69–82: 14 unique multiple-choice questions.

Being predominately comprising closed multiple-choice and matrix questions, the questionnaire favoured easy responses and the collection of a large volume of data within a short time.

TABLE 7.1

Sections of the Survey Questionnaire

Question No.	Sections and Question Descriptions
A: Background and socio demographics	
Q 2–5	Four closed questions on respondents' region, age, gender, and education.
B: Family structure	
Q 6–8	Three closed questions on respondents' family composition.
C: Employment	
Q 9–12	Four closed questions on respondents' job type, organisation, occupation, and income.
D: Car ownership and travel behaviour	
(No driverless car scenario)	
Q 13–18	13 closed questions on respondents' car ownership, driver's licence status,
Q 19–21,	transportation mode used for school/work/recreation, travel distance,
Q 23–26	travel time, etc.
Q 22	One matrix question on factors relating to respondents' choice of transportation mode.
E: Driverless car impact and opportunities	
Q 27–29	Three closed questions on respondents' familiarity with driving aids/smart
Q 30–34	tech and driverless car experience.
	Five matrix questions on respondents' perception of driverless car impacts/ benefits, social norms, and anxieties around using driverless cars.
F: Driverless car options and fuel source	
Q 35–41	Seven closed questions on respondents' choice of owning a driverless car, sharing a driverless car, riding on a driverless bus, or using a combination of these travel options.
G: Willingness to pay	
Driverless car ownership only + electric/diesel or petrol/hybrid	
Q 42–44	Three closed questions on respondents' choice on owning an electric/fossil fuel/hybrid driverless car.
H: Willingness to pay	
Driverless car sharing only + electric/diesel: petrol/hybrid	
Q 45–47	Three closed questions on respondents' choice on sharing an electric/fossil fuel/hybrid driverless car.
I: Willingness to pay	
Driverless bus riding only + electric/diesel: petrol/hybrid	
Q 48–50	Three closed questions on respondents' choice on riding on an electric/fossil fuel/hybrid driverless bus.
J: Willingness to pay	
Driverless car ownership and driverless car sharing + electric/diesel: petrol/hybrid	
Q 51–56	Six closed questions on respondents' choice on both owning and sharing an electric/fossil fuel/hybrid driverless car.
K: Willingness to pay	
Driverless car ownership and bus usage + electric/diesel: petrol/hybrid	
Q 57–62	Six closed questions on respondents' choice on both owning an electric/ fossil fuel/hybrid driverless car and riding on an electric/fossil fuel/ hybrid driverless bus.

(continued)

TABLE 7.1 (Continued)

Sections of the Survey Questionnaire

Question No.	Sections and Question Descriptions
L: Willingness to pay	
Driverless car sharing and bus usage + electric/diesel: petrol/hybrid	
Q 63–68	Six closed questions on respondents' choice on both sharing an electric/fossil fuel/hybrid driverless car and riding on an electric/fossil fuel/hybrid driverless bus.
M: Driverless car adoption and impact	
Q 69–71	14 closed questions on respondents' preference in adopting a driverless car,
Q 72–82	e.g., change in expected travel behaviour in terms of mode of transport they use, travel distance they cover, trips they make, car ownership, relocating residence.

The whole survey targeted data collection on metropolitan area consumers' choice of transportation mode, and the benefits, impacts, and
associated contributing factors of driverless car adoption. Representation of
survey respondents was from three Australian cities: Brisbane, Melbourne,
and Sydney. There were 2,608 total responses gathered. Of these, 1,108 were
connected to the adoption of driverless cars and this study, indicating that
roughly 43% of respondents are interested in their use within the next ten
years. A final dataset of 967 responses was chosen for the *driverless car adoption
modelling* after incomplete responses were screened out. This included
responses from 334 Brisbane participants, 233 Melbourne participants, and
400 Sydney participants. The sample size ratio for Brisbane, Melbourne,
and Sydney is 24:35:41, which is very close to the 2021 population ratio in
Brisbane: Melbourne: Sydney, 20:41:40 (Macrotrends, 2021). Therefore, there
were no weighting procedures carried out to make the sample more representative of the cities. Each respondent averaged 12 minutes per survey to
complete.

The next steps were taken to establish a high-quality set of data:

- Each respondent's responses to several pertinent questions were cross-
checked for any inconsistencies.
- Checked to see if any answers were missing from any questions.
- Checked to see if the age requirements (18+ years) match.
- If a question had multiple answer choices, those answers were reclassi-
fied by combining two answers to create a new answer.
- To prevent collinearity, all correlated variables were eliminated.

The survey sought to accumulate the views of respondents from varying
age groups, academic backgrounds, income levels, family structures, employment types, organisational groups, and occupations. The sample includes a

higher percentage of graduates relative to the population. Hence, this aspect is considered during the analysis and driving conclusions. Respondents were asked to provide their views on factors that might influence driverless car adoption, and these factors comprised most of the demographic and mobility behaviour factors as consolidated in the study conducted by Golbabaei et al. (2020). These factors are independent variables to *driverless car adoption modelling*.

The purpose of the study was to better understand factors that are likely to affect the timeline of driverless car adoption. The dependent variable, *driverless car adoption timeline*, was grouped into three time horizons to ensure each horizon had a decent number of observations. The descending order of the *adoption timeline* group was considered to reduce bias and variability of the estimated parameters for the ordered probit model (Ye et al., 2011). The groups were defined as follows:

T_0: *Quick* adoption: Uptake between zero and two years following market availability.

T_1: *Long way to go* adoption: Uptake five years following market availability.

T_2: *Hibernation* adoption: Uptake five to ten years following market availability.

Survey responses were collected against *driverless car adoption timeline* through the following survey question:

• When you are likely to adopt your preferred mode of driverless cars if they become available?

The data breakdown of the dependent variable *driverless car adoption timeline* is presented in Table 7.2.

To test for possible collinearity, the authors conducted Pearson's correlation test on 128 *independent explanatory variables/regressors*. Out of 128 variables, 24 regressors were highly correlated, with a correlation parameter of >0.50, in both the Brisbane and Melbourne data; 26 variables were highly correlated in the Sydney data. As a result, a total of 104 variables were tested in Brisbane and Melbourne models and 102 in the Sydney model. The independent variables were categorised into six groups: background and socio demographics, family structure, employment, travel behaviour, car ownership, driverless car impact, and opportunities. A summary of the statistics of key independent variables and dependent variable is presented in Table 7.2.

7.2.2 Methodology

Three *driverless car adoption time horizons* were considered: 'Quick', 'Long way to go' and 'Hibernation'. Two approaches, the ordered model and the unordered

TABLE 7.2

Descriptive Statistics of Variables Used in Driverless Car Adoption Modelling

Dependent Variable:

Variable Name	Description	Combined		Brisbane		Melbourne		Sydney	
		Number of Responses	%	Number of Responses	%	Number of Responses	%	Number of Responses	%
Driverless car adoption	Indicator for adoption configuration: 0 = Quick (T_0)	510	52.74	119	51.07	164	49.10	227	56.75
	1 = Long Way to Go (T_1)	350	36.19	92	39.48	130	38.92	128	32.00
	2 = Hibernation (T_2)	107	11.07	22	9.44	40	11.98	45	11.25
	Total	967	100	233	100	334	100	400	100

Independent Variable:

Factor	Variable Name	Combined			Brisbane			Melbourne			Sydney		
		%	Mean	SD.	%	Mean	SD.	%	Mean	SD.	%	Mean	SD.
Background and Socio demographics													
Gender	Female	50.57	0.506	0.500	55.36	0.554	0.498	48.20	0.482	0.500	49.75	0.498	0.501
	Male	49.43	0.494	0.500	44.64	0.446	0.498	51.80	0.518	0.500	50.25	0.503	0.501
Age	Age group: 18–25	8.27	0.083	0.276	7.73	0.077	0.268	9.58	0.096	0.295	7.5	0.075	0.264
	Age group: 25–35	19.44	0.194	0.396	18.03	0.180	0.385	20.36	0.204	0.403	19.5	0.195	0.397
	Age group: 35–55	36.30	0.363	0.481	30.90	0.309	0.463	35.93	0.359	0.481	39.75	0.398	0.490
	Age group: 55–65	13.96	0.140	0.347	15.45	0.155	0.362	13.17	0.132	0.339	13.75	0.138	0.345
	Age group: >65	22.03	0.220	0.415	27.90	0.279	0.449	20.96	0.210	0.408	19.5	0.195	0.397
Education	Academic qualification: primary	1.14	0.011	0.106	1.72	0.017	0.130	0.90	0.009	0.094	1	0.010	0.100

Academic qualification: secondary	17.89	0.179	0.383	30.90	0.309	0.463	17.07	0.171	0.377	11	0.110	0.313
Academic qualification: diploma	25.65	0.256	0.437	27.47	0.275	0.447	27.84	0.278	0.449	22.75	0.228	0.420
Academic qualification: graduate	55.33	0.553	0.497	39.91	0.399	0.491	54.19	0.542	0.499	65.25	0.653	0.477
Family Structure												
Family member — Household members: 1	18.72	0.187	0.390	19.74	0.197	0.399	19.76	0.198	0.399	17.25	0.173	0.378
Household members: 2	30.20	0.302	0.459	41.20	0.412	0.493	27.54	0.275	0.447	26	0.260	0.439
Household members: 3	21.92	0.219	0.414	18.03	0.180	0.385	21.86	0.219	0.414	24.25	0.243	0.429
Household members: >3	29.16	0.292	0.455	21.03	0.210	0.408	30.84	0.308	0.463	32.5	0.325	0.469
Employment												
Employed in HH — Household employment: 0	20.89	0.209	0.407	30.04	0.300	0.459	17.66	0.177	0.382	18.25	0.183	0.387
Household employment: 1	34.95	0.350	0.477	31.76	0.318	0.467	36.83	0.368	0.483	35.25	0.353	0.478
Household employment: 2	34.75	0.347	0.476	30.90	0.309	0.463	35.33	0.353	0.479	36.5	0.365	0.482
Household employment: 3	5.69	0.057	0.232	3.00	0.030	0.171	5.69	0.057	0.232	7.25	0.073	0.260
Household employment: >3	3.72	0.037	0.189	4.29	0.043	0.203	4.49	0.045	0.207	2.75	0.028	0.164
Occupation: transport planner/urban designer	13.44	0.134	0.341	12.45	0.124	0.331	13.47	0.135	0.342	14	0.140	0.347
Occupation: builder/developer/architect	3.52	0.035	0.184	2.15	0.021	0.145	4.49	0.045	0.207	3.5	0.035	0.184
Occupation: community group member	1.76	0.018	0.131	2.15	0.021	0.145	0.90	0.009	0.094	2.25	0.023	0.148

(continued)

TABLE 7.2 (Continued)

Descriptive Statistics of Variables Used in Driverless Car Adoption Modelling

Independent Variable:

Factor	Variable Name	Combined			Brisbane			Melbourne			Sydney		
		%	Mean	SD.	%	Mean	SD.	%	Mean	SD.	%	Mean	SD.
	Occupation: insurer/ legal adviser in automotive industry	1.03	0.010	0.101	0.43	0.004	0.066	1.50	0.015	0.122	1	0.010	0.100
	Occupation: research/ development in vehicle mfg/ride-share company	4.55	0.046	0.209	3.00	0.030	0.171	6.59	0.066	0.248	3.75	0.038	0.190
	Occupation: ride-share driver	2.69	0.027	0.162	1.72	0.017	0.130	4.49	0.045	0.207	1.75	0.018	0.131
	Occupation: not a professional	34.23	0.342	0.475	43.35	0.433	0.497	29.94	0.299	0.459	32.5	0.325	0.469
	Occupation: others	38.78	0.388	0.488	34.76	0.348	0.477	38.62	0.386	0.488	41.25	0.413	0.493
Organisation type	Organisation: not working	34.23	0.342	0.475	43.35	0.433	0.497	29.94	0.299	0.459	32.5	0.325	0.469
	Organisation: academician	1.55	0.016	0.124	1.72	0.017	0.130	1.80	0.018	0.133	1.25	0.013	0.111
	Organisation: federal government	5.27	0.053	0.224	4.29	0.043	0.203	7.78	0.078	0.268	3.75	0.038	0.190
	Organisation: state government	8.38	0.084	0.277	6.44	0.064	0.246	10.78	0.108	0.311	7.5	0.075	0.264
	Organisation: local government	6.00	0.060	0.238	3.86	0.039	0.193	7.49	0.075	0.264	6	0.060	0.238
	Organisation: others	44.57	0.446	0.497	40.34	0.403	0.492	42.22	0.422	0.495	49	0.490	0.501

Travel behaviour and car ownership

Mode of transport to work												
Mode of transport to work: active transport	1.55	0.016	0.124	1.72	0.017	0.130	0.60	0.006	0.077	2.25	0.023	0.148
Mode of transport to work: private car	41.37	0.414	0.493	36.05	0.361	0.481	50.90	0.509	0.501	36.5	0.365	0.482
Mode of transport to work: public transport	10.75	0.108	0.310	9.01	0.090	0.287	8.68	0.087	0.282	13.5	0.135	0.342
Mode of transport to work: shared car	1.34	0.013	0.115	1.72	0.017	0.130	1.50	0.015	0.122	1	0.010	0.100
Mode of transport to work: not working/ working from home	35.47	0.355	0.479	45.06	0.451	0.499	29.94	0.299	0.459	34.5	0.345	0.476
Mode of transport to work: private car+ public transport+ shared car	9.51	0.095	0.294	6.44	0.064	0.246	8.38	0.084	0.278	12.25	0.123	0.328
Mode of transport to school												
Mode of transport to school: active transport	2.90	0.029	0.168	3.43	0.034	0.182	2.10	0.021	0.143	3.25	0.033	0.178
Mode of transport to school: private car	41.37	0.414	0.493	38.63	0.386	0.488	45.21	0.452	0.498	39.75	0.398	0.490
Mode of transport to school: public transport	9.72	0.097	0.296	8.58	0.086	0.281	7.49	0.075	0.264	12.25	0.123	0.328
Mode of transport to school: shared car	2.38	0.024	0.152	0.86	0.009	0.092	2.69	0.027	0.162	3	0.030	0.171
Mode of transport to school: not a student/ not supporting student	34.75	0.347	0.476	39.91	0.399	0.491	32.93	0.329	0.471	33.25	0.333	0.472

(continued)

TABLE 7.2 (Continued)

Descriptive Statistics of Variables Used in Driverless Car Adoption Modelling

Independent Variable:

Factor	Variable Name	Combined			Brisbane			Melbourne			Sydney		
		%	Mean	SD.	%	Mean	SD.	%	Mean	SD.	%	Mean	SD.
	Mode of transport to school: private car+ public transport+ shared car	8.89	0.089	0.285	8.58	0.086	0.281	9.58	0.096	0.295	8.5	0.085	0.279
Mode of transport to social trip	Mode of transport to recreation/social trip: active transport	1.96	0.020	0.139	2.15	0.021	0.145	1.80	0.018	0.133	2	0.020	0.140
	Mode of transport to recreation/social trip: private car	69.08	0.691	0.462	69.10	0.691	0.463	70.36	0.704	0.457	68	0.680	0.467
	Mode of transport to recreation/social trip: public transport	10.24	0.102	0.303	8.15	0.082	0.274	9.28	0.093	0.291	12.25	0.123	0.328
	Mode of transport to recreation/social trip: shared car	4.03	0.040	0.197	5.15	0.052	0.221	4.49	0.045	0.207	3	0.030	0.171
	Mode of transport to recreation/social trip: not applicable	0.52	0.005	0.072	0.43	0.004	0.066	0.60	0.006	0.077	0.5	0.005	0.071
	Mode of transport to recreation/social trip: private car+public transport+shared car	14.17	0.142	0.349	15.02	0.150	0.358	13.47	0.135	0.342	14.25	0.143	0.350

Travel distance to works	Travel distance to work: <10 km	18.10	0.181	0.385	19.31	0.193	0.396	18.26	0.183	0.387	17.25	0.173	0.378
	Travel distance to work: 10–20 km	23.78	0.238	0.426	20.17	0.202	0.402	27.84	0.278	0.449	22.5	0.225	0.418
	Travel distance to work: 20–30 km	14.99	0.150	0.357	11.16	0.112	0.316	18.56	0.186	0.389	14.25	0.143	0.350
	Travel distance to work: >30 km	7.86	0.079	0.269	9.01	0.090	0.287	6.29	0.063	0.243	8.5	0.085	0.279
	Travel distance to work: working from home	8.38	0.084	0.277	5.15	0.052	0.221	5.99	0.060	0.238	12.25	0.123	0.328
	Travel distance to work: not applicable	26.89	0.269	0.444	35.19	0.352	0.479	23.05	0.231	0.422	25.25	0.253	0.435
Travel distance to school	Travel distance to school: <3 km	7.76	0.078	0.268	7.73	0.077	0.268	7.78	0.078	0.268	7.75	0.078	0.268
	Travel distance to school: 3–5 km	10.96	0.110	0.313	9.01	0.090	0.287	14.37	0.144	0.351	9.25	0.093	0.290
	Travel distance to school: 5–10 km	12.93	0.129	0.336	14.16	0.142	0.349	13.17	0.132	0.339	12	0.120	0.325
	Travel distance to school: >10 km	11.17	0.112	0.315	9.87	0.099	0.299	11.38	0.114	0.318	11.75	0.118	0.322
	Travel distance to school: nota a student or not supporting student	52.95	0.529	0.499	57.94	0.579	0.495	49.40	0.494	0.501	53	0.530	0.500
	Travel distance to school: online schooling	4.24	0.042	0.202	1.29	0.013	0.113	3.89	0.039	0.194	6.25	0.063	0.242

(continued)

TABLE 7.2 (Continued)

Descriptive Statistics of Variables Used in Driverless Car Adoption Modelling

Independent Variable:

Factor	Variable Name	Combined			Brisbane			Melbourne			Sydney		
		%	Mean	SD.	%	Mean	SD.	%	Mean	SD.	%	Mean	SD.
Travel distance to social trip	Travel distance to social trip: <10 km	27.92	0.279	0.449	26.61	0.266	0.443	20.66	0.207	0.405	34.75	0.348	0.477
	Travel distance to social trip: 10–25 km	34.44	0.344	0.475	35.19	0.352	0.479	38.32	0.383	0.487	30.75	0.308	0.462
	Travel distance to social trip: 25–50 km	21.30	0.213	0.410	23.18	0.232	0.423	23.95	0.240	0.427	18	0.180	0.385
	Travel distance to social trip: 50–100 km	7.14	0.071	0.258	6.87	0.069	0.253	8.08	0.081	0.273	6.5	0.065	0.247
	Travel distance to social trip: >100 km	2.69	0.027	0.162	1.72	0.017	0.130	3.89	0.039	0.194	2.25	0.023	0.148
	Travel distance to social trip: <25 km	4.34	0.043	0.204	3.00	0.030	0.171	3.89	0.039	0.194	5.5	0.055	0.228
	Travel distance to social trip: 10–50 km	2.17	0.022	0.146	3.43	0.034	0.182	1.20	0.012	0.109	2.25	0.023	0.148
Car ownership	Car ownership: 0	7.34	0.073	0.261	6.44	0.064	0.246	5.39	0.054	0.226	9.5	0.095	0.294
	Car ownership: 1	53.77	0.538	0.499	56.65	0.567	0.497	52.69	0.527	0.500	53	0.530	0.500
	Car ownership: 2	31.44	0.314	0.465	30.47	0.305	0.461	32.34	0.323	0.468	31.25	0.313	0.464
	Car ownership: >2	7.45	0.074	0.263	6.44	0.064	0.246	9.58	0.096	0.295	6.25	0.063	0.242
Car value	Car value: <$20k	43.12	0.431	0.496	50.64	0.506	0.501	40.72	0.407	0.492	40.75	0.408	0.492
	Car value: $20k–$30k	22.54	0.225	0.418	24.03	0.240	0.428	24.55	0.246	0.431	20	0.200	0.401
	Car value: $30k–$40k	16.13	0.161	0.368	13.30	0.133	0.340	17.66	0.177	0.382	16.5	0.165	0.372
	Car value: $40k–$50k	9.20	0.092	0.289	5.58	0.056	0.230	8.68	0.087	0.282	11.75	0.118	0.322
	Car value: >$50k	9.00	0.090	0.286	6.44	0.064	0.246	8.38	0.084	0.278	11	0.110	0.313

Driving licence duration	Driving licence duration: <1 year	4.65	0.047	0.211	4.29	0.043	0.203	5.69	0.057	0.232	4	0.040	0.196
	Driving licence duration: 1–3 years	14.89	0.149	0.356	14.59	0.146	0.354	15.27	0.153	0.360	14.75	0.148	0.355
	Driving licence duration: 3–5 years	12.20	0.122	0.327	9.44	0.094	0.293	12.28	0.123	0.329	13.75	0.138	0.345
	Driving licence duration: >5 years	61.43	0.614	0.487	63.95	0.639	0.481	62.28	0.623	0.485	59.25	0.593	0.492
	Driving licence duration: no driving licence	6.83	0.068	0.252	7.73	0.077	0.268	4.49	0.045	0.207	8.25	0.083	0.275
Household driving licence numbers	Household driving licence nos.: 0	20.68	0.207	0.405	22.32	0.223	0.417	20.66	0.207	0.405	19.75	0.198	0.399
	Household driving licence nos.: 1	41.68	0.417	0.493	45.06	0.451	0.499	41.02	0.410	0.493	40.25	0.403	0.491
	Household driving licence nos.: 2	27.09	0.271	0.445	23.61	0.236	0.426	27.84	0.278	0.449	28.5	0.285	0.452
	Household driving licence nos.: >2	10.55	0.105	0.307	9.01	0.090	0.287	10.48	0.105	0.307	11.5	0.115	0.319
Driverless car impact and opportunities													
Familiarity with smart tech	Smart tech familiarity type: 1	12.41	0.124	0.330	12.02	0.120	0.326	14.37	0.144	0.351	11	0.110	0.313
	Smart tech familiarity type: 2	9.20	0.092	0.289	6.44	0.064	0.246	9.28	0.093	0.291	10.75	0.108	0.310
	Smart tech familiarity type: 3	9.20	0.092	0.289	7.73	0.077	0.268	10.18	0.102	0.303	9.25	0.093	0.290

(continued)

TABLE 7.2 (Continued)

Descriptive Statistics of Variables Used in Driverless Car Adoption Modelling

Independent Variable:

Factor	Variable Name	Combined			Brisbane			Melbourne			Sydney		
		%	Mean	SD.	%	Mean	SD.	%	Mean	SD.	%	Mean	SD.
	Smart tech familiarity type: >3	67.63	0.676	0.468	72.96	0.730	0.445	64.97	0.650	0.478	66.75	0.668	0.472
	Smart tech familiarity type: no	1.55	0.016	0.124	0.86	0.009	0.092	1.20	0.012	0.109	2.25	0.023	0.148
Use of driving aid	Use of driving aid: 0	25.96	0.260	0.439	29.18	0.292	0.456	23.05	0.231	0.422	26.5	0.265	0.442
	Use of driving aid: 1–2	32.06	0.321	0.467	31.33	0.313	0.465	34.13	0.341	0.475	30.75	0.308	0.462
	Use of driving aid: 3–4	16.65	0.166	0.373	17.60	0.176	0.382	16.17	0.162	0.369	16.5	0.165	0.372
	Use of driving aid: >4	19.13	0.191	0.394	16.31	0.163	0.370	21.56	0.216	0.412	18.75	0.188	0.391
	Use of driving aid: not a driver	6.20	0.062	0.241	5.58	0.056	0.230	5.09	0.051	0.220	7.5	0.075	0.264
Driverless car riding experience	Driverless car riding experience: no	85.42	0.854	0.353	89.27	0.893	0.310	82.93	0.829	0.377	85.25	0.853	0.355
	Driverless car riding experience: yes	14.58	0.146	0.353	10.73	0.107	0.310	17.07	0.171	0.377	14.75	0.148	0.355
Driverless car trip purpose	Driverless car trip purpose when available: work	18.61	0.186	0.389	14.59	0.146	0.354	23.05	0.231	0.422	17.25	0.173	0.378
	Driverless car trip purpose when available: school	5.79	0.058	0.234	4.29	0.043	0.203	6.59	0.066	0.248	6	0.060	0.238
	Driverless car trip purpose when available: social	34.13	0.341	0.474	42.06	0.421	0.495	29.04	0.290	0.455	33.75	0.338	0.473

Driverless car trip purpose when available: work and school	3.10	0.031	0.173	5.15	0.052	0.221	2.40	0.024	0.153	2.5	0.025	0.156
Driverless car trip purpose when available: work and social	12.51	0.125	0.331	9.01	0.090	0.287	13.77	0.138	0.345	13.5	0.135	0.342
Driverless car trip purpose when available: school and social	1.34	0.013	0.115	0.86	0.009	0.092	1.20	0.012	0.109	1.75	0.018	0.131
Driverless car trip purpose when available: all	24.51	0.245	0.430	24.03	0.240	0.428	23.95	0.240	0.427	25.25	0.253	0.435

Note: All variables were in dummy coding (with values of 0 and 1). Therefore, the mean value can be interpreted as the proportion. For example, the mean value for the age group: 18–25 variable in the Brisbane dataset shows that 7.73% of the sample was age group: 18–25. These variables were subsequently examined in the driverless car adoption model specifications.

model, have been widely applied to examine the impact of contributing factors on similar data and modelling (Duncan et al., 1998; Khorashadi et al., 2005; Chang et al., 2013; Pahukula et al., 2015). Given that both ordered and unordered models have strengths as well as limitations, this study used the ordered probit model as it accounts for the indexed nature (McKelvey & Zavoina, 1975) of driverless car adoption levels/time horizons.

The ordered probit model is derived by introducing a latent variable, y^*, as a basis for modelling driverless car adoption of each observation, which can be defined as follows (Washington et al., 2020):

$$y^* = \beta'X + \varepsilon \tag{7.1}$$

where X is a vector of independent variables considered, β' is a vector of estimable parameters, and ε is a random error term assumed to be normally distributed across observations with a mean equal to 0 and a variance equal to 1.

Given Equation (7.1), the dependent variable, y, is defined by the unobserved variable, y^*, as follows:

$$y = \begin{cases} 2 \text{ if } y^* \geq \mu_1 \text{ (hibernation adoption)} \\ 1 \text{ if } \mu_0 < y^* \leq \mu_1 \text{ (long way to go adoption)} \\ 0 \text{ if } y^* \leq \mu_0 \text{ (quick adoption)} \end{cases} \tag{7.2}$$

where $\mu_0 = 0$, and μ_1 is threshold that is jointly estimated with β' parameters. Figure 7.1 illustrates the correspondence between latent, continuous underlying adoption variable, y^*, and the observed adoption time horizon, y (Kockelman & Kweon, 2002).

The probability of each adoption category for given variables can then be described on the distribution of random error, ε:

$$P(y = 2) = 1 - \Phi(\mu_1 - \beta'x)$$
$$P(y = 1) = \Phi(\mu_1 - \beta'x) - \Phi(\mu_0 - \beta'x) \tag{7.3}$$
$$P(y = 0) = \Phi(\mu_0 - \beta'x)$$

However, the standard probit model could lead to potential bias by treating the parameter, β', as a constant value across observations. This restricts each

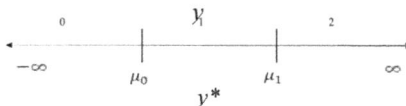

FIGURE 7.1
Relationship between latent and coded adoption variables.

variable to the same impact on every individual observation (Savolainen et al., 2011; Boes & Winkelmann, 2006; Christoforou et al., 2010). To account for this, a random parameter ordered probit model (ROPM) was developed to capture unobserved heterogeneity. This was achieved by adding a randomly distributed error term, φ (e.g., a normally distributed term with mean = 0 and variance = σ^2) (Greene, 2003):

$$\beta^* = \beta + \varphi \tag{7.4}$$

Since the interpretation of the estimated coefficient, β, on driverless car adoption is not straightforward, marginal effects were computed to measure the effect of one unit change in an independent variable on the probability of driverless car adoption. This is usually used to measure the influence of a variable within an adoption level while keeping all other variables constant. The marginal effects are calculated as follows (Washington et al., 2020):

$$
\begin{aligned}
\frac{\partial P(y=2)}{\partial X} &= \Phi(\mu_1 - \beta'X)\beta \\
\frac{\partial P(y=1)}{\partial X} &= \left[\Phi(\mu_0 - \beta'X) - (\mu_1 - \beta'X)\right]\beta \\
\frac{\partial P(y=0)}{\partial X} &= [\Phi(\mu_0 - \beta'X)\beta
\end{aligned}
\tag{7.5}
$$

7.2.3 Model Evaluation

A likelihood ratio (LR) test (Washington et al., 2020) was performed to determine whether separate models for Brisbane, Melbourne, and Sydney were warranted. The LR statistics are defined as follows (Ulfarsson & Mannering, 2004):

$$\chi^2 = -2\left[LL(\beta_{COMBINED}) - LL(\beta_{BRI}) - LL(\beta_{MEL}) - LL(\beta_{SYD})\right] \tag{7.6}$$

where $LL(\beta_{COMBINED})$, $LL(\beta_{BRI})LL(\beta_{MEL})$ and $LL(\beta_{SYD})$ are the log likelihoods at the convergence of the Combined data model, Brisbane, Melbourne, and Sydney model, respectively. The Log-LR statistic is a χ^2 distribution with the degree of freedom, d, equal to the sum of the number of parameters considered in each separate dataset minus the one in the joint dataset, which, in this case, is $d = K_{BRI} + K_{MEL} + K_{SYD} - K_{COMBINED}$. With a χ^2 value = 168.838 and 60 degrees of freedom, a confidence level of over 99.99% was obtained. This indicated

TABLE 7.3

Likelihood Ratio Test between Fixed and Random Parameters Models

Dataset	Combined (All)	Brisbane	Melbourne	Sydney
χ2 value	8.694	15.229	17.2951	20.184
Degrees of freedom	3	7	8	4
p value	0.0336	0.0331	0.0271	0.0004

that modelling Brisbane, Melbourne, and Sydney data separately was more likely to present a better fit than the joint dataset (Combined) model.

Another LR test was conducted to test the null hypothesis that the fixed parameter model is statistically equivalent to the random parameters model, i.e., to compare the differences between the random parameters model and their fixed parameters model using the test statistics below (Washington et al., 2020):

$$LR = -2\left[LL\left(\beta_{fixed}\right) - LL\left(\beta_{random}\right) \right]$$
(7.7)

where $LL\left(\beta_{fixed}\right)$ and $LL\left(\beta_{random}\right)$ are the log likelihoods at the convergence of FOPM and ROPM estimated using the same dataset (Combined, Sydney, Melbourne, and Brisbane), respectively. This LR statistic is a χ^2 distribution with the degree of freedom equal to the difference in the number of parameters between the two models. A list of χ^2 values and the degrees of freedom for each dataset are presented in Table 7.3.

The test results were significant for all four (Combined, Brisbane, Melbourne, and Sydney) driverless car adoption datasets. The p-values were 0.033 for Combined, 0.033 for Brisbane, 0.027 for Melbourne models, and 0.0004 for Sydney, showing with more than 95.00% confidence for Combined, Brisbane, and Melbourne models, and more than 99.99% confidence for Sydney model that the random parameters ordered probit models outperformed the corresponding fixed models. This means that the null hypothesis, that the random parameter model not being better than the fixed parameter model, is rejected for the Combined, Brisbane, Melbourne, and Sydney models.

7.3 Analysis and Results

The ROPM was estimated through simulated maximum likelihood 200 Halton draws, which has been demonstrated to be an efficient method for producing accurate results for discrete choice models with low dimensionality of integration (Bhat, 2003). In addition, Halton draws provide better simulation performance than random draws due to their dramatic speed gains with no degradation. Normal distribution was chosen for random parameters over other options, including lognormal, triangular, and uniform distribution, since normal distribution almost always outperforms the others (Islam et al., 2013; Naik et al., 2016; Fountas & Anastasopoulos, 2017).

Fixed and random parameters model estimations for Brisbane, Melbourne, and Sydney driverless car adoptions are presented in Tables 7.4–7.6, respectively. The marginal effects of the random parameters model for these three datasets are also provided. Backward selection was performed to select the best subsets of the independent variables with a criterion of p>0.1. Hence, all estimated parameters included in the final model were statistically significant at a confidence level of 90%—i.e., at a significance level of 10%. The results were plausible, as discussed below.

It is interesting that any two models only share a small number of significant variables. There are only four common significant variables between Brisbane and Melbourne models; there are two between Melbourne and Sydney and four between Sydney and Brisbane. Notably, the explanatory variables 'age group: 35–55 years' and "driving licence duration: less than 1 and 1–3 years" yielded similar results on the Brisbane and Melbourne driverless car adoption models, while the variable 'age group: 55–65 years' yielded results that were essentially the opposite of each other.

In the Melbourne and Sydney models, one explanatory variable—'use of driving aids: no'—yielded similar outcomes, whereas another—"travel distance to school: >10 km"—produced results that were the opposite. In the Sydney and Brisbane models, one explanatory variable—"driving distance to school: less than 3 km"— yielded similar results, but three others— 'income range: $40–60k', "travel distance to a social trip: 50–100 km," and "number of smart tech use: 3 types"—produced outcomes that were the complete opposite. The random parameter model shows a better fit than its fixed parameters counterpart. Here, the random parameters (all normally distributed) suggest that the effects varied across the observations.

The increase in the goodness of fit in the case of a ROPM over the respective FOPM is relatively small. Even a relatively small increase in the goodness of fit of a ROPM over a FOPM may indicate the presence of unobserved heterogeneity and associated insights. The ROPM captured random parameters associated with the following explanatory variables: (a) six for the Brisbane model including male, age group (35–55 years), household members (2) employment status (homemaker), and driving licence duration (1–3 and 3–5 years); (b) seven for the Melbourne model including age group (35–55) and (55–65), level of education (secondary), travel to school (more than 10 km), social trip distance (more than 100 km and less than 25 km), and car ownership (2); (c) four for the Sydney model including income range

TABLE 7.4

Fixed and Random Parameter Ordered Probit Model Results: Brisbane Driverless Car Adoption

Explanatory Variables	Fixed Parameters Ordered Probit Model (FOPM)				Random Parameter Ordered Probit Model (ROPM)				Marginal Effects (ROPM)		
	Coefficient (β)	Standard Error	z-stat	Prob (p) |z|>Z*	Coefficient (β)	Standard Error	z-stat	Prob (p) |z|>Z*	Quick Adoption (y=0)	Long Way to Go Adoption (y=1)	Hibernation Adoption (y=2)
Constant	0.406**	0.188	2.170	0.030	0.637***	0.240	2.650	0.008			
• *Standard deviation of parameter density function*					*1.151***	*0.236*	*4.880*	*0.000*			
Background and socio demographics											
Male	0.314*	0.174	1.810	0.071	0.810***	0.236	3.430	0.001	-0.178	0.135	0.043
• *Standard deviation of parameter density function*					*1.957***	*0.284*	*6.890*	*0.000*			
Age group: 35–55 years	-0.772***	0.212	-3.640	0.000	-1.231***	0.272	-4.530	0.000	0.261	-0.209	-0.053
• *Standard deviation of parameter density function*					*1.015***	*0.282*	*3.600*	*0.000*			
Age group: 55–65 years	-0.684***	0.251	-2.730	0.006	-1.515***	0.350	-4.330	0.000	0.302	-0.250	-0.052
Level of education: primary	1.578**	0.640	2.470	0.014	3.234***	1.091	2.960	0.003	-0.475	-0.044	0.519
Household members: 2	-0.312*	0.173	-1.810	0.071	-0.754***	0.239	-3.160	0.002	0.164	-0.128	-0.036
• *Standard deviation of parameter density function*					*1.638***	*0.284*	*5.760*	*0.000*			

Employment

Employment status: home maker	1.236***	0.432	2.860	0.004	1.530**	0.725	2.110	0.035	−0.304	0.151	0.154
• *Standard deviation of parameter density function*					*4.428**	*1.731*	*2.560*	*0.011*			
Work organisation type: academia	1.269**	0.592	2.140	0.032	1.631*	0.890	1.830	0.067	−0.317	0.142	0.175
Work organisation type: state government	0.774**	0.329	2.350	0.019	1.029**	0.431	2.390	0.017	−0.215	0.133	0.083
Income range: $40k–$60k	0.560**	0.266	2.110	0.035	1.218***	0.330	3.690	0.000	−0.255	0.157	0.098
Income range: $60k–$80k	0.446*	0.247	1.810	0.071	1.025***	0.324	3.170	0.002	−0.212	0.133	0.079
Income range: $80k–$100k	0.774**	0.366	2.120	0.034	1.267**	0.497	2.550	0.011	−0.254	0.139	0.115

Travel behaviour and car ownership

Travel distance to school: <3 km	−1.080***	0.413	−2.610	0.009	−2.269***	0.681	−3.330	0.001	0.402	−0.360	−0.042
Travel distance for social purpose: 50–100 km	0.656**	0.335	1.960	0.050	1.782***	0.382	4.670	0.000	−0.344	0.155	0.189
Driving licence duration: <1 year	−1.297***	0.485	−2.670	0.008	−2.875***	0.830	−3.460	0.001	0.453	−0.409	−0.044
Driving licence duration: 1–3 years	−1.229***	0.279	−4.400	0.000	−2.224***	0.429	−5.190	0.000	0.431	−0.379	−0.052
• *Standard deviation of parameter density function*					*1.700***	*0.439*	*3.870*	*0.000*			

(continued)

TABLE 7.4 (Continued)

Fixed and Random Parameter Ordered Probit Model Results: Brisbane Driverless Car Adoption

Explanatory Variables	Fixed Parameters Ordered Probit Model (FOPM)				Random Parameter Ordered Probit Model (ROPM)				Marginal Effects (ROPM)		
	Coefficient (β)	Standard Error	z-stat	Prob (p) |z|>Z*	Coefficient (β)	Standard Error	z-stat	Prob (p) |z|>Z*	Quick Adoption (y=0)	Long Way to Go Adoption (y=1)	Hibernation Adoption (y=2)
Driving licence duration: 3–5 years	−1.011***	0.331	−3.050	0.002	−2.551***	0.681	−3.750	0.000	0.436	−0.387	−0.048
• Standard deviation of parameter density function					1.554**	0.707	2.200	0.028			
Driverless car impact and opportunities											
Number of smart tech use: 3	0.694**	0.331	2.100	0.036	0.968*	0.495	1.950	0.051	−0.204	0.127	0.077
Driverless car trip purpose: work and school	−1.157**	0.500	−2.310	0.021	−2.293***	0.652	−3.520	0.000	0.407	−0.365	−0.042
Driverless car feature preference: 3	−0.518**	0.259	−2.000	0.046	−1.048***	0.370	−2.830	0.005	0.230	−0.196	−0.034
Threshold											
Mu (01)	1.634***	0.150	10.870	0.000	2.813***	0.297	9.460	0.000			

Goodness of fit statistics6

Number of observations	233	233
Restricted log likelihood or log likelihood at 0 (LL_0)	-217.369	-217.369
Log-likelihood function or log likelihood at convergence (LL_B)	-177.962	-170.348
Akaike information criteria (AIC)	397.9	396.7
AIC/N	1.708	1.703
Chi-squared (dof = 19)	78.813	
Significance level	<0.001	
Halton draws		200
K	21	28

***, **, * ==> Significance at 1%, 5%, 10% level

TABLE 7.5

Fixed and Random Parameter Ordered Probit Model Results: Melbourne Driverless Car Adoption

Explanatory Variable	Fixed Parameters Ordered Probit Model (FOPM)				Random Parameter Ordered Probit Model (ROPM)				Marginal Effects (ROPM)		
	Coefficient (β)	Standard Error	z-stat	Prob (p) |z|>Z*	Coefficient (β)	Standard Error	z-stat	Prob (p) |z|>Z*	Quick Adoption (y=0)	Long Way to Go Adoption (y=1)	Hibernation Adoption (y=2)
Constant	0.0177	0.145	0.120	0.903	-0.106	0.192	-0.56	0.579			
• Standard deviation of parameter density function					1.788***	0.214	8.37	0.000			
Background and socio demographics											
Age group: 35–55 years	-0.433***	0.158	-2.740	0.006	-0.477**	0.200	-2.39	0.017	0.120	-0.089	-0.031
• Standard deviation of parameter density function					0.710***	0.239	2.97	0.003			
Age group: 55–65 years	0.434**	0.203	2.140	0.032	0.830***	0.252	3.30	0.001	-0.206	0.128	0.078
• Standard deviation of parameter density function					2.478***	0.316	7.85	0.000			
Level of education: secondary	0.588***	0.192	3.060	0.002	1.312***	0.244	5.38	0.000	-0.310	0.172	0.138
• Standard deviation of parameter density function					1.779***	0.266	6.69	0.000			
Level of education: diploma	0.548***	0.156	3.510	0.001	0.917***	0.201	4.56	0.000	-0.226	0.146	0.080

Employment

Employed in household: 3 members	−0.920***	0.330	−2.790	0.005	−1.507***	0.481	−3.13	0.002	0.337	−0.282	−0.055
Occupation: transport planner/urban designer	−0.554**	0.236	−2.360	0.019	−0.856***	0.315	−2.71	0.007	0.215	−0.174	−0.041
Occupation: builder	0.637**	0.324	1.970	0.049	0.866**	0.413	2.10	0.036	−0.204	0.112	0.092

Travel behaviour and car ownership

Travel distance to work: 20–30 km	−0.535***	0.200	−2.670	0.008	−0.727***	0.280	−2.60	0.009	0.182	−0.142	−0.040
Travel distance to school: >10 km	0.477**	0.228	2.090	0.037	0.977***	0.283	3.46	0.001	−0.224	0.119	0.105
• *Standard deviation of parameter density function*					2.729***	0.361	7.56	0.000			
Travel distance for social purpose: >100 km	1.027***	0.351	2.920	0.004	1.572***	0.484	3.25	0.001	−0.335	0.122	0.212
• *Standard deviation of parameter density function*					2.721***	0.579	4.70	0.000			
Travel distance for social purpose: <25 km	0.820**	0.339	2.420	0.015	1.487***	0.447	3.33	0.001	−0.324	0.128	0.196
• *Standard deviation of parameter density function*					2.196***	0.547	4.01	0.000			
Car ownership: 2 nos.	0.524***	0.148	3.540	0.000	0.739***	0.196	3.78	0.000	−0.180	0.118	0.062

(continued)

TABLE 7.5 (Continued)

Fixed and Random Parameter Ordered Probit Model Results: Melbourne Driverless Car Adoption

Explanatory Variable	Fixed Parameters Ordered Probit Model (FOPM)				Random Parameter Ordered Probit Model (ROPM)				Marginal Effects (ROPM)		
	Coefficient (β)	Standard Error	z-stat	Prob (p) |z|>Z*	Coefficient (β)	Standard Error	z-stat	Prob (p) |z|>Z*	Quick Adoption (y=0)	Long Way to Go Adoption (y=1)	Hibernation Adoption (y=2)
• *Standard deviation of parameter density function*					1.135***	0.233	4.87	0.000			
Driving licence duration: <1 year	-1.175***	0.347	-3.390	0.001	-1.617***	0.406	-3.98	0.000	0.358	-0.303	-0.055
Driving licence duration: 1–3 years	-0.769***	0.219	-3.510	0.000	-1.419***	0.298	-4.76	0.000	0.345	-0.290	-0.054
Driverless car impact and opportunities											
Use of driving aids: >4 types	-0.305*	0.170	-1.800	0.071	-0.737***	0.219	-3.370	0.001	0.183	-0.140	-0.043
Threshold											
Mu (01)	1.499***	0.115	13.080	0.000	2.569***	0.220	11.690	0.000			
Goodness of fit statistics											
Number of observations	334				334						
Restricted log likelihood or log likelihood at 0 (LL_0)	-324.208				-324.208						

Log-likelihood function or log likelihood at convergence (LL_B)	−270.255	−261.607
Akaike information criteria (AIC)	574.5	573.200
AIC/N	1.720	1.716
Chi-squared (dof = 15)	107.907	
Significance level	<0.001	
Halton draws		200
K	17	25

***, **, * ==> Significance at 1%, 5%, 10% level

TABLE 7.6

Fixed Parameter Ordered Probit Model Results: Sydney Driverless Car Adoption

Explanatory Variable	Fixed Parameters Ordered Probit Model (FOPM)				Random Parameter Ordered Probit Model (ROPM)				Marginal Effects (ROPM)		
	Coefficient (β)	Standard Error	z-stat	Prob (p) |z|>Z*	Coefficient (β)	Standard Error	z-stat	Prob (p) |z|>Z*	Quick Adoption (y=0)	Long Way to Go Adoption (y=1)	Hibernation Adoption (y=2)
Background and socio demographics											
Age group: 25–35 years	−0.489***	0.175	−2.8	0.005	−0.615***	0.197	−3.12	0.002	0.173	−0.127	−0.046
Employment											
Employment status: unemployed	−0.433*	0.242	−1.8	0.073	−0.963***	0.303	−3.17	0.002	0.250	−0.191	−0.059
Income range: $40k–$60k	−0.519**	0.218	−2.38	0.017	−1.257***	0.345	−3.65	0.000	0.315	−0.247	−0.068
• Standard deviation of parameter density function					2.491***	0.440	5.66	0.000			
Income range: >$100k	−0.389**	0.167	−2.33	0.020	−0.508***	0.188	−2.7	0.007	0.143	−0.103	−0.041
Travel behaviour and car ownership											
Mode of transport to work: private car, public transport, and shared car	−0.551**	0.232	−2.37	0.018	−0.807***	0.290	−2.79	0.005	0.217	−0.164	−0.053

Mode of transport to school: shared car	0.683*	0.372	1.84	0.066	0.891**	0.442	2.02	0.044	-0.244	0.110	0.134
Mode of transport to social event: private car, public transport, and shared car	0.467**	0.189	2.48	0.013	0.747***	0.226	3.3	0.001	-0.211	0.113	0.098
Travel distance to work: <10 km	0.481***	0.178	2.71	0.007	0.759***	0.215	3.54	0.000	-0.211	0.113	0.098
Travel distance to work: >30 km	0.675***	0.248	2.73	0.006	1.075***	0.288	3.73	0.000	-0.287	0.120	0.168
Travel distance to school: <3 km	-0.622**	0.259	-2.4	0.017	-0.828***	0.313	-2.64	0.008	0.218	-0.164	-0.053
Travel distance to school: 3–5 km	-0.646***	0.242	-2.67	0.008	-0.716***	0.258	-2.78	0.006	0.195	-0.146	-0.048
Travel distance to school: 5–10 km	-0.734***	0.216	-3.4	0.001	-0.981***	0.251	-3.9	0.000	0.258	-0.199	-0.060
Travel distance to school: >10 km	-0.606***	0.224	-2.71	0.007	-0.646**	0.258	-2.51	0.012	0.176	-0.130	-0.046
Travel distance to social trip: <10 km	-0.320**	0.137	-2.34	0.019	-0.530***	0.163	-3.26	0.001	0.148	-0.100	-0.048
• *Standard deviation of parameter density function*					*0.525***	*0.173*	*3.04*	*0.002*			

(continued)

TABLE 7.6 (Continued)

Fixed Parameter Ordered Probit Model Results: Sydney Driverless Car Adoption

Explanatory Variable	Fixed Parameters Ordered Probit Model (FOPM)				Random Parameter Ordered Probit Model (ROPM)				Marginal Effects (ROPM)		
	Coefficient (β)	Standard Error	z-stat	Prob (p) |z|>Z*	Coefficient (β)	Standard Error	z-stat	Prob (p) |z|>Z*	Quick Adoption (y=0)	Long Way to Go Adoption (y=1)	Hibernation Adoption (y=2)
Travel distance to social trip: 50–100 km	−0.685**	0.303	−2.27	0.024	−0.778*	0.441	−1.77	0.077	0.207	−0.158	−0.050
Car value: $40k–$50k	−0.863***	0.253	−3.41	0.001	−1.172***	0.320	−3.67	0.000	0.299	−0.236	−0.063
Driving licence duration: >5 years	0.317**	0.137	2.32	0.020	0.430***	0.152	2.84	0.005	−0.123	0.085	0.039
Driverless car impact and opportunities											
Number of smart tech use: 3	−0.491**	0.237	−2.07	0.039	−0.628**	0.262	−2.4	0.016	0.172	−0.128	−0.044
Use of driving aids: no	0.608**	0.247	2.47	0.014	0.615*	0.370	1.66	0.096	−0.174	0.095	0.079
• *Standard deviation of parameter density function*					4.520***	0.807	5.6	0.000			
Driverless car riding experience: no	0.276*	0.146	1.9	0.058	0.402**	0.164	2.45	0.014	−0.113	0.081	0.032
• *Standard deviation of parameter density function*					0.545***	0.101	5.39	0.000			

Threshold								
Mu (01)	1.260***	0.097	13.06	0.000	1.783***	0.143	12.48	0.000
Goodness of fit statistics								
Number of observations	400				400			
Restricted log likelihood or log likelihood at 0 (LL₀)	−372.762				−372.762			
Log-likelihood function or log likelihood at convergence (LL_B)	−320.769				−310.677			
Akaike information criteria (AIC)	683.5				671.4			
AIC/N	1.709				1.679			
Chi-squared (dof = 19)	103.987							
Significance level	<0.001							
Halton draws	200				200			
K	21				25			

***, **, * ==> Significance at 1%, 5%, 10% level

($40k–$60k), travel distance to a social trip (less than 10 km), familiarity with driving aids (No), and driverless car experience (No); and (d) two for the Combined model including age group (35–55 years) and driving licence duration (3–5 years). These random parameters captured the associated high variances in the driverless car adoption timelines.

As an example, the Melbourne model shows that the observed sample associated an early adoption with the age group 35–55. Further, the estimated variances for the age group 35–55 capture varying levels in the adoption timeline of customers—i.e., 75% of the responses are associated with early adoption, and the remaining 25% are linked to late adoption. This implies that while a significant portion of the sample perceived an early adoption, the remaining sample perceived a belated adoption in this age group.

The heterogeneities identified through the ROPM have a degree of implications for driverless car manufacturers, organisations involved in policy and regulation formulation, and research organisations. For instance, manufacturers cannot assume naively that Melbourne's population of 35–55-year-olds will be the early adopters. They must consider the 25% of people in this Melbourne age group who are considered late adopters. To gain the trust of this 25% of the population, either they need to educate them about the functionality, advantages, freedom, flexibility, and safety that a driverless car can offer, or they need to pinpoint any weaknesses in the privacy, security, privacy, and safety features of driverless cars. To accomplish this, manufacturers can conduct the necessary research and consult with the target audience, research organisations, and pertinent governmental bodies.

The following section elaborates the findings in more detail and addresses the issue of why the same variable produces different results in different cities.

7.4 Findings and Discussion

The timeliness of adoption is reflected in the scale of parameter results of a significant variable: the more positive the results, the later the adoption compared to non-significant variables of a particular factor; the more negative, the earlier the adoption. In total, 19 variables had a significant impact on the adoption of driverless cars in Brisbane. Six of the 19 significant variables were random parameters and, thus, had a variable effect on driverless car adoption outcome probabilities (Table 7.4). Within the Melbourne data, 15 variables had a significant impact on driverless car adoption. Seven of the 15 significant variables were determined to be random parameters (Table 7.5). For Sydney, 20 variables were found significant. Four of the 20 significant variables were determined to be random parameters (Table 7.6).

7.4.1 Interpretation of Significant Variables in the Model

7.4.1.1 Background and Socio Demographics

Several background and socio-demographic factors were assessed. Age was a significant factor in all three models, namely Brisbane, Melbourne, and Sydney. With respect to other age groups, the 25–35 age group in Sydney showed a preference for early adoption compared to other age groups (coefficient = –0.615, z = –3.12, p = 0.002) and led to an increase in the probabilities of *quick* adoption by 17.3% (Table 7.6). In Melbourne, the 35–55 age group was more inclined to adopt driverless cars earlier than other age groups (coefficient = –0.477, z = –2.39, p = 0.017) and led to an increase in the probabilities of *quick* adoption by 12.00% (Table 7.5). However, the coefficient for this group was found to be a random parameter (mean: –0.477; standard deviation: 0.710). This indicates that the estimated coefficient (β) associated with the group indicator is positive for 25% of the responses and negative for the rest 75%. Like Melbourne, in Brisbane, the 35–55 age group was more inclined to adopt driverless cars earlier than other age groups (coefficient = –1.231, z = –4.53, p = 0.000) and led to an increase in the probably of *quick* adoption by 26.1%, age group 55–65 showed even faster adoption and more increased probability towards *quick* adoption (Table 7.4). The coefficient of the Brisbane age group 35–55 was found to be random (mean: –1.231; standard deviation: 1.015). This indicates that the estimated coefficient (β) for the age group 35–55 is positive for 11% of the responses and negative for the rest 89%. Brisbane results suggest that mid-aged adults (35–55 and 55–65) would be the early adopters than other younger (18–25 and 25–35) and older (65+) age groups, though findings from other studies (Rödel et al., 2014; Schoettle et al., 2014; Haboucha et al., 2017; Nordhoff et al., 2018; Butler et al., 2021; Othman, 2021) confirmed that younger people had a positive mindset towards driverless car adoption. On the contrary, another France study showed that elder individuals are more willing to accept such technology, but less willing to pay for it (Payre et al., 2014).

The Melbourne age group of 55–65 had a positive coefficient value (β = 0.830), predicting belated adoption compared to other age groups, and this led to an increase in the likelihood of *long way to go* and *hibernation* adoptions and a decrease in the likelihood of *quick* adoption (Table 7.5). The estimate β was a random parameter, with a mean of 0.830 and a standard deviation of 2.478. This indicates that the indicator was negative for 37% of the responses and positive for 63%. This suggests that the 55–65 age group in Melbourne has a high propensity to be early adopters (36.87%), like the findings from Brisbane.

Among the considered background factors, academic qualification had a significant impact on driverless car adoption across Brisbane and Melbourne cities. People with diploma qualifications were late adopters in Melbourne compared to other qualifications. In Melbourne, people with secondary education qualifications were identified as far late adopters with respect to Melbourne diplomas. In Brisbane, people with primary qualifications showed the slowest adoption preference among all other significant qualification groups (Tables 7.4–7.6).

The Melbourne secondary qualification group indicator was found to be a random parameter, with a mean of 1.312 and a standard deviation of 1.779. This indicates that the parameter associated with the group indicator is negative for 23% of the responses and positive for 77%. The probability increases for *long way to go* and *hibernation* adoption were 17.21% and 13.83%, respectively (Table 7.5).

A common factor emerged across Brisbane and Melbourne models regarding academic qualification; people with lower qualifications—primary, secondary, or diploma—indicated comparatively later adoption of driverless cars compared to those with graduate or higher qualifications. This might reflect a level of knowledge and awareness of the safety features, benefits, ongoing trials, and policy support towards this new technology. Or this cohort may not be aware of current progress with developments in the field of driverless car technology. Specifically, people with secondary qualifications in Melbourne were less eager to adopt driverless cars compared to those qualified to diploma level. According to studies in the US, UK, Australia, and France (Piao et al., 2016; Schoettle et al., 2014), having more education was related to having a more positive attitude towards driverless cars, even though this research did not categorise people as early or late adopters based on their academic background. Another US study discovered that the intention to use driverless cars and educational attainment were unrelated (Zmud et al., 2016).

Gender was identified as a significant factor in the Brisbane model only, which implies that males are late adopters with respect to females. The probability increases for *long way to go* and *hibernation* adoption were 13.53% and 4.26%, respectively, and decrease for *quick* adoption by 17.80% (Table 7.4). According to other studies in the US, UK, Australia, Austria, Finland, and France (Payre et al., 2014, Piao et al., 2016; Schoettle et al., 2014; König & Neumayr, 2017; Hulse et al., 2018; Liljamo et al., 2018), males are more likely than females to use driverless cars, or males are more in favour of their adoption. However, these studies did not identify how quickly or slowly men will adopt driverless cars over women.

7.4.1.2 Family Structure

Family structure was found to affect driverless car adoption in Brisbane metropolitan area (Table 7.4). Small family size had a significantly positive effect on early adoption. Brisbane results revealed that families with two members are the early adopters among all other household sizes, and they led to an increase in the likelihood of *quick* adoption by 16.40% (Table 7.4).

Brisbane result may reflect the fact that small families might want to switch from public transport to driverless car to leverage their office or university commute, or they might want to add a convenient first- and last-mile connectivity to their ongoing public transport trips. This is supported by the results of Brisbane office commuters' travel behaviour because office commuters who the use private car, public transport, and shared car are the early adopters.

7.4.1.3 Employment

The number of people employed in a household had a different impact on driverless car adoption in Brisbane, Melbourne, and Sydney metropolitan areas. Future selection of a driverless car in Brisbane and Sydney is unaffected by the number of employed individuals in a household. On the other hand, Melbourne families with three employed were the early adopters compared to families with other numbers of employed members and led to an increase in the likelihood of *quick* adoption by 33.70% (Table 7.5). The Melbourne families with three employed members may be early adopters of shared driverless car rides to avail the opportunity of working on wheels without being late to the workplace while the car is routing to drop-off the riders one by one.

For the early adopters, a shared driverless car can also emerge as a convenient first/last mile solution to reduce cost per traveller and vehicle kilometres travelled (Fagnant & Kockelman, 2018; Levin, 2017; Martinez & Crist, 2015; Zhu et al., 2016), provide the opportunity of travel cost reduction by downsizing the privately owned conventional human-driven car(s) ownership (Golbabaei et al., 2021), or offer the services on less frequent public transport routes (Liang et al., 2016; Levine et al., 2018). Melbourne families with more than three numbers of employed were not early adopters compared to the families with three members employed in a household. One possible reason behind their reluctance could be a belief that they will not gain benefit from a driverless car if household members' workplaces are spread across multiple locations.

Employment factors also strongly influenced driverless car adoption in Brisbane, Melbourne, and Sydney (Tables 7.4–7.6). In Brisbane, people who worked in academia were the strongest late adopters compared to other organisation types, and they led to a big increase in the likelihood of *hibernation* (0.175) and *long way to go* (0.142) and a decrease in the *quick* adoption (–0.317), and people who worked for state government organisations were the strong late adopters compared to other organisation types, and they led to a big increase in the likelihood of *long way to go* (0.133) and decrease in the *quick* adoption (–0.215) (Table 7.4). Federal and state organisations are more involved in the formulation of policy, regulations, standards, and operational safety documents. It was expected that these groups would have shown an indication of being early adopters of driverless cars. Considering this, it seems the Brisbane model does not reflect a cross-section of state government employees who work in the transport sector. Similarly, results reflect the same fact for the academia people. Sydney and Melbourne data on employment organisation type (academic, federal, state, local government) showed no significant differences to each other in terms of driverless car adoption timeframe, which was unexpected and very interesting.

In terms of occupation, Melbourne transport/urban planners were the early adopters compared to other occupations, and they led to a big increase in the likelihood of *quick* (0.215) and a decrease in the *long way to go* adoption (–0.174) (Table 7.5). This is likely attributable to obvious future changes in built and natural environments linked with increasing driverless car technology that these two occupational groups might foresee, translating into quicker adoption of driverless cars compared to other groups.

In terms of income, the Sydney income group: $40k–$60k were the strongest early adopters, then income group: >$100k were the stronger early adopters compared to other income groups, and income groups: $40k–$60k and >$100k led to an increase in the likelihood of *quick* adoptions by 31.5% and 14.3%, respectively (Table 7.6). Almost a similar trend was seen for Brisbane income groups: except for income group $20k–$40k is early adopter along with >$100k group with respect to other mid-ranged income groups (Table 7.4). This means the highest income ranges (>$100k) were the early adopters in both models. Results showing the low-end income groups: $20k–$40k and $40k–$60k as early adopters are different to the results of other studies that indicate middle to higher-income groups are more likely to have a positive attitude towards AV (Wang et al., 2020; Yuen et al., 2020).

7.4.1.4 Travel Behaviour and Car Ownership

Among the factors associated with travel behaviour, travel distance to school and social trips and driver's licence duration had a significant influence on driverless car adoption in Brisbane (Table 7.4). In Melbourne, travel distance to work, school, and social trips, car ownership and driver's licence duration were identified as significant factors in influencing driverless car adoption (Table 7.5). Travel mode to work, school, and social trips, travel distance to work, school, and social trips, driver's licence duration and car value were the significant factors in Sydney (Table 7.6).

In comparison to other working classes that use other modes of transportation, those in the Sydney working class who use public transportation and shared transportation in addition to a privately owned car are more likely to adopt driverless cars as *quick* adopters. In the early stages of the deployment of driverless cars, this trend may have been influenced by a desire to find a suitable and effective replacement for the privately owned car, the current first- and last-mile connectivity. On the other hand, this early adopter group may completely switch from their current mode of transportation to their place of employment to a driverless car because it is a preferred alternative

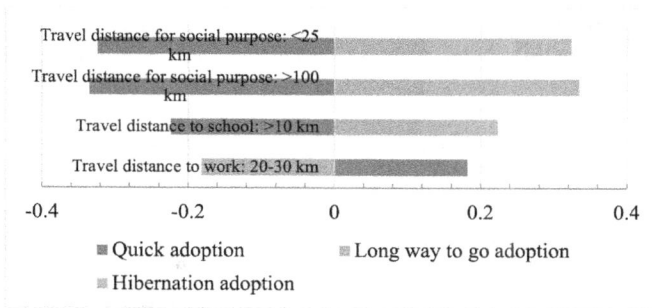

FIGURE 7.2
Average marginal effects of travel distance to work/school/social trips (Melbourne).

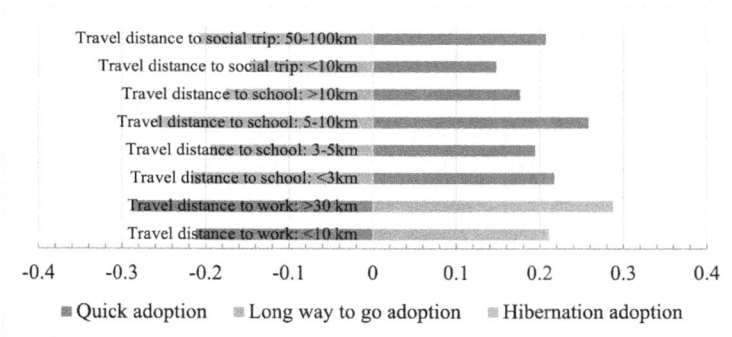

FIGURE 7.3
Average marginal effects of travel distance to work/school/social trips (Sydney).

for those who use public transportation and commuters who feel unsafe in it (Butler et al., 2021).

Sydney parents and students who commute to school in a shared vehicle demonstrated an increase in the likelihood of adopting driverless cars as *long way to go* and *hibernation* adopters, making them the late adopters compared to other parent and student groups who use other modes of transportation. This outcome might be explained by the fact that the driverless school trip is not significantly superior to the shared car school trip.

People in Brisbane who commute to work from varying distances are indifferent to the driverless choice timeframe (Table 7.4). This could be due to their current level of comfort and convenience in commuting with a human-driven car, and the fact that they are not yet confident in driverless technology and infrastructure readiness. On the contrary, Melbourne workers who commute to work in the range of 20–30 km led to an increase in the likelihood of *quick* adoption by 18.17% and a decrease in the *hibernation* adoption by 3.98%, and this group is the early adopter in comparison to other low- and high-range commuting Melbourne working classes (Table 7.5, Figure 7.2). It is quite understandable that commuters travelling 20–30 km would be early adopters with respect to less than 20 km commuters, non-working individuals and those who work from home. However, it is unclear why this group favours early adoption over the commuters who travel more than 30 km.

One of the reasons might be, with such long commutes (more than 30 km), driverless cars may not be as appealing in the early days of deployment to support multiple work trips or, work and school trips through a single driverless car during morning and afternoon pickup and drop off hours that people might be interested in. One of the other reasons might be those with the slightly longer commute are yet to be convinced by any visible steps in infrastructure readiness. When compared to the other commuting groups in Sydney, people who commuted for distances of less than 10 km and more than 30 km were found to be late adopters. It is not entirely clear, though, why Sydney commuters who travel less than 10 km are the late adopters when compared to those who are unemployed and those who work remotely. It's possible that Sydney commuters who travel more than 30 km are late adopters for the same reasons Melbourne commuters who travel more than 30 km are. Such Sydney work trips are associated with an increase in the *long way to go* and *hibernation* adoption.

Brisbane parents who travel less than 3 km for school trips are the early adopters in comparison to other school travellers and they led to an increase in the *quick* adoption (Table 7.4). Less-than-3-km-school-trip makers are one of the strongest early adopters by possessing a bigger parameter value (–2.269). Within the short school trip duration, Brisbane parents might do household work or get ready for work while a driverless car drops off their kids and comes back to pick them up for their commute. Melbourne parents who travelled more than 10 km for school trips were statistically significant with a random parameter with a mean of 0.977 and standard deviation of 2.729, indicating the parameter was negative for 36% of the responses and positive for 64%. They are the late adopters among all other Melbourne school travellers, and such school trips led to an increase in the probability of *long way to go* and *hibernation* adoption. This trend was likely observed since for such long school trips, driverless cars might not be appropriate in terms of time to complete school drop-off, then come back to pickup the working parents during peak school/office hours. Schools in Sydney follow the same trends as those in Brisbane and Melbourne, except that the 5–10 km commuters are the strongest early adopters, and all four commuter groups are significant early adopters.

Long-distance social trips of 50–100 km will not be appealing when driverless cars are first deployed, according to data from Brisbane. However, this group contributed to an increase in *long way to go* adoption— adoption in the first five years. In contrast, Sydney long-distance social trips: 50–100 km show these trips will be appealing, leading to an increase

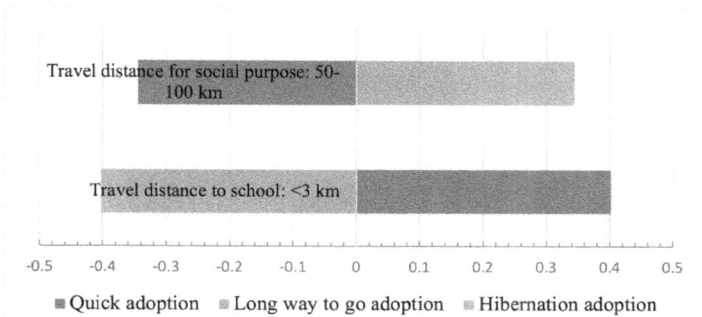

FIGURE 7.4
Average marginal effects of travel distance to school/social trips (Brisbane).

in *quick* adoption by 20.70% (Figure 7.3). Results from Melbourne residents who made social trips of more than 100 km were statistically significant with a normally distributed random parameter with a mean of 1.572 and a standard deviation of 2.721, indicating the parameter was negative for 28.17% of the responses and positive for 71.83%. These high-range social trip makers in Melbourne are the late adopters among all other ranges of social trip makers and led to an increase in the probability of *long way to go* and *hibernation* adoption (Table 7.5). The statistical significance of the Melbourne social trip group: less than 25 km was supported by a normally distributed random parameter with a mean of 1.487 and standard deviation of 2.196, indicating that the parameter was negative for 25% of the responses and positive for 75%. This Melbourne trend was likely observed because driverless cars cannot be relied upon at early stages of their deployment for short- or long-range social trips considering that infrastructure and legalities may not be completely ready within complex urban environments, i.e., within central business districts (CBDs) or far from the metropolitan centres. Again, one of the other reasons might be—for those making social trips within 25 km of home, active transport or public transport may seem more appealing. Or they might want to choose the thrill of manual driving for shorter recreation trips. Interestingly, Sydney short-distance travellers who travel less than 10 km are more likely to be early adopters. One of the plausible causes could be the willingness to remain active on the wheels, even over short distances in urban traffic jams. The statistical significance of these Sydney social trip group: <10 km results were supported by a normally distributed random parameter with a mean of −0.530 and standard deviation of 0.525, indicating that the parameter was negative for 84% of the responses and positive for 16% (Table 7.6).

For better interpretation as well as comparing the effects of "travel distance to work/school/social trips" on different adoption time horizons, marginal effects of travel distance are presented below in bar diagram (Figures 7.2–7.4) in addition to the data presented in Tables 7.4–7.6.

Results from Brisbane drivers who owned a driver's licence for a period of <1 year, 1–3 years and 3–5 years were found to be statistically significant, and the early adopters in comparison to other groups of licence holders. Among these three driver groups, the <1 year-group is the most aggressive early adopter, and the other two groups have normally distributed random parameters, indicating that parameter values are not negative for all responses, i.e., for the 1–3-year group, 10% of responses are positive and for the 3–5-year group, 5% are positive. All these three drivers group led to an increase in the likelihood of *quick* adoption. There is a possibility that the 1-, 1–3-, and 3–5-year groups consist of novice to mature young drivers who are interested in experimenting with the new technology before other licence-holder groups. Melbourne and Sydney results are very similar to that of Brisbane, only major exception is in Melbourne, where the 3–5-year group is not an early adopter. It means that Melbourne drivers with licences that are at least three years old are more likely to continue driving conventional vehicles for a few more years before accepting driverless vehicles (Table 7.5). The reasoning regarding Sydney drivers is comparable to that presented for Brisbane drivers.

Results from Sydney respondents who own a car valued between $40k and $50k were statistically significant as early adopters compared to other car value ranges (coefficient = −1.172, z = −3.67, p = 0.0002). These car owners had 29.90% probability increases for *quick* adoption. This means, current above-average value ($40k–$50k) car owners may be able to afford more expensive, cutting-edge technology driverless cars when first available.

7.4.1.5 Familiarity with Technology, Driverless Car Features, and Driverless Car Experience

Of the factors linked with driverless car technology, familiarity, and smart technology adoptability, driverless car trip purpose and driverless car feature preference had a strong influence on driverless car adoption in Brisbane (Table 7.4). Whether or not a Brisbane respondent has ridden in a driverless vehicle has no bearing on his or her preference for driverless vehicles. In the Sydney model, driverless car riding experience had a good influence on driverless car adoption (Table 7.6). Results from Sydney respondents who had not experienced riding in a driverless car were found to be statistically significant and late adopters compared to the people who had experienced riding on driverless cars, with a parameter mean of 0.402. Those with no driverless car experiences led to a decrease in the likelihood of *quick* adoption and an increase in the likelihood of *long way to go* and *hibernation* adoptions. It is logical that users who had not experienced riding in a driverless car would show hesitation around early adoption of a driverless car.

Brisbane users who planned to use driverless cars for both work and school-trip purposes were strong early adopters of driverless cars compared to other user groups with a parameter value of –2.293 and led to an increase in the likelihood of *quick* adoption by 40.70% and decrease in the likelihood of *long way to go* adoption by 36.50%. This finding is like the research findings of Irannezhad and Mahadevan (2022). Their research revealed that Australian regional residents with tech-centric attitudes and metropolitan area residents with anti-driving attitude are more likely to use AV-based services for all trips purposes. Research by Yigitcanlar et al. (2020) also revealed similar findings.

Expertise in using driving aids had strong influence on driverless car adoption both in Sydney and Melbourne. The results revealed that the greater a person's familiarity with driving aids, the more likely they are to be an early adopter. But there is no correlation between a Melbourne respondent's familiarity with smart technology and his or her preference for autonomous vehicles, though it was expected that respondents conversant with a greater number of smart technologies will be the early adopters (Table 7.5). In the case of Sydney, those who are accustomed to three types of smart technology are the early adopters, as opposed to those who use no smart technology or more than three types. This indicates that it is not always the case that tech-savvy individuals are the first to adopt new technologies such as driverless cars. In contrast, in the Brisbane scenario, three types of smart technology users are late adopters compared to less than three or more than three types of technology users. All three scenarios for Brisbane, Melbourne, and Sydney demonstrate that tech-friendliness is not the best indicator of an individual's early adoption of driverless cars. This result is comparable to research findings of Irannezhad and Mahadevan (2022) to some extent.

In the metropolitan areas of Melbourne and Sydney, driverless car feature preference is not a significant factor in determining early or late adoption of driverless cars. Most early adopters in Brisbane, however, are those who prefer three driverless car features as opposed to fewer or additional features. Considering the preferences for driverless car features in the three metropolitan areas, it is impossible to determine definitively which feature preference groups are early or late adopters of driverless cars.

7.4.1.6 Recommendations

The following study findings offer insights for future planning and policy direction related to driverless car adoption in Australia. These findings might also inform policymaking in other country contexts with similar characteristics. Ideas have been offered for turning late adopters into early adopters that may aid in the promotion of autonomous vehicles.

The findings from *Brisbane* provide key insights regarding driverless car adoption behaviour. These include the following:

- The 35–55 and 55–65 age groups were the earliest adopters and demonstrated a higher likelihood of *quick* adoption, indicating that mid- to older-generation customers should be the target customer groups to persuade them of the advantages of driverless cars.
- Primary-level learners were more reluctant to adopt driverless cars, indicating that less academically accomplished individuals require more exposure to and education about the features and advantages of driverless car technology.
- Income groups: $20k–$40k and >$100k were early adopters compared to other income groups, suggesting a survey is required to capture the mindset barrier of the mid-ranged income groups in between, then targeted promotional marketing activities emphasising driverless cars' ease of use, reliability, safety, and capacity for multitasking will be required to change their minds and make them early adopters.
- Less-than-3-km-school-trip-takers are the strongest early adopters among all school-trip-takers, suggesting that parents of long-trip-takers must be made aware of the potential benefit of office on the wheels that a driverless car will be able to offer from the day one of deployment for them to change their minds.
- Long-distance social trips will be unattractive when driverless cars are first introduced, suggesting that infrastructure be made driverless-car-friendly in rural or remote areas.
- Stronger early adopters were found to have held their licences for between less than one and five years, indicating that these demographics will be influential and proponents of driverless car technology in the early stages of its availability.
- Compared to <3 or >3 types of technology users, smart technology user: three types are late adopters. This indicates that future driverless car sellers in Brisbane should prioritise a variety of other factors besides tech-friendliness when considering their initial customer base.
- One commuter group that uses driverless cars for both work and school trips was discovered to be an early adopter, indicating that this group will make another early-stage breakthrough in commercialisation of driverless cars.

The findings from *Melbourne* provided key insights regarding driverless car adoption behaviour. These include the following:

- Those aged 35–55 were early adopters of driverless cars, suggesting the core target age group for popularising driverless cars in the early days. Those aged 55–65 being delayed adopters suggested the necessity of marketing campaigns tailored specifically to this demographic.

- Both secondary and diploma-qualified groups were slower to adopt the driverless car, suggesting that low- to middle-educated groups need more information about the features and benefits of driverless car technology in comparison to those with higher level of educations.

- Early adopters were households with three employed individuals. This is logical in the sense that they will likely realise the benefits of driverless cars sooner due to their multiuse of the vehicle's idle time. Consequently, it may not be worthwhile to promote driverless cars to families with fewer than three employed members in the early stages of their deployment.

- Transport/urban planners were early adopters compared to other occupations, suggesting that urban planners can act as a motivator in promoting driverless cars' positive impact on the built and natural environment, e.g., reduction of CBD parking, reduction in lane width, elimination of traffic signal visual clutter, promotion of ex-urban living, exclusion of grade-separated footpath in urban area, etc., which will inevitably attract more people to driverless cars in the early foreseeable future.

- In comparison to other low- and high-range commuting Melbourne working classes, the 20–30 km commuter class was the earliest adopter. To entice the high-distance commuter workforce group to adopt driverless cars in the early days, state and local governments must ensure the visibility of their road infrastructure readiness on the periphery. Initially, it will be difficult to persuade workers with short commutes to adopt driverless cars. Still, the ability of driverless cars to provide effective first- and last-mile connectivity, as well as their immediate use by other household members, can be promoted to customers with short commutes.

- The delayed adopters were those who commuted more than 10 km to school, as well as less than 25 and more than 100 km for social outings. Getting those who commute more than 10 km to school to adopt early technology may be extremely difficult. One possible way to attract this group, especially parents of students, would be to add dressing facilities, food and beverage preparation options, and virtual meeting options inside the car. This might entice parents to bring their kids along on the longer trip. If driverless car can be proven as an appealing alternative to active or public transportation, people who frequently make social trips within 25 km of their homes may be drawn to adopting them in early going. Infrastructure and enforcement readiness at remote and regional areas might be the key for encouraging early adoption for social trips longer than 100 km. Both the groups, therefore, represent a lower-tier targets for marketers in early days of deployment.

- In comparison to other car owner groups, two-car owners were the slower to adopt driverless cars, indicating that this group needs better driverless car advertising campaigns.
- Insights for licence holders less than a year and 1–3 years were the same as that of Melbourne.
- Those who use more than four driving aids to control dynamic driving tasks in human-driven cars are predicted to be early promoters and target customers in the early days of deployment.

The findings from *Sydney* provide key insights regarding driverless car adoption behaviour. These include the following:

- Insights for age group 25–35 were the same as Melbourne age group 35–55, income group: $40k–$60k and >$100k were the same as income groups $20k–$40k and >$100k in Brisbane, social trip makers: <10 km and 50–100 km were the opposite to 50–100 km group in Brisbane, school commuters travelling in the range of <3 km–>10 km were same as <3 km in Brisbane and use of driving aids: 'no' was the opposite to use of driving aids: >4 types in Melbourne.
- People using the private car and public transport in conjunction with shared transport for work commute were the early adopters of driverless cars, indicating that driverless cars will be a threat to the transit-oriented multimodal office modes at the early stage of its deployment.
- Those who commute to work in the range of <10 km or >30 km were the delayed adopters, suggesting that this short- and long-distance commute group may need to be 'sold' on how they can use their car for pre- and post-office activities or convert their car to an office on wheels.
- All ranges of school-trip makers, except non-students/online school group are the early adopters with a minor difference in their parameter values, suggesting that households having any number of students will be most informed and prospective early-day customers of driverless cars where future car manufacturers can blindly sell their driverless cars.
- Less-than-10-km and 50–100-km social trip makers were the earliest adopters. Other social trip makers, being the later adopters, can be encouraged to use driverless cars in the early days of deployment by providing the same facilities that were suggested for the late social trip adopters in Melbourne.
- Those who currently possess the cars valued at $40k–$50k were the early adopter compared to other car value ranges, suggesting that members of this customer group could prove promising in paving the path to driverless car adoption. That is why target group-oriented promotional or marketing activities need to be initiated by the original equipment

manufacturers (OEMs) among low- to medium- and high-value car owners at the early days of driverless car deployment.

- Compared to less than or greater than three types of technology users, three types of users of smart technology are early adopters. This indicates that future driverless car sellers in Sydney should prioritise a variety of other factors besides tech-friendliness when considering their initial customer base.

7.5 Conclusion

This chapter analysed the adoption timeline for future driverless car users using a random parameter ordered probit model estimated with data from three Australian metropolitan cities—Brisbane, Melbourne, and Sydney. To the best of our knowledge, this is the first attempt in Australian context to incorporate random parameters to analyse driverless car users' behaviour in terms of adoption timeframe.

The proposed quantitative behavioural model of driverless car adoption timeline can assist OEMs, urban planners, ride-sharing businesses, and government agencies in reaching out to early adoption groups and addressing potential barriers that might be repelling prospective driverless car users from being early adopters. Researchers in the field of driverless car simulations may also find this study useful. This study provides a broad overview by considering all age groups and working classes that may benefit from the technology. This identified specific groups that may be investigated in more detail in future studies to work out group-specific car models and features, maximise sales and marketing strategies, and offer group-specific driverless car services. A similar study can also be conducted to expand data collection to other Australian metropolitan and regional cities so that OEMs, transport and passenger service providers, and government agencies can align their strategies and programmes accordingly.

In general, the results showed that higher levels of education, being young or middle-aged, living in a small household, earning a middle-to-high income, taking any range of school trips, taking small- and large-scale social trips, commuting for 20–30 km, possessing a licence valid for up to five years, driving a car worth $40k–$50k, using driverless cars for multiple purposes, having used more than four conventional driving aids, having experience with driverless cars before, and being tech-friendly increase the probability of being an early adopter. As indicated before, the sample includes a higher percentage of graduates relative to the population. Although the results are consistent with published literature (Dennis et al., 2021; Haboucha et al.,

2017; Liljamo et al., 2018), further research is recommended to access any potential bias because of many educated people in the sample.

Due to study limitations, driverless car adoption modelling was performed for only three Australian metropolitan areas. In this study, psychological factors of, willingness to pay and mode and fuel preference for driverless car adoption were not considered to keep the study scope within a reasonable and manageable limit, but these factors should be considered in a follow-up study. Similarly, this study tested multiple specifications and determine that a random parameter ordered probit model better captured the observed behaviour. Further research considering an extensive hypothesis testing is recommend exploring opportunities to gain additional insights from the data (Paz et al., 2019; Beeramoole et al., 2023).

Acknowledgements

This chapter, with permission from the copyright holder, is a reproduced version of the following journal article: Faisal, A., Yigitcanlar, T., & Paz, A. (2023). Understanding driverless car adoption: random parameters ordered probit model for Brisbane, Melbourne, and Sydney. *Journal of Transport Geography*, 110(1), 103633.

References

ADVI (2021). Australia and New Zealand driverless vehicle initiative. www.ccat.org.au

Beeramoole, P., Arteaga, C., Haque, M., Pinz, A., & Paz, A. (2023). Extensive hypothesis testing for estimation of mixed-logit models. *Journal of Choice Modeling*, 47, 100409.

Bhat, C. (2001). Quasi-random maximum simulated likelihood estimation of the mixed multinomial logit model. *Transportation Research Part B*, 35(7), 677–693.

Bhat, C. (2003). Simulation estimation of mixed discrete choice models using randomized and scrambled Halton sequences. *Transportation Research Part B*, 37(9), 837–855.

Boes, S., & Winkelmann, R. (2006). Ordered response models. *Allgemeines Statistisches Archiv*, 90, 167–181.

Butler, L., Yigitcanlar, T., Areed, W., & Paz, A., (2022). How can smart mobility bridge the first/last mile gap? Empirical evidence on public attitudes from Australia. *Journal of Transport Geography*, 104, 103452.

Butler, L., Yigitcanlar, T., & Paz, A., (2020a). Smart urban mobility innovations: A comprehensive review and evaluation. *IEEE Access*, 8, 196034–196049.

Butler, L., Yigitcanlar, T., & Paz, A., (2020b). How can smart mobility contribute to alleviate transportation disadvantage? Assembling a conceptual framework through a systematic review. *Applied Sciences*, 10(18), 6306.

Butler, L., Yigitcanlar, T., & Paz, A., (2021). Factors influencing public awareness of autonomous vehicles: empirical evidence from Brisbane. *Transportation Research Part F*, 82, 256–267.

Chang, L., & Chien, J. (2013). Analysis of driver injury severity in truck-involved accidents using a non-parametric classification tree model. *Safety Science*, 51, 17–22.

Chen, F., Song, M., & Ma, X. (2019). Investigation on the injury severity of drivers in rear-end collisions between cars using a random parameters bivariate ordered probit model. *International Journal of Environmental Research and Public Health*, 16(14), 2632.

Christoforou, Z., Cohen, S., & Karlaftis, M. (2010). Vehicle occupant injury severity on highways: An empirical investigation. *Accident Analysis & Prevention*, 42(6), 1606–1620.

Cloke, P., Cook, I., Crang, P., Goodwin, M., Painter, J., & Philo, C. (2004). Talking to people. In: *Practising Human Geography*. London: Sage.

Dennis, S., Paz, A., & Yigitcanlar, T., (2021). Perceptions and attitudes towards the deployment of autonomous and connected vehicles: insights from Las Vegas, Nevada. *Journal of Urban Technology*, 28(3–4), 75–95.

Duncan, C., Khattak, A., & Council, F. (1998). Applying the ordered probit model to injury severity in truck-passenger car rear-end collisions. *Transportation Research Record*, 1635, 63–71.

Fagnant, D., & Kockelman, K. (2018). Dynamic ride-sharing and fleet sizing for a system of shared autonomous vehicles in Austin, Texas. *Transportation*, 45, 143–158.

Faisal, A., Yigitcanlar, T., Kamruzzaman, M., & Paz, A. (2021). Mapping two decades of autonomous vehicle research: A systematic scientometric analysis. *Journal of Urban Technology*, 28(3–4), 45–74.

Fountas, G., & Anastasopoulos, P. (2017). A random thresholds random parameters hierarchical ordered probit analysis of highway accident injury-severities. *Analytic Methods in Accident Research*, 15, 1–16.

Golbabaei, F., Yigitcanlar, T., & Bunker, J., (2021). The role of shared autonomous vehicle systems in delivering smart urban mobility: A systematic review of the literature. *International Journal of Sustainable Transportation*, 15(10), 731–748.

Golbabaei, F., Yigitcanlar, T., Paz, A., & Bunker, J., (2020). Individual predictors of autonomous vehicle public acceptance and intention to use: A systematic literature review. *Journal of Open Innovation*, 6(4), 106.

Greene, W. (2003). *Econometric Analysis*. Mumbai: Pearson Education India.

Haboucha, C., Ishaq, R., & Shiftan, Y. (2017). User preferences regarding autonomous vehicles. *Transportation Research Part C*, 78, 37–49.

Hulse, L., Xie, H., & Galea, E. (2018). Perceptions of autonomous vehicles: Relationships with road users, risk, gender and age. *Safety Science*, 102, 1–13.

Irannezhad, E., & Mahadevan, R. (2022). Examining factors influencing the adoption of solo, pooling and autonomous ride-hailing services in Australia. *Transportation Research Part C*, 136, 103524.

Islam, M., & Hernandez, S. (2013). Large truck – Involved crashes: Exploratory injury severity analysis. *Journal of Transportation Engineering*, 139(6), 596–604.

Jared, C. (2022). Advantages and disadvantages of online surveys. www.proprofssur vey.com

Khorashadi, A., Niemeier, D., Shankar, V., & Mannering, F. (2005). Differences in rural and urban driver-injury severities in accidents involving large-trucks: An exploratory analysis. *Accident Analysis & Prevention*, 37(5), 910–921.

Kockelman, K., & Kweon, Y. (2002). Driver injury severity: An application of ordered probit models. *Accident Analysis & Prevention*, 34(3), 313–321.

König, M., & Neumayr, L. (2017). Users' resistance towards radical innovations: The case of the self-driving car. *Transportation Research Part F*, 44, 42–52.

Levin, M. (2017). Congestion-aware system optimal route choice for shared autonomous vehicles. *Transportation Research Part C*, 82, 229–247.

Levine, J., Zellner, M., Arquero de Alarcon, M., Shiftan, Y., & Massey, D. (2018). The impact of automated transit, pedestrian, and bicycling facilities on urban travel patterns. *Transportation Planning and Technology*, 41(5), 463–480.

Liang, X., de Almeida Correia, G., & van Arem, B. (2016). Optimizing the service area and trip selection of an electric automated taxi system used for the last mile of train trips. *Transportation Research Part E*, 93, 115–129.

Liljamo, T., Liimatainen, H., & Pöllänen, M. (2018). Attitudes and concerns on automated vehicles. *Transportation Research Part F*, 59, 24–44.

Macrotrends (2021). Australia metro area populations 1950–2023. www.macrotre nds.net.

Martinez, L., & Crist, P. (2015). Urban mobility system upgrade: How shared self-driving cars could change city traffic. In: International Transport Forum.

McKelvey, R., & Zavoina, W. (1975). A statistical model for the analysis of ordinal level dependent variables. *Journal of Mathematical Sociology*, 4, 103–120.

Naik, B., Tung, L., Zhao, S., & Khattak, A. (2016). Weather impacts on single-vehicle truck crash injury severity. *Journal of Safety Research*, 58, 57–65.

Nastjuk, I., Herrenkind, B., Marrone, M., Brendel, A., & Kolbe, L. (2020). What drives the acceptance of autonomous driving? An investigation of acceptance factors from an end-user's perspective. *Technological Forecasting and Social Change*, 161, 120319.

Nordhoff, S., De Winter, J., Kyriakidis, M., Van Arem, B., & Happee, R. (2018). Acceptance of driverless vehicles: Results from a large cross-national questionnaire study. *Journal of Advanced Transportation*, 2018, 5382192.

NTC (2019). In-service safety for automated vehicles. www.ntc.gov.au

NTC (2022). The regulatory framework for automated vehicles in Australia. www.ntc.gov.au

Othman, K. (2021). Public acceptance and perception of autonomous vehicles: A comprehensive review. *AI and Ethics*, 1(3), 355–387.

Pahukula, J., Hernandez, S., & Unnikrishnan, A. (2015). A time-of-day analysis of crashes involving large trucks in urban areas. *Accident Analysis & Prevention*, 75, 155–163.

Parfitt, J. (2013). Questionnaire design and sampling. In: *Methods in Human Geography*. London: Routledge.

Payre, W., Cestac, J., & Delhomme, P. (2014). Intention to use a fully automated car: Attitudes and a priori acceptability. *Transportation Research Part F*, 27, 252–263.

Paz, A., Arteaga, C., & Cobos, C. (2019). Specification of mixed logit models assisted by an optimization framework. *Journal of Choice Modelling*, 30, 50–60.

Perveen, S., Kamruzzaman, M., & Yigitcanlar, T. (2017). Developing policy scenarios for sustainable urban growth management: A Delphi approach. *Sustainability*, 9(10), 1787.

Piao, J., McDonald, M., Hounsell, N., Graindorge, M., Graindorge, T., & Malhene, N. (2016). Public views towards implementation of automated vehicles in urban areas. *Transportation Research Procedia*, 14, 2168–2177.

Revelt, D., & Train, K. (1998). Mixed logit with repeated choices: Households' choices of appliance efficiency level. *Review of Economics and Statistics*, 80(4), 647–657.

Rödel, C., Stadler, S., Meschtscherjakov, A., & Tscheligi, M. (2014). Towards autonomous cars: The effect of autonomy levels on acceptance and user experience. In: International Conference on Automotive User Interfaces and Interactive Vehicular Applications.

Savolainen, P., Mannering, F., Lord, D., & Quddus, M. (2011). The statistical analysis of highway crash-injury severities: A review and assessment of methodological alternatives. *Accident Analysis & Prevention*, 43(5), 1666–1676.

Sayer, A. (1992). Problems of explanation and the aims of social science. In: *Method in Social Science*. London: Routledge.

Schoettle, B., & Sivak, M. (2014). A survey of public opinion about autonomous and self-driving vehicles in the US, the UK, and Australia. https://deepblue.lib.umich.edu/handle/2027.42/108384

Ulfarsson, G., & Mannering, F. (2004). Differences in male and female injury severities in sport-utility vehicle, minivan, pickup and passenger car accidents. *Accident Analysis & Prevention*, 36(2), 135–147.

Wang, S., Jiang, Z., Noland, R., & Mondschein, A. (2020). Attitudes towards privately-owned and shared autonomous vehicles. *Transportation Research Part F*, 72, 297–306.

Washington, S., Karlaftis, M., Mannering, F., & Anastasopoulos, P. (2020). *Statistical and Econometric Methods for Transportation Data Analysis*. London: CRC Press.

Ye, F., & Lord, D. (2011). Investigation of effects of underreporting crash data on three commonly used traffic crash severity models: Multinomial logit, ordered probit, and mixed logit. *Transportation Research Record*, 2241(1), 51–58.

Yigitcanlar, T., Dodson, J., Gleeson, B., & Sipe, N. (2007). Travel self-containment in master planned estates: Analysis of recent Australian trends. *Urban Policy and Research*, 25(1), 129–149.

Yigitcanlar, T., Kankanamge, N., Regona, M., Ruiz Maldonado, A., Rowan, B., Ryu, A., … , & Li, R. (2020). Artificial intelligence technologies and related urban planning and development concepts: how are they perceived and utilized in Australia? *Journal of Open Innovation*, 6(4), 187.

Yuen, K., Huyen, D., Wang, X., & Qi, G. (2020). Factors influencing the adoption of shared autonomous vehicles. *International Journal of Environmental Research and Public Health*, 17(13), 4868.

Zhu, M., Liu, X., Tang, F., Qiu, M., Shen, R., Shu, W., & Wu, M. (2016). Public vehicles for future urban transportation. *IEEE Transactions on Intelligent Transportation Systems*, 17(12), 3344–3353.

Zmud, J., Sener, I., & Wagner, J. (2016). Consumer acceptance and travel behavior: impacts of automated vehicles. https://static.tti.tamu.edu/tti.tamu.edu/documents/PRC-15-49-F.pdf

Appendix

TABLE 7A.1

Fixed and Random Parameter Ordered Probit Model Results: Combined (Brisbane, Melbourne, and Sydney) Driverless Car Adoption

Explanatory Variable	Fixed Parameters Ordered Probit Model (FOPM)				Random Parameter Ordered Probit Model (ROPM)				Marginal Effects (ROPM)		
	Coefficient (β)	Standard Error	z-stat	Prob (p) |z|>Z*	Coefficient (β)	Standard Error	z-stat	Prob (p) |z|>Z*	Quick Adoption (y=0)	Long Way to Go Adoption (y=1)	Hibernation Adoption (y=2)
Constant	0.434***	0.073	5.950	0.000	1.015***	0.109	9.340	0.000			
• *Standard deviation of parameter density function*					2.038***	0.116	17.600	0.000			
Background and socio demographics											
Age group: 35–55 years	-0.226***	0.085	-2.680	0.007	-0.687***	0.126	-5.470	0.000	0.180	-0.169	-0.012
• *Standard deviation of parameter density function*					0.508***	0.152	3.340	0.001			
Travel behaviour and car ownership											
Mode of transport to work: private car, public transport and shared car	-0.390***	0.146	-2.680	0.007	-0.832***	0.206	-4.040	0.000	0.212	-0.202	-0.010
Mode of transport to school: private car	-0.290***	0.085	-3.400	0.001	-0.647***	0.116	-5.560	0.000	0.169	-0.157	-0.012

Mode of transport to school: public transport	–0.466***	0.141	–3.320	0.001	–1.138***	0.193	–5.890	0.000	0.285	–0.273	–0.012
Car ownership: 2 nos.	0.161*	0.084	1.920	0.055	0.328***	0.114	2.880	0.004	–0.082	0.074	0.009
Car value: $40k–$50k	–0.521***	0.147	–3.560	0.000	–1.133***	0.206	–5.490	0.000	0.283	–0.272	–0.011
Driving licence age: <1 year	–0.792***	0.212	–3.750	0.000	–1.814***	0.313	–5.800	0.000	0.402	–0.390	–0.011
Driving licence age: 1–3 years	–0.686***	0.124	–5.550	0.000	–1.612***	0.185	–8.740	0.000	0.402	–0.390	–0.012
Driving licence age: 3–5 years	–0.565***	0.134	–4.230	0.000	–0.757***	0.189	–4.000	0.000	0.195	–0.186	–0.010
• *Standard deviation of parameter density function*					*3.499****	*0.237*	*14.760*	*0.000*			
Driverless car impact and opportunities											
Number of smart tech use: no	0.616**	0.302	2.040	0.041	1.374***	0.400	3.440	0.001	–0.300	0.201	0.099
Driverless car riding experience: yes	–0.406***	0.134	–3.040	0.002	–0.879***	0.191	–4.610	0.000	0.229	–0.219	–0.010
Driverless car trip purpose: work and school	–0.783***	0.289	–2.710	0.007	–2.015***	0.471	–4.280	0.000	0.418	–0.407	–0.011
Driverless car feature preference over human-driven car: 4 nos.	0.244**	0.111	2.200	0.028	0.544***	0.147	3.700	0.000	–0.135	0.117	0.018

(*continued*)

TABLE 7A.1 (Continued)

Fixed and Random Parameter Ordered Probit Model Results: Combined (Brisbane, Melbourne, and Sydney) Driverless Car Adoption

Explanatory Variable	Fixed Parameters Ordered Probit Model (FOPM)				Random Parameter Ordered Probit Model (ROPM)				Marginal Effects (ROPM)		
	Coefficient (β)	Standard Error	z-stat	Prob (p) $|z|>Z^*$	Coefficient (β)	Standard Error	z-stat	Prob (p) $|z|>Z^*$	Quick Adoption (y=0)	Long Way to Go Adoption (y=1)	Hibernation Adoption (y=2)
Threshold											
Mu (01)	1.293***	0.060	21.580	0.000							
Goodness of Fit statistics											
Number of observations (N)	967				967						
Restricted log likelihood or log likelihood at 0 (LL_0)	−917.531				−917.531						
Log-likelihood function or log likelihood at convergence (LL_B)	−831.397				−827.050						
Akaike information criteria (AIC)	1692.800				1690.100						
AIC/N	1.751				1.748						
Chi-squared (dof = 13)	172.268										
Significance level	0.000										
Halton draws					100						
K	15				18						

***, **, * ==> Significance at 1%, 5%, 10% level

8

Autonomous Vehicle Adoption in Developing Nations

8.1 Introduction

The fields of information and communication technology (ICT) have advanced significantly in recent years, resulting in innovations like artificial intelligence (AI), robotics, quantum computing, Internet-of-Things (IoT), and fifth-generation wireless technologies (5G) (Yigitcanlar et al., 2020). As a result of the obvious effects that these technologies will have on urban areas and other facets of human life, the concept of smart cities is gaining popularity (Manfreda et al., 2021). In the field or transportation, autonomous vehicles (AVs) are viewed as an essential component of the transportation system in smart cities and have been a contentious subject in recent years (Chen et al., 2020). This is such a critical issue that some governments have incorporated it into their macro plans and produced short-term and long-term strategies. In this respect, several developed countries, including Australia, Canada, the UK, the US, and Germany, have been investing in research and development (R&D) of AV technology.

There is no doubt that AVs would have a significant impact and vast market in developing countries, even though they are more likely to be adopted in developed countries initially. However, the findings from research conducted in developed countries may not be transferrable to developing countries due to significant differences in urbanisation and development levels, cultural norms, and transportation rationales (Moody et al., 2020). There are serious drawbacks associated with increasing urbanisation in developing nations' large metropolises. The increased demand for transportation brought on by rapid urbanisation exacerbates existing infrastructure constraints. Number of factors has contributed to this crisis, including a lack of investment in transportation infrastructure, an inadequate supply of public transit options, ineffective traffic management, and a general lack of technical understanding and expertise (Wang et al., 2021).

AVs and other forms of cutting-edge technology are being used in metropolitan areas to alleviate the traffic congestion that comes along with rapid urbanisation. Here, an AI-based traffic system may govern traffic flow in the city and reduce congestion by collecting and analysing real-time and historical traffic data from linked traffic systems. Less time spent idling translates to greater fuel economy and a smaller carbon impact for cars powered by fossil fuels. Analysing data from the infrastructure itself is another use of AI and its subsets in traffic efficiency. In addition to improving safety, deep learning algorithms may help cities save money by predicting when and where repair is required on transportation infrastructure (Soltani & Ivaki, 2011). Therefore, progress in technology might be seen as a potential solution to urbanisation's challenges. At present, these issues are more pressing in developing nations; yet the adoption of these technologies in these regions is hindered by low-quality infrastructures.

Thus, this research hopes to shed light on the most important elements influencing AV adoption in the context of developing nations. Furthermore, the current study addresses the following key research question: What are the most significant factors to AV use in developing countries, with reference to Iran?

Iran is a nation in West Asia with an economy that the World Bank classifies as lower-middle-income. Services contributed the most to gross domestic product (GDP), followed by industry (mining and manufacturing) and agriculture (Soltani & Ivaki, 2011). The population of Iran is above 80 million, and approximately 74.9% of the population lived in urban areas, and the nation is home to eight megacities with populations of a million or more each (Statistics Center of Iran, 2019). These cities' problems stem in part from the high population densities that characterise them, but also from the additional challenges that this phenomenon presents to city governments.

Urban transportation has been plagued by issues that have repercussions for urban residents, including traffic congestion, pollution, noise, mortality from vehicle accidents, rising car ownership rates, and more (Soltani, 2017; Soltani & Askari, 2017; Abdi & Soltani, 2022). AV technology appears to be able to help urban management tackle these issues, as it has the potential to drastically cut down on things like pollution, traffic congestion, transportation costs (including vehicle depreciation, energy, and fuel consumption), infrastructure maintenance, wasted parking space, and accidents.

Due to the state of the existing scientific platform and infrastructure pertinent to the area of AV technology, including road infrastructure, communication infrastructure, legal infrastructure, and so on, Iran is still in the early stages of R&D in this industry. The greatest way to plan forwards and aspire for future architecture under these conditions is to make sure you do not get left behind in terms of technological progress. Since Iran is a developing country, learning about the obstacles it faced while trying to implement AV technology might be instructive for similar nations. They have a comparable degree of urbanisation and level of development, as well as cultural norms and transportation rationale (Villacorta et al., 2014).

Dealing with the future via forecasting and trend analysis has often led to innumerable difficulties in programme implementation (Puglisi & Marvin, 2002). These issues arise when people fail to consider how new technology will influence their daily lives, or when they fail to give enough attention to the forces and circumstances that will ultimately facilitate future difficulties or development challenges. If vital forces and drives are ignored throughout development, they will diminish over time, which will harm the system.

The structure of the chapter is as follows. Following this introduction first, a review of the theoretical literature on AVs is presented. Then, the techniques for data collection and analysis are detailed. Afterward, the research results are presented in three sections, followed by a discussion and conclusion.

8.2 Literature Background

8.2.1 Autonomous Vehicles in a Nutshell

AVs have been defined as systems that can perform all driving activities without the participation of a person. These smart vehicles utilise a combination of AI, computers, software, and hardware sensors that can interact with one another (Collingwood, 2017). Due to the time of adaptation and acceptance of prior technologies such as automatic transmissions and hybrid vehicles, it is anticipated that by 2040, AVs would account for half of the vehicle sales, 30% of vehicle registrations, and 40% of total automobile journeys (Litman, 2022). As a result, it is critical to understand the possible concerns and obstacles to assist a country in planning for the usage of AVs in the future. Here below a concise review of the literature highlights the issues and developments in the field in a nutshell.

For instance, according to Manivasakan et al. (2021), alternative measures and arrangements may be considered for various locations. Kröger et al. (2019) analysed the influence of AVs on German and American travel behaviour and found that the National policy's underlying components impact the acceptability of AVs and the shift in travel demand.

In another study, Faisal et al. (2021) using the Scientometrics method showed that educational institutions performed 87.7% of AV research. Europe is the most productive continent in audiovisual research, accounting for 35.9% of all publications, while with 41.1% of citations, North America is the most significant region in AV research. However, AV research in many settings is still in its infancy, and collaboration and transfer of information between academia and business are quite limited (Faisal et al., 2021).

In an empirical study Butler et al. (2021), by investigating factors affecting public awareness of AVs in Brisbane, Australia, found that public awareness is positively correlated with age and public transport usage, while increasing the number of household vehicles has a negative correlation with AV acceptance. This is to say that AV technology adoption is strongly correlated with technology acceptance (Butler et al., 2021).

A review study conducted by Golbabaei et al. (2021) found that the introduction of dynamic sharing AV (SAV) services may lead to considerable reductions in traffic congestion, travel costs, parking demand, car ownership, and glasshouse gas emissions. Additionally, the good environmental consequences may be reinforced by completely electrifying the SAV fleet by charging with renewable energy. As a smart urban mobility system with dynamic sharing services, SAV integration may enhance sustainability, social fairness, and transportation uptake (Golbabaei et al., 2021).

Additionally, some AV studies were also carried out in undeveloped nations, with an emphasis on public perception and societal acceptability of AVs. For example, in research from Pakistan, Shafique et al. (2021) investigated the adoption of AV technology from the standpoint of Pakistani youth. A questionnaire was issued to 356 students from three Lahore Universities, including Universities of Central Punjab, Engineering and Technology, Management and Technology. The findings indicated that, while AVs have not yet been deployed in Pakistan, most participants are aware of the technology. Furthermore, respondents expressed considerable reservations about going by autonomous public transportation and seeing autonomous trucks on the road, despite admitting most of the benefits associated with AVs. Furthermore, the findings revealed respondents' scepticism about the usage of driverless vehicles by youngsters (Shafique et al., 2021).

Likewise, Ackaah et al. (2021) investigated popular attitudes towards AVs. A poll with 417 participants was undertaken for this aim. According to the findings, most respondents (66.4%) were familiar with AVs. The majority (55.4%) had a favourable attitude towards AVs, and the far majority (78.2%) were eager to test them out. However, respondents still wanted to retain some control over their vehicles if they had prior experience with AVs (Automation Level 3 and below). In addition, respondents voiced anxiety about the safety of their AVs. Finally, the researchers presented several recommendations to the Ghanaian government to secure public acceptance of AVs (Ackaah et al., 2021).

Moreover, a study of the development and deployment of AVs in Singapore showed that formalising safety evaluations and publishing technical standards are among the most consistent and innovative strategies permitted in Singapore for managing unmanned aerial vehicles. In addition, Singapore employs a proactive and responsive adaptive method for AVs. In addition, it reveals the concurrent adoption of two opposing prescriptive-experimentalist approaches to the acceptance of AVs (Tan & Taeihagh, 2021). A summary of the background of the literature is presented in the form of Table 8.1.

Lastly, the review of the literature on AVs has undercover the key factors affecting the use of AVs in the following areas: (a) Social acceptance;

TABLE 8.1

A Summary of the Background Literature

Title	References	Study Field
Infrastructure requirement for autonomous vehicle integration for future urban and suburban roads: current practice and a case study of Melbourne, Australia	(Manivasakan et al., 2021)	Transport infrastructure
Does context matter? A comparative study modelling autonomous vehicle impact on travel behaviour in Germany and the USA	(Kröger et al., 2019)	Travel behavior
Mapping two decades of autonomous vehicle research: a systematic scientometric analysis	(Faisal et al., 2021)	Autonomous driving
Factors influencing public awareness of autonomous vehicles: empirical evidence from Brisbane	(Butler et al., 2021)	Transport technology acceptance
The role of shared autonomous vehicle systems in delivering smart urban mobility: a systematic review of the literature	(Golbabaei et al., 2021)	Smart urban mobility
Public perception regarding autonomous vehicles in developing countries: a case study of Pakistan	(Shafique et al., 2021)	Transport technology acceptance
Perception of autonomous vehicles: a Ghanaian perspective	(Ackaah et al., 2021)	Transport technology acceptance
Adaptive governance of autonomous vehicles: accelerating the adoption of disruptive technologies in Singapore	(Tan & Taeihagh, 2021)	Transport technology adoption

(b) Infrastructure; (c) Policy and legislation, and; (d) Technology and innovation. These areas are elaborated on in detail in the next sections.

8.2.2 Social Acceptance

The adoption of new technology has always been influenced by how individuals think and act. The public's acceptance of AVs is essential for their rapid and broad adoption (Liljamo et al., 2018; Faisal et al., 2019). In general, specialists and the public have a favourable view of AVs, although they have serious reservations. Because there are no human drivers, certain members of society may be sceptical about the safety of AVs in their early phases of development (Schoettle & Sivak, 2014). However, human mistakes cause at least 90% of all vehicle accidents; therefore, the arrival of AVs on the road can remove or diminish the major source of accidents, i.e., human error. In addition, these vehicles are faster in making decisions and conducting driving-related tasks than human drivers. However, the manufacture and sale of AVs will introduce additional safety concerns (Collingwood, 2017). The deadly accident involving Tesla's AVs in 2016 demonstrated the lack of safety in AVs as well as the incapacity of technology to avert mishaps in certain conditions (Coca-Vila, 2018; Butler et al., 2020). On the other hand, packed streets with heavy traffic and the presence of people will provide difficulty for AVs (Raza, 2018). As a result, public adoption of this technology may be hampered by concerns about collisions between AVs and human-centred vehicles in mixed traffic, as well as pedestrians (Nikitas et al., 2019).

Although the safety features of AVs are a priority, the issue of security has received little attention. Indeed, another fear that might dramatically impede AV acceptability is the technology's vulnerability to cyber and privacy threats (Parkinson et al., 2017). According to a 2015 poll of 5,000 individuals in 109 countries, AV users with all degrees of autonomy were more concerned about hacking and exploiting vehicle software (Kyriakidis et al., 2015). The California Consumer Watchdog organisation has expressed privacy issues for AV users. This problem poses five data-related concerns: Who should have ownership of or control over vehicle data? What sort of information will be saved? Who will have access to this dataset? How will such information be made available? And how will this information be used? (Fagnant & Kockelman, 2015)].

Another issue with societal acceptance of AVs is 'responsibility'. There are numerous uncertainties surrounding AVs right now, such as who is accountable in the event of an accident (Bichiou & Rakha, 2019). Because of these uncertainties, people have been hesitant to adopt this new technology. Further explanations on 'cyber security', 'privacy', and 'responsibility' will be provided in Section 8.2.4 (policy and legislation). In other circumstances, the navigation system of AVs is sensitive to severe weather conditions, e.g., fog, rain, and snow, and dark regions, e.g., the entry or departure of a road tunnel, limiting people's adoption of this technology (Chehri & Mouftah, 2019; Golbabaei et al., 2020). It is critical to have appropriate knowledge about new technology and the benefits of AVs to enhance the adoption of AVs. The individual and social advantages of utilising AVs have a beneficial impact on the technology's acceptability. Users are typically attracted to benefits that save them time and money (Whittle et al., 2019), such as shorter travel times, less congestion, and greater fuel economy.

Furthermore, the widespread use of AVs has benefits not only for individuals but for society, such as reducing emissions (Bansal et al., 2016), including lower CO_2 emissions, and reducing energy consumption and reliance on fossil fuels (Nikitas et al., 2019), leading to a reduction in environmental footprint, as well as improving living conditions (Millard-Ball, 2019). Other benefits include increased independence of vulnerable social groups such as the elderly, children, and the disabled (Nikitas et al., 2019), as well as improved safety and more efficient traffic flow (Papadoulis et al., 2019). AVs lower the need for parking space, especially since the introduction of AVs, which allow customers to enter and exit a location without having to search for and pay for parking space (Millard-Ball, 2019). In addition, consumers can avoid the high cost of purchasing, maintaining a vehicle, and paying only the annual fee or payment per use (Haboucha et al., 2017). It is anticipated that the cost per kilometre of shared AVs would be not only less than taxi fares, but also less than the cost of individual vehicle ownership (Xu & Fan, 2019). Shared autonomous public transportation is cited by researchers as a possible replacement for conventional modes of transportation (Ribeiro et al., 2022). This is both ecologically friendly and economical for households (Wang & Zhao, 2019).

In general, the benefits experienced from a personal or societal standpoint have a favourable influence on AV acceptability. A good view of AV technology enhances trust in the technology while decreasing risk perception among potential customers (Kassens-Noor et al., 2021).

Trust is the strongest determinant of performance hope and essential to reducing risk perception (Xu & Fan, 2019). Therefore, researchers have suggested trust as a determining factor for the acceptance of AVs (Wang & Zhao, 2019; Ribeiro et al., 2022). Some researchers argue that AVs are not yet commercialised and that most end users are unfamiliar with this technology. As a result, media and public advertising play an important role in moulding end-user perceptions about AVs (Anania et al., 2018). Another concern is the degree of individual readiness, which is affected by a variety of characteristics such as adult literacy, digital literacy, the number of Internet users, and the number of active social media users, as indicated in many studies. Residents of major cities are more inclined to adopt this technology since living in big cities is closely tied to access to facilities. As a result, the social acceptability of new technology will grow in these cities (Threlfall, 2020; Dennis et al., 2021).

8.2.3 Infrastructure

The first area is to concentrate on the geometric design of roadways and adjust their cross-sectional dimensions in preparation for the introduction of AVs. The reduction in needed lane width is one of the projected changes in road layout with the advent of AVs (Engholm et al., 2018). This is due to AVs' enhanced lateral control as compared to conventional vehicles, which means that the safe cross-section for AVs can be deemed less than that of conventional vehicles. Based on current regulations, the cross-sectional area of highways is 3.5 m, which is high for the movement of AVs. According to studies, construction plans and standards should be re-evaluated, improved, and updated to guarantee that they can respond to altered loading patterns and higher loads associated with AVs (Manivasakan et al., 2021). Regardless of the type of vehicle, an AV can drive at faster speeds than a conventional vehicle. This may result in higher traffic flow and greater pavement loads. As a result of increased traffic loads and vehicle movement at high speeds, the construction and materials of roads, such as road pavements and bridges, necessitate new requirements (Butler et al., 2020; Chen et al., 2020).

The proper operation of AVs necessitates the use of road signs and markers. Clear and consistent markers are required for AVs to prevent misunderstanding and for sensors to interpret road information (Lee et al., 2022). As a result, cities should have regular markings and quality road signs that are not readily worn out and can be seen properly (Khan et al., 2019). Furthermore, various infrastructure adjustments are suggested, such as the placement of a transmitter on the road surface, which allows for a more effective diagnosis to avoid misinterpretation, for example, the identification of a painted white line and a genuine line. In recent years, a digital alternative known as Vehicle to Everything Technology (V2X) has gained popularity (Kuutti et al., 2018). This technology enables digital mapping to include information such as road signs and lines, road conditions, and traffic (Rubin et al., 2019). This technology includes communications such as vehicle-to-vehicle (V2V), vehicle-to-infrastructure (V2I), and vehicle-to-pedestrian (V2P) (Rubin et al., 2019).

Facility maintenance is critical to ensuring that road infrastructure remains useable for the duration of its design life. Over time, road markings and pavements deteriorate, resulting in the removal of lines, pits, and vague signs. Any deficiency in road infrastructure can lead to inconsistency. As a result, it may impair AVs' capacity to navigate, perhaps leading to accidents (Carreras et al., 2018). Parking facilities and road lines that are suitable for AVs should also be created. Furthermore, the provision of specialised places for AV transportation aids in infrastructure preparation because AVs act differently than human drivers, and having dedicated spaces makes them easier to utilise. High-tech roadside amenities are among the other necessary road constructions. Because most AVs are likely to be electric, a considerable number of high-tech roadside amenities, such as electric vehicle charging stations, are necessary (Khan et al., 2019; Butler et al., 2021).

The availability of an integrated Internet for the successful transmission of AV-related data is an important part of communication infrastructure readiness. AVs may interact with the infrastructure network by utilising the coverage of the fifth generation of 5G wireless technologies. They can successfully communicate and receive important safety information to and from a centralised mobile control centre thanks to their high-speed Internet connection (Khan et al., 2019). The data transfer speed with 5G will be 100 times quicker than with 4G. As a result, because AVs generate and consume a large amount of data, reaction times will be substantially faster (Chehri & Mouftah, 2019).

Cities should have local data centres for efficient processing and analysis of such massive amounts of big data. A data centre of this type may act as a command-and-control centre for all intelligent transportation system (ITS)-related operations, as well as aid AVs while driving (Khan et al., 2019).

The ITS is a network-based system in which the vehicle serves as a router, transmitter, and receiver (Zeadally et al., 2012). As a result, developing an acceptable ITS system in cities, such as putting cones and lights coupled with sensors, is crucial.

8.2.4 Policy and Legislation

The future orientation of government in each country is defined based on average measures in the fields of policy stability, government accountability for change, adaptability of the legal framework to change, and long-term government perspective (Threlfall, 2020), which plays a critical and effective role in the maturity of new technologies and government readiness. It demonstrates that national policies have a significant impact on the acceptability of AVs (Kröger et al., 2019), and the efficacy of law is a difficult problem that must be addressed. Indeed, legal difficulties are a key worry for the usage of AVs, since duties should be explicit in the event of a system failure (López-Lambas, 2017). Responsibility is an important aspect since it is intertwined with insurance. According to some experts, the purpose of such a regulation is to resolve the question of who is liable for the costs of accidents. Is it the victim's fault, the fault of another actor, or the fault of both (Schellekens, 2015)? The driver oversees the vehicle in most accidents involving regular automobiles.

As a result, he bears primary responsibility for the vehicle's fate, although individuals in an AV no longer have control over the vehicle. As a result, it appears that manufacturers and third parties engaged in the design of AV safety systems may face increased liability in associated cases (Collingwood, 2017). As a result, automakers are more at risk of tarnishing their brand because of accidents caused by design and manufacturing flaws in their vehicles (Collingwood, 2017). There is currently no legislative structure in place to determine how responsibility for the design and manufacturing of AV systems should be divided among third parties such as the manufacturer, component supplier, software developer, or software user. This makes identifying and distinguishing the many components and circumstances that have caused the vehicle to malfunction challenging (Collingwood, 2017). Another unresolved problem is the transfer of liability to the insurer and its impact on insurance premiums (Hevelke & Nida-Rümelin, 2015).

Hackers are drawn to AVs because of their capacity to store and send data. Because the computer controls most of the vehicle, such vehicles are more vulnerable to cyber-attacks than standard automobiles, and the driver has less ability to interfere during the assault. Without adequate security in the communication routes between these vehicles, such vehicles are vulnerable to hacking, which can result in serious collisions. Entering phony messages and tampering with such vehicles' worldwide satellite navigation systems are the most serious risks that AVs may face because tampering with this satellite data might impair the vehicle's vital safety functions (Bagloee et al., 2016; Butler et al., 2022). Other cyber dangers include the manipulation of vehicle sensors to mislead decision-making systems, the shutdown of cameras, and the creation of radar news to conceal the obstructions in the path of vehicle cameras (West, 2016).

The performance of AVs is dependent on cameras, sensors, precise maps, and other equipment that collect data about the vehicle. AI in the vehicle system then optimises this information to assure the vehicle's performance (West, 2016). However, there are significant worries regarding how to regulate and use this information. Some concerns about information confidentiality remain unresolved. The reasons for collecting such information, the categories of information obtained, third-party access to this information, and the duration permitted to retain the gathered information have not yet been stated. The capacity of AVs to interact with one another allows for the exchange of information acquired by each vehicle for the safe operation of such vehicles. However, it exposes the vehicle's movements and position to detection by external networks, allowing third parties to get access to the user or occupant of the AVs (Glancy, 2012). The use of black boxes to explain the cause of an accident is another source of concern, since this information might be sold to other parties such as insurance companies and used against drivers.

Other concerns associated with information confidentiality include the possibility of exploiting information obtained by the vehicle system to harass AV users via marketing and advertising, stealing users' identities and profile images, and so on (Glancy, 2012). Another problem is the use of camera monitoring in AVs, such as those employed in self-driving taxis. Because the occupants of AVs are not the proprietors, it is unclear whether these vehicles are deemed public spaces and whether public monitoring of them can be tolerated (Schoonmaker et al., 2016).

Governments should collaborate with manufacturers and research institutions to welcome this new mobility while ensuring that related regulations are as safe as feasible (Ruggeri et al., 2018). To build a legislative framework that provides minimal safety, large-scale AV experiments on public roads must be encouraged. Furthermore, this testing may result in technological advancements and legal laws (de Bruyne & Werbrouck, 2018). Thus, investing in AV infrastructure is a vital aspect of promoting the deployment of driverless vehicles, as technological restrictions now prevent AVs from fully functioning on ordinary roads, requiring them to be driven in specialised infrastructure. This issue necessitates the establishment of specific test facilities for such vehicles, where complete external assistance for their operation may be supplied (Khan et al., 2019).

Furthermore, because of the specialised and extremely complicated repair, maintenance, and operation, specialist crew training is critical. Thus, encouraging colleges to perform research on AVs and perhaps including this topic in university curricula might be a beneficial step forward. Motor vehicle ownership and use legislation, as well as traffic laws, should be amended including the development of laws and regulations which prioritise the use of AVs (Raza, 2018). One of these regulations may permit AVs to operate high-occupancy vehicles (HOV) even if only one person is present. Similarly, roadways in downtown regions are typically congested, and allowing only AVs to access these routes will encourage individuals to utilise such vehicles. Aside from such laws, local administration can provide AVs the ability to park in private spaces, even in congested locations, or offer toll savings in contrast to conventional vehicles (Khan et al., 2019; Yigitcanlar et al., 2022).

8.2.5 Technology and Innovation

KPMG, one of the world's leading consulting, financial consulting, tax, and auditing organisations, has released statistics on the preparedness of countries to deploy AVs, taking the following technological and innovative factors into account: Industrial relationships between automakers and suppliers of automotive technology. Indeed, the rapid and disruptive nature of AV technology has led to the formation of several partnerships between automakers and technology suppliers. In addition to the number of AV technology companies that are engaged in investing and testing for this new technology. The number of patents associated with AVs, the amount of industrial investment in AVs, the cybersecurity of infrastructure, the allocation of a portion of the automobile market to electric vehicles, cloud computing, AI, and the Internet of Things are being evaluated (Threlfall, 2020).

Humans have long envisioned 'thinking' devices that operate without human assistance. In the late 1950s, the introduction of the first computers amplified this interest and made it possible for the project to focus on the emerging field of AI. By developing three-dimensional maps of their environs, AVs gain knowledge from their surroundings. Such maps enable AVs to have a complete understanding of their position in real-time. Using radar and lidar together, three-dimensional pictures are captured. After the photographs have been captured and filtered, it is the responsibility of the computer system to appropriately analyse their meaning and assess the worth of the gathered data. AI arrives at this point (Chehri & Mouftah, 2019).

Such vehicles need AI to comprehend, learn, and navigate the actual world. AI should get data via many channels and make decisions based on analysis. Consequently, this procedure can be divided into four phases: environmental comprehension, decision-making, route design, control, and navigation. Cloud computing is heavily used in AV decision-making. In addition, they leverage cloud computing on IoT to handle traffic data, meteorological information, maps, and road conditions. This offers a deeper awareness of the situation and the surrounding environment, allowing for more effective decision-making. Due to the complexity and difficulty of urban environments, particular research is needed in the fields of big data, sensor technologies, IoT, cloud computing, and AI to build the necessary infrastructure for their administration (Villagra et al., 2017).

8.2.6 Conceptual Framework

Considering the review, this study formed a conceptual framework for AV adoption with reference to Iran. Based on this conceptual framework, the variables affecting the usage of AV in the four categories of social acceptability, infrastructure, policy and legislation, and technology and innovation have been explored (Figure 8.1). Each of these constructs contains number of specific components that may be assessed using an expert opinion survey. For instance, five measures are stated as follows for social acceptance: (a) Users' trust; (b) Media and publicity; (c) Cost-effective use of AVs; (d) Individual readiness, and; (e) Social benefits of using AVs.

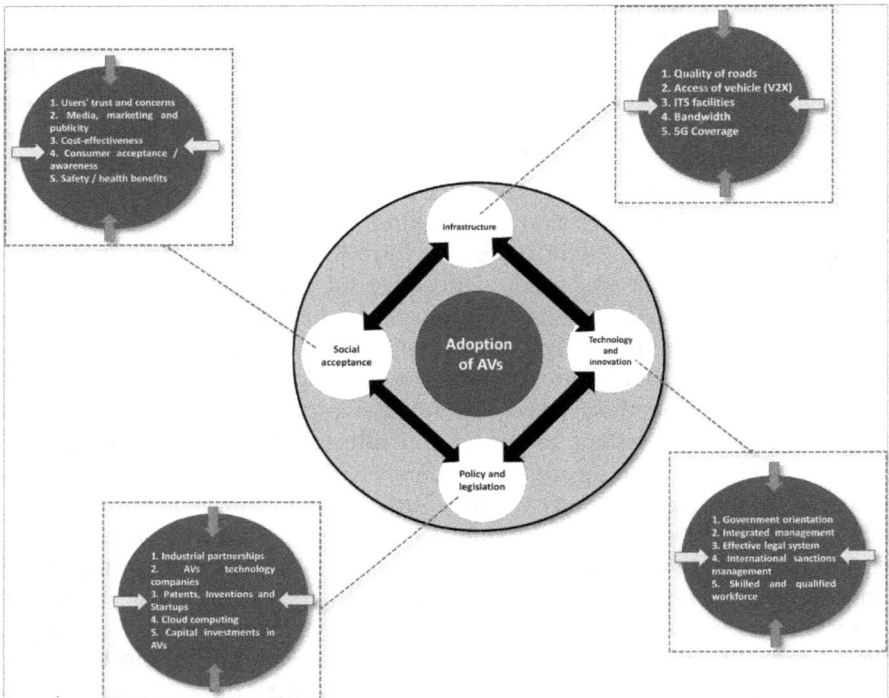

FIGURE 8.1
Conceptual framework of AV adoption.

8.3 Research Design

This research used a mixed methodological strategy involving both quantitative and qualitative analyses. To achieve the study objective, structural analysis was used. There is a network of interrelationships between variables in this system, and their examination is crucial to comprehending the future development of the system. The significant outcome of structural analysis is the discovery of the essential factors that regulate the development of AVs (Nematpour & Faraji, 2019). The structural analysis consists of three steps: creating an inventory matrix of factors, describing the relationships amongst the factors, and then identifying key factors (Godet et al., 2013).

8.3.1 Step 1: Identifying Factors

Based on the literature, there are four major categories of societal acceptability, infrastructure (roads and communication), policy and legislation, and technology and innovation, as well as around 54 long-listed factors. The Fuzzy Delphi approach was then utilised in two phases to find additional factors and assess the initial 54 factors. Delphi methodology is a way of acquiring collective knowledge that has been used for over half a century. Although experts' mental talents and abilities are employed for comparisons in traditional Delphi methods, the quantification of expert judgements cannot fully capture the human way of thinking. Using fuzzy sets is more consistent with human language and often imprecise descriptions, and it facilitates better decision-making in the actual world. Using fuzzy sets, the error rate is minimised (Soltani et al., 2018). The questionnaire was distributed to all experts at each phase as a combination of closed and open questions. Closed questions use the Likert scale to judge the importance of the long list of factors. Expert-suggested factors are the subject of open questions. The long list of factors that did not score in the first stage of Delphi was deleted for the second stage.

Eight of the first 54 long-listed factors were eliminated during the first round of Delphi, and the first 3 were recommended by experts. More details of the panel are provided in Table 8A.1. The second round authorised 49 long-listed factors in four targeted sectors. In this study, 25 experts from different fields of expertise were members of the Delphi panel. All 25 experts have participated in two Delphi rounds. More details of the panel are provided in Figure 8A.1.

To choose appropriate experts, a non-probabilistic snowball sampling method was used. The snowball sampling strategy guides the researcher to additional individuals who may be able to contribute to the study topic until the researcher reaches a point of repetitive responses (Handcock & Gile, 2011).

8.3.2 Step 2: Describing Relationships among Factors

A two-dimensional matrix known as the 'Structural Analysis Matrix' was utilized in structural analysis to determine the correlations between factors. This matrix should ideally be populated by experts who have already participated in the first stage. The filling of the matrix was a qualitative procedure. The following question was presented for each pair of factors: Is there a direct connection between factors i and j? The factors in the row influence the factors in the column. The degree of association is represented by a number between 0 and 4. Number 'zero' denotes 'no effect', number 'one' denotes 'weak effect', number 'two' denotes 'moderate impact', number 'three' denotes 'strong effect', and lastly number 'four', even if no association exists now, may exist in the future (Godet et al., 2013).

As a result, if the number of recognised factors is n, an n × n matrix containing the influence of the factors on each other will be generated. In this work, a matrix having 49 × 49 dimensions was developed. Only five volunteer experts from the previous phase were employed for scoring due to the high dimension of the matrix, the time needed to fill in the matrix, as well as to simplify the computations. According to Clayton (Habibi et al., 2015), between five and ten people are enough if a mix of specialists with various expertise is utilised. The degree of filling the matrix was 84.4%. Based on statistical indicators, the data had 100% usefulness and optimisation, showing the questionnaire's excellent dependability (Table 8.2).

8.3.3 Step 3: Identification of Key Factors

This stage entails finding and re-ranking the critical factors, i.e., the factors required for the AV's development. The key factors were extracted using MICMAC (Matrix of Crossed Impact Multiplications Applied to a Classification) (Godet et al., 2013). The software output includes the Matrix of

TABLE 8.2

Details on the Dimensions of the Structural Analysis Matrix

Indicator	Value
Matrix size	49
Number of iterations	2
Number of twos	637
Number of threes	632
Number of P or four	0
Number of zeros	374
Number of ones	758
Total	2027
Response rate	84.4%

Direct Influences (MDI) and its associated graphs, and the Matrix of Indirect Influences (MI) and its related graphs. If the structural analysis matrix reveals possible links between factors, the programme generates a Matrix of Potential Direct Influences (MPDI) and a Matrix of Potential Indirect Influences (MPII), as well as accompanying graphics. The amount of the impact and influence of each factor may be displayed in a two-dimensional graph with the *x*-axis representing influence and the *y*-axis representing impact (Figure 8.2). In addition to finding the most influential factors in the system, the different mappings of such factors may be studied (Godet et al., 2013).

Their placements illustrate the various roles performed by system factors (input factors, intermediate factors, resultant factors, excluded factors, and clustered factors). These maps illustrate the current and future participants' perceptions of the system's potentials (factors with high influence and dependency capacity), opportunities (factors with moderate influence and dependence capacity), and constraints (factors that cannot be affected) for change. By creating influence graphs, the networks or loops of connected components are formed (Delgado-Serrano et al., 2016) (a description is given in Table 8A.2).

And finally, node connection analysis has been done using Gephi software version 0.9.0 to better display the relationships between factors.

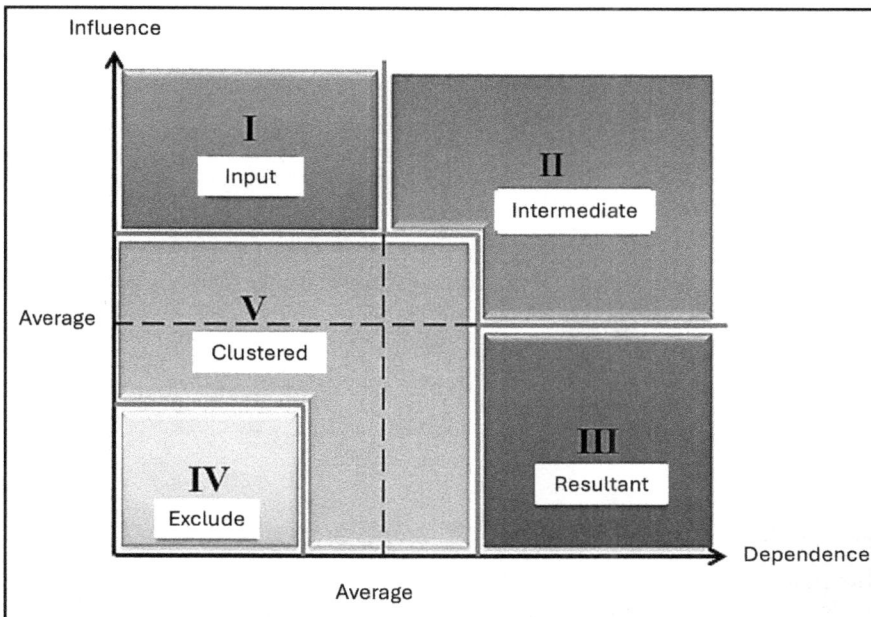

FIGURE 8.2
Different types of factors on the matrix with axes of influence and dependence.

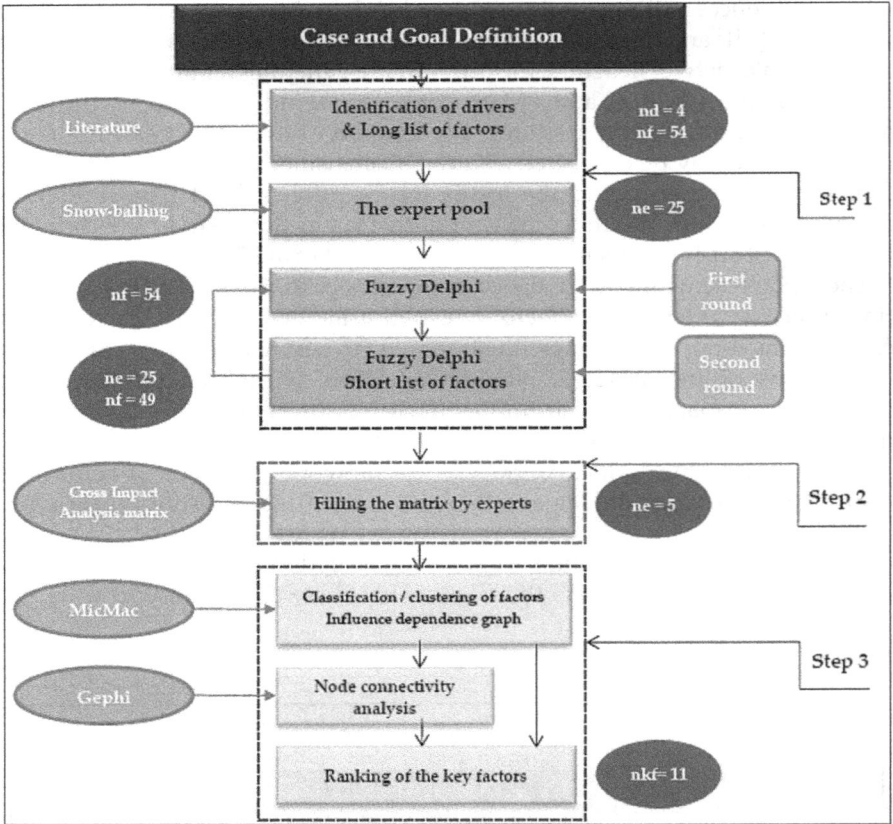

FIGURE 8.3
The research process (nd= number of drivers, n e= number of experts, nf= number of factors, nkf= number of key factors).

Based on the provided explanations, the research process is presented in Figure 8.3.

8.4 Results

8.4.1 Step 1: Identifying Factors

This phase entails developing a list of factors that is as complete as feasible. The long list can be divided into four categories: policy and legislation, technology and innovation, infrastructure (communication and roads), and societal (public) acceptability. A total of 49 factors were discovered (Table 8.3).

TABLE 8.3

Factors Affecting the Development of AVs, Fuzzy Delphi Results

Drivers	Factors	Abbreviation	Delphi Value of Each Factor	Percentage of Consensus	Amount of Weight	Rank
Social acceptance	Sufficient technical knowledge of AV technology	A1	3.438	56.67	0.020	30
	Users' trust in the companies responsible for introducing and selling AVs	A2	3.260	60	0.019	42
	Media and marketing	A3	4.013	70	0.023	6
	Resistance to lifestyle change among potential customers	A4	3.708	53.33	0.021	16
	Residence in larger cities (cosmopolitans)	A5	3.222	53.33	0.018	43
	Concerns related to cybersecurity and endangering privacy and personal information	A6	3.030	63.33	0.017	49
	Concerns about AVs accident with other vehicles (autonomous—ordinary)	A7	3.554	53.33	0.020	24
	Concerns about AVs accidents with pedestrians	A8	3.364	63.33	0.019	36
	Social acceptance of new technologies	A9	3.871	60	0.022	9
	Cost-effectiveness compared to human-driven vehicles	A10	4.250	70	0.024	3
	Individual readiness	A11	3.459	62.07	0.020	27
	Awareness of the environmental benefits of using AVs	A12	3.0634	56.67	0.017	47
	Awareness of the social benefits of using AVs	A13	3.165	51.72	0.018	45
	Legal and technical gaps related to AVs	A14	3.298	56.67	0.019	41
	Feasibility of using AVs in collective and shared modes	A15	3.069	60	0.017	46

(continued)

TABLE 8.3 (Continued)

Factors Affecting the Development of AVs, Fuzzy Delphi Results

Drivers	Factors	Abbreviation	Delphi Value of Each Factor	Percentage of Consensus	Amount of Weight	Rank
Infrastructure	Quality of streets and roads (quality of markings and signs, quality of asphalt, etc.)	B1	3.666	74.19	0.021	17
	Revision in the geometric design of roads	B2	3.413	53.33	0.019	32
	Revision in road paving standards	B3	3.730	51.72	0.021	15
	Revision in the design and standards of bridges	B4	3.0399	51.72	0.017	48
	Revision in the design of parking spaces	B5	3.221	51.72	0.018	44
	Backup equipment, data clouding, and maintenance	B6	3.575	62.07	0.020	21
	Collecting and transmission of online data on road conditions	B7	3.321	60	0.019	40
	Supplying roadside service stations (charging and repair stations)	B8	3.597	63.33	0.020	20
	5G Internet coverage	B9	3.839	58.62	0.022	10
	V2X (passing information from a vehicle to any entity)	B10	3.989	55.17	0.023	7
	Availability of local data centres	B11	3.781	53.33	0.022	13
	ITS facilities	B12	3.624	66.67	0.021	19
	Bandwidth	B13	3.824	56.67	0.022	11
Policy and legislation	Developing transparent supportive laws/regulations for the use of AVs in the field of priority and right of way, insurance, and certification	C1	3.444	58.62	0.020	29
	Developing legislation in the field of criminals, including responsibility for road accidents, privacy, and cybersecurity	C2	3.555	62.07	0.020	23
	Supporting R&D activities	C3	3.637	68.97	0.021	18
	Funding mega projects on AVs	C4	3.475	63.33	0.020	26

Category	Description	Code				
	Providing incentives to attract private-sector investment	C5	3.4156	60	0.019	31
	Reduction of tariffs for importing AVs vehicles, systems, and technologies	C6	3.458	53.33	0.020	28
	Encouraging PPT (public and private partnership) in developing AVs systems	C7	3.331	53.33	0.019	39
	Supporting skilled and qualified workforce (in the areas of manufacturing, repair, maintenance, and operation)	C8	3.358	56.67	0.019	37
	Funding support of research/education activities	C9	3.764	60	0.021	14
	Managing the international sanctions for foreign investment	C10	3.368	63.33	0.019	35
	future orientation of government (stability and accountability of policies, the long-term vision of the government for developing AV industry)	C11	3.568	63.33	0.020	22
	Supporting transparency and accountability in the AV market	C12	3.818	66.67	0.022	12
	Establishing a coordination plan among stakeholders and influential parties	C13	3.983	73.33	0.023	8
	Refining and developing the legal system in addressing the AVs regulations	C14	3.399	53.33	0.0194	33
Technology and innovation	Tendency to mass production and industrial partnerships	D1	4.277	66.67	0.0244	2
	Number of patents, innovations, and products related to AVs	D3	3.340	63.33	0.0194	38
	The volume of investment in AVs	D4	4.143	56.67	0.024	5
	Development of high-level cybersecurity protection	D5	3.487	60	0.020	25
	Development of IoT	D6	4.328	76.67	0.025	1
	Number and quality of research publications/outputs and training/education activities	D7	3.391	65.52	0.019	34

8.4.2 Step 2: Describing Relationships among Factors

The total of the row values for each factor in the structural analysis matrix reflects the degree of influence, and the column sum of each factor reveals the degree of dependence on other factors. According to the analytical results of this matrix, the influence of 'Policy and legislation' and 'Social acceptability' among the recommended factors is significantly more than their dependence and has a significant influence on the system. On the contrary, 'infrastructure (communication and road network)' and 'technology and innovation' are positioned differently from the other two drivers. In other words, their dependence much outweighs their influence. The numerical disparity between the influence and dependence of the 'Legal and policy' drivers is more obvious among the other drivers (Table 8.4).

8.4.3 Step 3: Identification of Key Factors

The factors distribution reveals the degree of stability or instability of the researched system. In general, there are two sorts of distributions: stable and unstable systems. The distribution of components in stable systems is as L in English; that is, some factors have a high influence, and some have high dependence; However, in unstable systems, the issue is more complicated than in stable ones. The factors in this system are dispersed along the diagonal axis and frequently represent an intermediate state of influence and dependence, making it extremely difficult to evaluate the key factors. However, Figure 8.2 is offered to identify key system factors (Zali et al., 2018). The map of scattering the factors which affect the development of AVs in Iran showed the instability state of the system (Figure 8.4). Since the results of direct and indirect influences of the factors were the same, only the results obtained from the MDI were presented. Figure 8.4 presents the distribution map of long-listed factors and their position in the axis of influence-dependence.

Gephi software version 0.9.0 has been used to better display the relationships between factors. Figure 8.5, which is called the 'weighted-out-degree' graph, shows the degree of influence of the factors on each other. Nodes that have a larger weighted out-degree value have more influence than other nodes. Weighted out-degree is a measure of the system-wide influence that a node has (Hua et al., 2019). Based on Figure 8.5 and Table 8.4, the future orientation of government (stability and accountability of policies, the long-term vision of the government for developing AV industry) (C11), Managing the international sanctions for foreign investment (C10), Funding mega projects on AVs (C4) have the highest influence on other factors.

Figure 8.6, which is called a 'weighted-in-degree' graph, shows the dependence of each factor on other factors. Nodes with high in-degree are in the centre which can affect others easily. The weighted in-degree is a measure of the system-wide influence that a particular node has (Hua et al., 2019). Based on Figure 8.6 and Table 8.4, the feasibility of using AVs in collective and

TABLE 8.4

Ranking of Long-Listed Factors Based on the Degree of Influence and Dependence

Drivers	Factors	Influence	Rank	Dependence	Rank
Social acceptance	A1	75	23	65	41
	A2	63	36	110	3
	A3	111	10	91	14
	A4	67	32	105	7
	A5	103	12	4	49
	A6	62	37	81	28
	A7	82	17	97	12
	A8	82	18	96	13
	A9	71	28	110	4
	A10	115	8	106	6
	A11	55	42	66	40
	A12	74	24	75	36
	A13	71	29	75	37
	A14	66	35	36	45
	A15	113	9	130	1
Infrastructure	B1	59	40	98	16
	B2	53	43	89	20
	B3	53	44	89	21
	B4	53	45	89	22
	B5	49	48	90	16
	B6	67	33	102	9
	B7	59	41	90	17
	B8	51	47	90	18
	B9	68	31	88	23
	B10	118	6	87	24
	B11	60	39	90	19
	B12	119	4	87	25
	B13	67	34	79	33
Policy and	C1	117	7	80	31
legislation	C2	119	5	83	26
	C3	84	16	79	34
	C4	126	3	78	35
	C5	86	14	80	32
	C6	47	49	74	38
	C7	81	19	81	29
	C8	74	25	82	27
	C9	78	20	81	30
	C10	132	2	7	48
	C11	139	1	42	43
	C12	53	46	37	44
	C13	76	22	32	47
	C14	62	38	33	46
Technology and	D1	77	21	105	8
innovation	D2	85	15	101	10
	D3	72	27	91	15
	D4	87	13	107	5
	D5	70	30	52	42
	D6	104	11	73	39
	D7	73	26	115	2

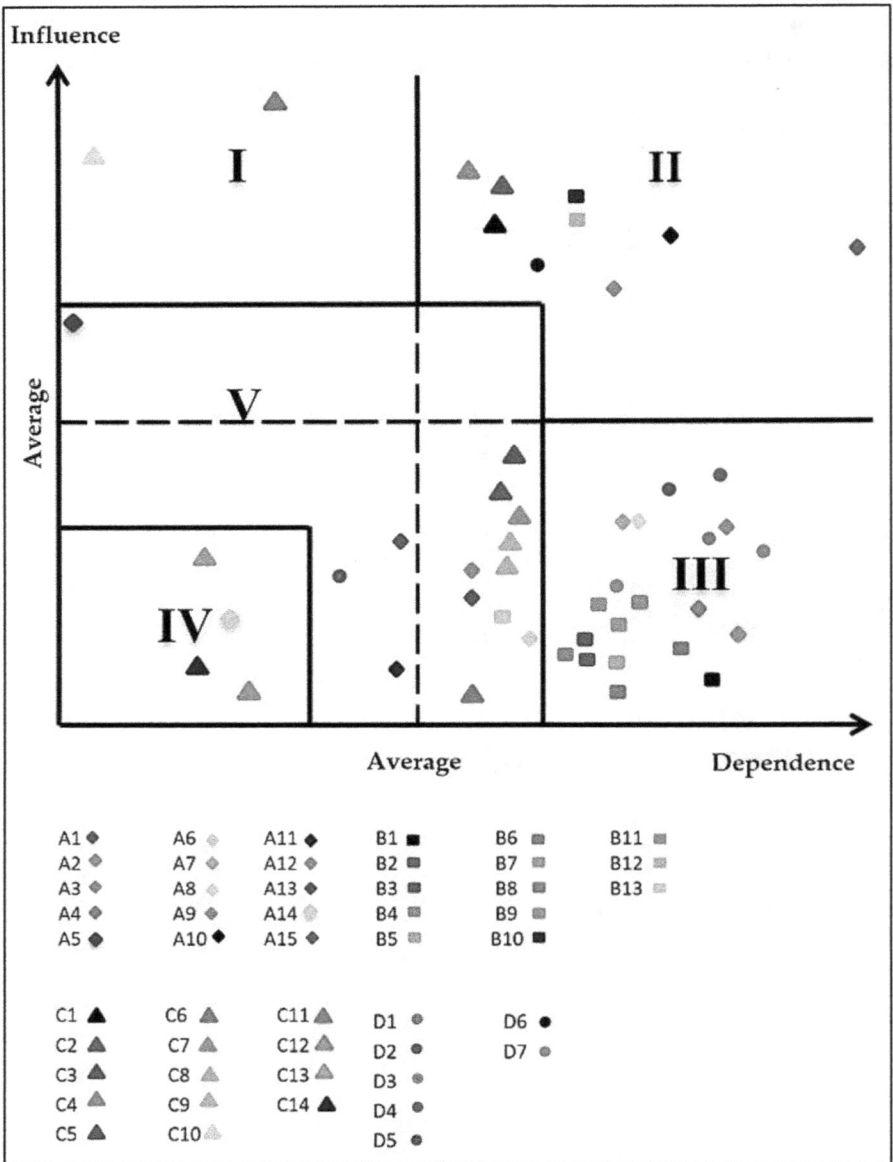

FIGURE 8.4
Distribution map of long-listed factors and their position in the axis of influence-dependence.

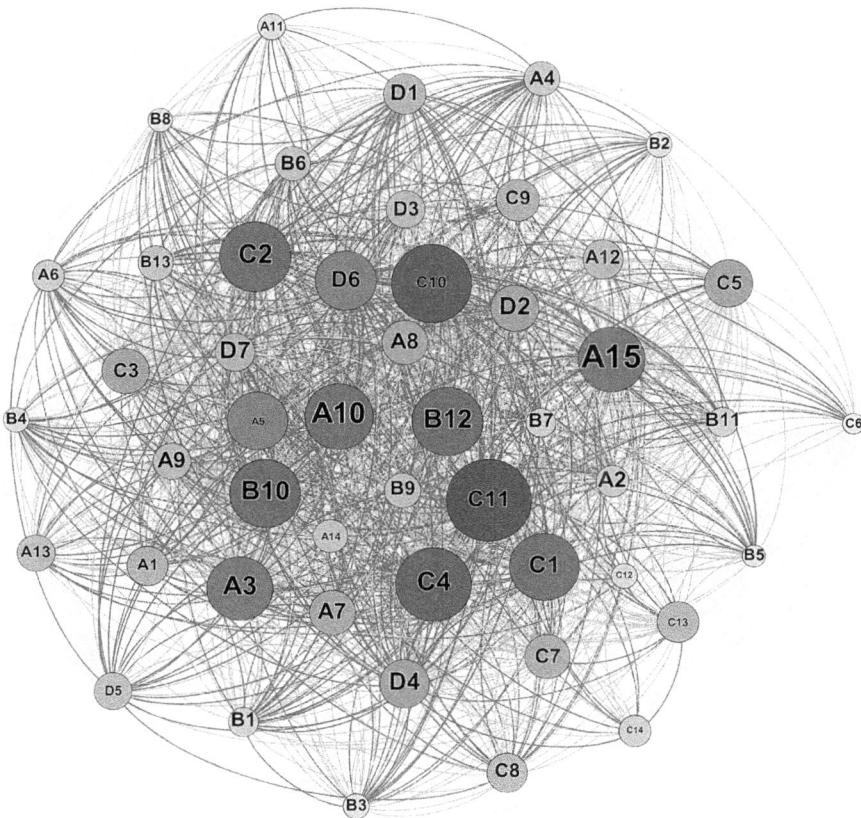

FIGURE 8.5
Weighted-out-degree.

shared modes (A15) and the number and quality of research publications/ outputs and training/education activities (D7) receive the highest influence from other factors.

Thus, the status of each factor according to their position in Figure 8.4 can be identified in the following (Table 8.5).

Based on Table 8.5 and the explanations provided in the appendix, the influence and rank of the key AV adoption factors are as shown in Figure 8.7.

8.5 Findings and Discussion

In this chapter, the critical factors and barriers impacting the development of AVs in Iran are identified: Societal acceptance, Infrastructure (roads and

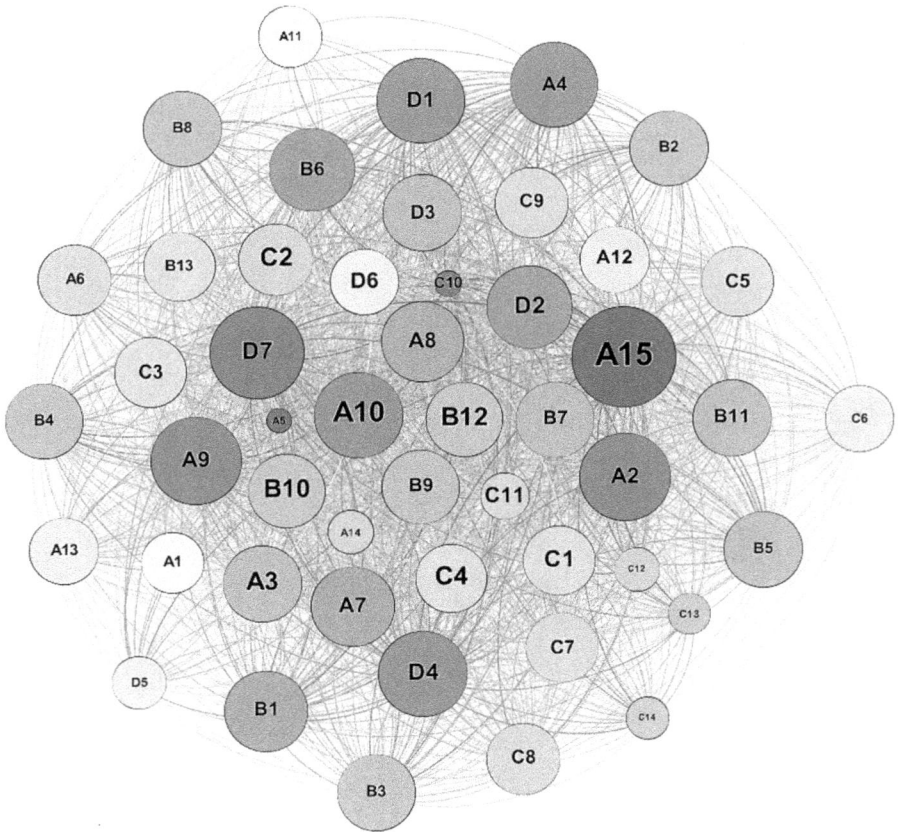

FIGURE 8.6
Weighted-in-degree.

TABLE 8.5

The Position of Each Factor in the Influence-Dependence Map

Position	Factors
Input factors (1)	C10, C11
Intermediate factors (2)	C1, C2, C4, A3, A10, A15, D6, B10, B12
Resultant factors (3)	D1, D2, D3, D4, D7, B1, B2, B3, B4, B5, B6, B7, B8, B9, B11, A2, A4, A7, A8, A9, C8
Excluded factors (4)	C12, C13, C14, A14
Clustered factors (5)	A1, A5, A6, A11, A12, A13, D5, C3, C5, C6, C7, C9, B13

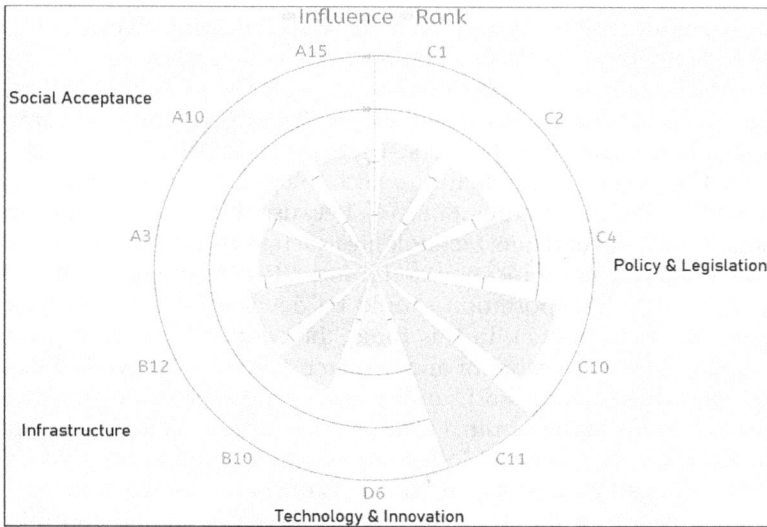

FIGURE 8.7
Influence and rank of the key AV adoption factors.

communication), Policy and legislation, and Technology and innovation. The findings showed that in the field of 'Policy and legislation', the factors of 'future orientation of government', "Managing the international sanctions for foreign investment", 'Funding mega projects on AVs', "Developing legislation in the field of criminals, including responsibility for road accidents, privacy, and cyber security", and "Developing of transparent supportive laws/regulations for the use of AVs in the field of priority and right of way, insurance, and certification", appeared as key factors.

International sanctions imposed on Iran's automotive and manufacturing sectors, along with the departure of multinational manufacturers, have effectively prevented the Iranian car market from receiving cutting-edge technology. This limitation has affected not just ordinary automotive manufacturing but has also caused Iran to lag far behind in the development of emerging technologies such as AVs. Furthermore, due to current uncertainty in the sphere of criminal laws, such as culpability for road accidents, privacy, cybersecurity, as well as traffic rules, insurance laws, and AV licensing, legislation efficiency is a significant barrier. On the other hand, it may imperil a country's critical infrastructure, disrupt key service delivery, and harm society's general well-being (Lim & Taeihagh, 2018). As a result, a roadmap for collaboration between the public and commercial sectors, as well as research institutes, is necessary to build a clear legislative framework. The elements 'Media and marketing', "Cost-effectiveness compared to human-driven cars", and "Feasibility of using AVs in collective and shared modes" emerged as major considerations in the realm of 'social acceptance'. Based on cultural norms and socio-demographic environment, social acceptability would be closely associated with the client's lifestyle and personal preferences. The public would appreciate it if they were more aware of the genuine benefits, such as reduced travel time and cost savings. Media, advertising, and campaigns have a significant impact on how end-users perceive AVs.

The most important variable in the 'technology and innovation' area was determined as the 'development of IoT'. Because the issue of smart vehicles is entangled with several new technologies such as cloud computing, AI, and IoT, a roadmap in communication and information technology connected to intelligent vehicle transportation should be devised (electric vehicles, AVs, passages, smart signs, etc.). In this topic, the country's status in each field is assessed to define investment and research priorities, as well as decision-making intentions towards R&D. In the field of 'Infrastructure', the factors of "V2X, passing information from a vehicle to any entity", and 'ITS equipment' were proposed as key factors. Modern cities are expected to supply local data centres for efficient processing and analysis of such massive amounts of big data. A data centre of this type may act as a command-and-control centre for all ITS-related operations, as well as aid AVs while driving. An ITS is an advanced application that aims to provide innovative services relating to different modes of transport and traffic management and enable users to be better informed and make safer, more coordinated, and 'smarter' use of transport networks. Some of these technologies include calling for emergency services when an accident occurs and using cameras to enforce traffic laws or signs that mark speed limit changes depending on conditions.

This study showed that the important element with the highest effect was 'Policymaking and Legislation', demonstrating that the government's intervention would remove a large share of the impediments. Furthermore, the government's investment in developing infrastructure (road and communication) and technology essential for the introduction of AVs into the country is highly effective. All planning and decision-making procedures in Iran are determined by the role and origin of the government. State control and management of economic concerns is a sovereign responsibility enshrined in national legislation.

The study's findings are in line with earlier works reported in the literature. For instance, Talebian and Mishra (2018) discovered that the adoption of AVs among American academics (in the case of Memphis University) would be greatly reliant on the social network among individuals, which is afterward impacted by media advertisement and marketing. Likewise, an Australian study discovered that gender would influence whether tertiary students use AVs, even though students are worried about cybersecurity and the failure of AVs (Soltani et al., 2021). Similarly, in terms of urban and transport infrastructure development involving advanced technologies, Yigitcanlar and Bulu (2015) highlighted various challenges in the context of a developing country, Turkey, that align well with the findings of the study at hand. Additionally, another study revealed, in the context of Brazil, that stimulating technological innovation through government incentives is critical (Yigitcanlar et al., 2019).

Although in this research the variables influencing the development of AVs in Iran were examined, the findings may be applied to other emerging nations with similar economic and social situations. As most of the existing study on AVs focuses on the public (Soltani et al., 2021), it is advised that the viability of implementing AVs for passengers with diverse mobility and communication demands be investigated (Etminani-Ghasrodashti et al., 2021). This is particularly crucial for Iran, a nation with a large proportion of elderly (Soltani et al., 2018) and handicapped citizens who have difficulty using the standard public transportation networks now in place (Panahi et al., 2022). This research may go further by investigating transport priorities such as safety, capacity, and accessibility of the services, particularly if they are linked with other public transportation systems (Miller et al., 2022). One of the main constraints of this study was finding the most pertinent specialists to take part in it because the technology of AVs in Iran is still in the early phases of study and research. This research may potentially be enhanced by recruiting a broader pool of stakeholders and especially conducting focus groups with potential AV clients. Additionally, it is beneficial to go deeper into the technical technique and undertake a cross-impact balance analysis for scenario analysis, as is customary in futurist research.

8.6 Conclusion

In conclusion, this chapter underscores the intricate yet promising path for AV adoption in developing nations, with Iran as a focal case. The findings illustrate that successful AV integration relies on a synergistic interplay of societal acceptance, infrastructure, policy and legislation, and technology and innovation. Key challenges in infrastructure and legislative frameworks reveal the need for cohesive, supportive policies and substantial investment to enable AV compatibility with current urban landscapes. Moreover, societal acceptance will be crucial, with public awareness, cybersecurity assurances, and reliable infrastructure poised to shape user trust and AV adoption. For developing nations, addressing these factors within unique economic and social contexts is vital to leveraging AVs' potential benefits in reducing traffic congestion, environmental impact, and enhancing urban mobility. Future studies could expand these insights by involving a broader range of stakeholders and focusing on adaptive strategies to better integrate AVs within existing transportation systems.

To facilitate AV adoption, developing nations should prioritise a phased approach that integrates AVs within existing infrastructure, coupled with supportive policy and public education initiatives. Governments can start by investing in foundational infrastructure, such as upgrading road markings, establishing data centres for real-time traffic management, and expanding high-speed Internet coverage, especially in urban areas. Implementing clear, adaptable regulatory frameworks is essential for addressing liability, cybersecurity, and safety standards, providing a stable foundation for AV deployment. Additionally, public awareness campaigns focusing on the environmental and economic benefits of AVs can foster acceptance and trust, which are crucial for adoption. By encouraging public-private partnerships, countries can mobilise investments, drive technological innovation, and facilitate skills development, ensuring that AV deployment aligns with local needs and societal values.

Acknowledgements

This chapter, with permission from the copyright holder, is a reproduced version of the following journal article: Zali, N., Amiri, S., Yigitcanlar, T., & Soltani, A., (2022). Autonomous vehicle adoption in developing countries: Futurist insights. *Energies*, 15(22), 8464.

References

Abdi, M.H., & Soltani, A. (2022). Which fabric/scale is better for transit-oriented urban design: Case studies in a developing country. *Sustainability*, 14, 7338. https://doi.org/10.3390/su14127338

Ackaah, W., Leslie, V.L.D., & Osei, K.K. (2021). Perception of autonomous vehicles— A Ghanaian perspective. *Transportation Research Interdisciplinary Perspectives*, 11, 100437. https://doi.org/10.1016/j.trip.2021.100437

Anania, E.C., Rice, S., Walters, N.W., Pierce, M., Winter, S.R., & Milner, M.N. (2018). The effects of positive and negative information on consumers' willingness to ride in a driverless vehicle. *Transport Policy*, 72, 218–224. https://doi.org/10.1016/j.tranpol.2018.04.002

Bagloee, S.A., Tavana, M., Asadi, M., & Oliver, T. (2016). Autonomous vehicles: Challenges, opportunities, and future implications for transportation policies. *Journal of Modern Transportation*, 24, 284–303. https://doi.org/10.1007/s40534-016-0117-3

Bansal, P., Kockelman, K.M., & Singh, A. (2016). Assessing public opinions of and interest in new vehicle technologies: An Austin perspective. *Transportation Research Part C: Emerging Technologies*, 67, 1–14. https://doi.org/10.1016/j.trc.2016.01.019

Bichiou, Y., & Rakha, H.A. (2019). Developing an optimal intersection control system for automated connected vehicles. *IEEE Transactions on Intelligent Transportation Systems*, 20, 1908–1916. https://doi.org/10.1109/TITS.2018.2850335

Butler, L., Yigitcanlar, T., & Paz, A. (2020). How can smart mobility innovations alleviate transportation disadvantage? Assembling a conceptual framework through a systematic review. *Applied Sciences*, 10, 6306. https://doi.org/10.3390/APP1 0186306

Butler, L., Yigitcanlar, T., & Paz, A. (2020). Smart urban mobility innovations: A comprehensive review and evaluation. *IEEE Access*, 8. 196034–196049. https://doi.org/10.1109/ACCESS.2020.3034596

Butler, L., Yigitcanlar, T., & Paz, A. (2021). Barriers and risks of mobility-as-a-service (MaaS) adoption in cities: A systematic review of the literature. *Cities*, 109, 103036. https://doi.org/10.1016/j.cities.2020.103036

Butler, L., Yigitcanlar, T., & Paz, A. (2021). Factors influencing public awareness of autonomous vehicles: Empirical evidence from Brisbane. *Transportation Research Part F: Traffic Psychology and Behaviour*, 82, 256–267. https://doi.org/10.1016/j.trf.2021.08.016

Butler, L., Yigitcanlar, T., & Paz, A. (2022). How can smart mobility bridge the first/last mile gap? Empirical evidence on public attitudes from Australia. *SSRN Electronic Journal*, 104, 103452. https://doi.org/10.2139/ssrn.4068716

Carreras, X.D.A., Erhart, J., & Ruehrup, S. (17–21 September 2018). Road infrastructure support levels for automated driving. In Proceedings of the 25th ITS World Congress, Copenhagen, Denmark, pp. 12–20. Available: www.inframix.eu/wp-content/uploads/ITSWC2018-ASF-AAE-Final-paper_v4.pdf (accessed on).

Chehri, A., & Mouftah, H.T. (2019). Autonomous vehicles in the sustainable cities, the beginning of a green adventure. *Sustainable Cities and Society*, 51, 101751. https://doi.org/10.1016/j.scs.2019.101751

Chen, B., Sun, D., Zhou, J., Wong, W., & Ding, Z. (2020). A future intelligent traffic system with mixed autonomous vehicles and human-driven vehicles. *Information Sciences*, 529, 59–72. https://doi.org/10.1016/j.ins.2020.02.009

Chen, F., Song, M., & Ma, X. (2020). A lateral control scheme of autonomous vehicles considering pavement sustainability. *Journal of Cleaner Production*, 256, 120669. https://doi.org/10.1016/j.jclepro.2020.120669

Coca-Vila, I. (2018). Self-driving cars in dilemmatic situations: An approach based on the theory of justification in criminal law. *Criminal Law and Philosophy*, 12, 59–82. https://doi.org/10.1007/s11572-017-9411-3

Collingwood, L. (2017). Privacy implications and liability issues of autonomous vehicles. *Information & Communications Technology Law*, 26, 32–45. https://doi.org/10.1080/13600834.2017.1269871

de Bruyne, J., & Werbrouck, J. (2018). Merging self-driving cars with the law. *Computer Law & Security Review*, 34, 1150–1153. https://doi.org/10.1016/j.clsr.2018.02.008

Delgado-Serrano, M.d., Vanwildemeersch, P., London, S., Ortiz-Guerrero, C.E., Semerena, R.E., & Rojas, M. (2016). Adapting prospective structural analysis to strengthen sustainable management and capacity building in community-based natural resource management contexts. *Ecology and Society.*, 21, 36. https://doi.org/10.5751/ES-08505-210236

Dennis, S., Paz, A., & Yigitcanlar, T. (2021). Perceptions and attitudes towards the deployment of autonomous and connected vehicles: insights from Las Vegas,

Nevada. *Journal of Urban Technology*, 28, 75–95. https://doi.org/10.1080/10630 732.2021.1879606

Engholm, A., Pernestål, A., & Kristoffersson, I. (2018). System-Level Impacts of Self-Driving Vehicles: Terminology, Impact Frameworks and Existing Literature Syntheses; Swedish National Road and Transport Research Institute (VTI): Linköping, Sweden.

Etminani-Ghasrodashti, R., Patel, R., Kermanshachi, S., Rosenberger, J., Weinreich, D., & Foss, A. (2021). Integration of shared autonomous vehicles (SAVs) into existing transportation services: A focus group study. *Transportation Research Interdisciplinary Perspectives*, 12, 100481. https://doi.org/10.1016/j.trip.2021.100481

Fagnant, D.J., & Kockelman, K. (2015). Preparing a nation for autonomous vehicles: Opportunities, barriers and policy recommendations. *Transportation Research Part A: Policy and Practice*, 77, 167–181. https://doi.org/10.1016/j.tra.2015.04.003

Faisal, A., Yigitcanlar, T., Kamruzzaman, M., & Currie, G. (2019). Understanding autonomous vehicles: A systematic literature review on capability, impact, planning and policy. *Journal of Transport and Land Use*, 12, 45–72. https://doi.org/10.5198/jtlu.2019.1405

Faisal, A., Yigitcanlar, T., Kamruzzaman, M., & Paz, A. (2021). Mapping two decades of autonomous vehicle research: A systematic scientometric analysis. *Journal of Urban Technology*, 28, 45–74. https://doi.org/10.1080/10630732.2020.1780868

Glancy, D.J. (2012). Privacy in autonomous vehicles. *Santa Clara Law Review*, 52, 1171.

Godet, M., Durance, P., & Gerber, A. (2013). Strategic foresight La prospective use and misuse of scenario building. *Circle of Future Entrepreneurs*, 65, 421.

Golbabaei, F., Yigitcanlar, T., & Bunker, J. (2021). The role of shared autonomous vehicle systems in delivering smart urban mobility: A systematic review of the literature. *International Journal of Sustainable Transportation*, 15, 731–748. https://doi.org/10.1080/15568318.2020.1798571

Golbabaei, F., Yigitcanlar, T., Paz, A., & Bunker, J. (2020). Individual predictors of autonomous vehicle public acceptance and intention to use: A systematic review of the literature. *Journal of Open Innovation: Technology, Market, and Complexity*, 6, 106. https://doi.org/10.3390/joitmc6040106

Habibi, A., Jahantigh, F.F., & Sarafrazi, A. (2015). Fuzzy Delphi technique for forecasting and screening items. *Asian Journal of Economics and Business Management*, 5, 130. https://doi.org/10.5958/2249-7307.2015.00036.5.

Haboucha, C.J., Ishaq, R., & Shiftan, Y. (2017). User preferences regarding autonomous vehicles. *Transportation Research Part C: Emerging Technologies*, 78, 37–49. https://doi.org/10.1016/j.trc.2017.01.010

Handcock, M.S., & Gile, K.J. (2011). Comment: On the concept of snowball sampling. *Sociological Methodology*, 41, 367–371. https://doi.org/10.1111/j.1467-9531.2011.01243.x

Hevelke, A., & Nida-Rümelin, J. (2015). Responsibility for crashes of autonomous vehicles: An ethical analysis. *Science and Engineering Ethics*, 21, 619–630. https://doi.org/10.1007/s11948-014-9565-5

Hua, J., Huang, M.L., Huang, W., & Zhao, C. (2019). Applying graph centrality metrics in visual analytics of scientific standard datasets. *Symmetry*, 11, 30. https://doi.org/10.3390/sym11010030

Kassens-Noor, E., Wilson, M., & Yigitcanlar, T. (2021). Where are autonomous vehicles taking us? *Journal of Urban Technology*, 28, 1–4. https://doi.org/10.1080/10630 732.2021.1985318

Khan, J.A., Wang, L., Jacobs, E., Talebian, A., Mishra, S., Santo, C.A., Golias, M., & Astorne-Figari, C. (10–12 September 2019). Smart cities connected and autonomous vehicles readiness index. In Proceedings of the 2nd ACM/EIGSCC Symposium on Smart Cities and Communities, Portland, OR, USA. https://doi. org/10.1145/3357492.3358631

Kröger, L., Kuhnimhof, T., & Trommer, S. (2019). Does context matter? A comparative study modelling autonomous vehicle impact on travel behaviour for Germany and the USA. *Transportation Research Part A: Policy and Practice*, 122, 146–161. https://doi.org/10.1016/j.tra.2018.03.033

Kuutti, S., Fallah, S., Katsaros, K., Dianati, M., Mccullough, F., & Mouzakitis, A. (2018). A survey of the state-of-the-art localization techniques and their potentials for autonomous vehicle applications. *IEEE Internet of Things Journal*, 5, 829–846. https://doi.org/10.1109/JIOT.2018.2812300

Kyriakidis, M., Happee, R., & de Winter, J.C.F. (2015). Public opinion on automated driving: Results of an international questionnaire among 5000 respondents. *Transportation Research Part F: Traffic Psychology and Behaviour*, 32, 127–140. https://doi.org/10.1016/j.trf.2015.04.014

Lee, S., Jang, K.M., Kang, N., Kim, J., Oh, M., & Kim, Y. (2022). Redesigning urban elements and structures considering autonomous vehicles: Preparing design strategies for wide implementation in cities. *Cities*, 123, 103595. https://doi.org/ 10.1016/j.cities.2022.103595

Liljamo, T., Liimatainen, H., & Pöllänen, M. (2018). Attitudes and concerns on automated vehicles. *Transportation Research Part F: Traffic Psychology and Behaviour*, 59, 24–44. https://doi.org/10.1016/j.trf.2018.08.010

Lim, H.S.M., & Taeihagh, A. (2018). Autonomous vehicles for smart and sustainable cities: An in-depth exploration of privacy and cybersecurity implications. *Energies*, 11, 1062. https://doi.org/10.3390/en11051062

Litman, T. (6 November 2022). Autonomous vehicle implementation predictions: Implications for transport planning. In Proceedings of the Transportation Research Board (TRB) Annual Meeting. Victoria Transport Policy Institute, Victoria, British Columbia Canada. www.vtpi.org/avip.pdf

López-Lambas, M.E. (2017). The socioeconomic impact of the intelligent vehicles: Implementation strategies. In *Intelligent Vehicles: Enabling Technologies and Future Developments*; Elsevier: Amsterdam, pp. 437–453. https://doi.org/ 10.1016/B978-0-12-812800-8.00011-4

Manfreda, A., Ljubi, K., & Groznik, A. (2021). Autonomous vehicles in the smart city era: An empirical study of adoption factors important for millennials. *International Journal of Information Management*, 58, 102050. https://doi.org/ 10.1016/j.ijinfomgt.2019.102050

Manivasakan, H., Kalra, R., O'Hern, S., Fang, Y., Xi, Y., & Zheng, N. (2021). Infrastructure requirement for autonomous vehicle integration for future urban and suburban roads–Current practice and a case study of Melbourne, Australia. *Transportation Research Part A: Policy and Practice*, 152, 36–53. https://doi.org/ 10.1016/j.tra.2021.07.012

Millard-Ball, A. (2019). The autonomous vehicle parking problem. *Transport Policy*, 75, 99–108. https://doi.org/10.1016/j.tranpol.2019.01.003

Miller, K., Chng, S., & Cheah, L. (2022). Understanding acceptance of shared autonomous vehicles among people with different mobility and communication needs. *Travel Behaviour and Society*, 29, 200–210. https://doi.org/10.1016/j.tbs.2022.06.007

Moody, J., Bailey, N., & Zhao, J. (2020). Public perceptions of autonomous vehicle safety: An international comparison. *Safety Science*, 121, 634–650. https://doi.org/10.1016/j.ssci.2019.07.022

Nematpour, M., & Faraji, A. (2019). Structural analysis of the tourism impacts in the form of future study in developing countries (case study: Iran). *Journal of Tourism Futures*, 5, 259–282. https://doi.org/10.1108/JTF-05-2018-0028

Nikitas, A., Njoya, E.T., & Dani, S. (2019). Examining the myths of connected and autonomous vehicles: Analysing the pathway to a driverless mobility paradigm. *International Journal of Automotive Technology and Management*, 19, 10–30. https://doi.org/10.1504/IJATM.2019.098513

Panahi, N., Pourjafar, M.R., Ranjbar, E., & Soltani, A. (2022). Examining older adults' attitudes towards different mobility modes in Iran. *Journal of Transport and Health*, 26, 101413. https://doi.org/10.1016/j.jth.2022.101413

Papadoulis, A., Quddus, M., & Imprialou, M. (2019). Evaluating the safety impact of connected and autonomous vehicles on motorways. *Accident Analysis & Prevention*, 124, 12–22. https://doi.org/10.1016/j.aap.2018.12.019

Parkinson, S., Ward, P., Wilson, K., & Miller, J. (2017). Cyber threats facing autonomous and connected vehicles: Future challenges. *IEEE Transactions on Intelligent Transportation Systems*, 18, 2898–2915. https://doi.org/10.1109/TITS.2017.2665968

Puglisi, M., & Marvin, S. (2002). Developing urban and regional foresight: Exploring capacities and identifying needs in the North West. *Futures*, 34, 761–777. https://doi.org/10.1016/S0016-3287(02)00019-8

Raza, M. (2018). Autonomous vehicles: Levels, technologies, impacts and concerns. *International Journal of Applied Engineering Research*, 13(16), 12710–12714. www.ripublication.com/ijaer18/ijaerv13n16_42.pdf

Ribeiro, M.A., Gursoy, D., & Chi, O.H. (2022). Customer acceptance of autonomous vehicles in travel and tourism. *Journal of Travel Research*, 61, 620–636. https://doi.org/10.1177/0047287521993578

Rubin, I., Baiocchi, A., Sunyoto, Y., & Turcanu, I. (2019). Traffic management and networking for autonomous vehicular highway systems. *Ad Hoc Networks*, 83, 125–148. https://doi.org/10.1016/j.adhoc.2018.08.018

Ruggeri, K., Kácha, O., Menezes, I.G., Kos, M., Franklin, M., Parma, L., Langdon, P., Matthews, B., & Miles, J. (2018). In with the new? Generational differences shape population technology adoption patterns in the age of self-driving vehicles. *Journal of Engineering and Technology Management (JET-M)*, 50, 39–44. https://doi.org/10.1016/j.jengtecman.2018.09.001

Schellekens, M. (2015). Self-driving cars and the chilling effect of liability law. *Computer Law & Security Review*, 31, 506–517. https://doi.org/10.1016/j.clsr.2015.05.012

Schoettle, B., & Sivak, M. (2014). *A Survey of Public Opinion about Autonomous and Self-Driving Vehicles in the US, UK and Australia*. University of Michigan, Ann Arbor, Transportation Research Institute: Ann Arbor, MI, pp. 1–38.

Schoonmaker, J. (2016). Proactive privacy for a driverless age. *Information & Communications Technology Law*, 25, 96–128. https://doi.org/10.1080/13600834.2016.1184456

Shafique, M.A., Afzal, M.S., Riaz, N., & Ahmed, A. (2021). Public perception regarding autonomous vehicles in developing countries: A case study of Pakistan. *Pakistan Journal of Engineering and Applied Sciences*, 28, 1–6.

Soltani, A. (2017). Social and urban form determinants of vehicle ownership; evidence from a developing country. *Transportation Research Part A: Policy and Practice*, 96, 90–100. https://doi.org/10.1016/j.tra.2016.12.010

Soltani, A., & Askari, S. (2017). Exploring spatial autocorrelation of traffic crashes based on severity. *Injury*, 48, 637–647. https://doi.org/10.1016/j.injury.2017.01.032

Soltani, A., & Ivaki, Y.E. (2011). The influence of urban physical form on trip generation, evidence from metropolitan Shiraz, Iran. *Indian Journal of Science and Technology*, 4, 1168–1174.

Soltani, A., Ananda, D., & Rith, M. (2021). University students' perspectives on autonomous vehicle adoption: Adelaide case study. *Case Studies on Transport Policy*, 9, 1956–1964. https://doi.org/10.1016/j.cstp.2021.11.004

Soltani, A., Pojani, D., Askari, S., & Masoumi, H.E. (2018). Socio-demographic and built environment determinants of car use among older adults in Iran. *Journal of Transport Geography*, 68, 109–117. https://doi.org/10.1016/j.jtrangeo.2018.03.001

Statistics Center of Iran (2019). *Statistical Yearbook of the Country*; Statistics Center of Iran: Tehran, Iran.

Talebian, A., & Mishra, S. (2018). Predicting the adoption of connected autonomous vehicles: A new approach based on the theory of diffusion of innovations. *Transportation Research Part C: Emerging Technologies*, 95, 363–380. https://doi.org/10.1016/j.trc.2018.06.005

Tan, S.Y., & Taeihagh, A. (2021). Adaptive governance of autonomous vehicles: Accelerating the adoption of disruptive technologies in Singapore. *Government Information Quarterly*, 38, 101546. https://doi.org/10.1016/j.giq.2020.101546

Threlfall, R. (2020). Autonomous Vehicles Readiness Index; Kpmg: Swiss, 2022; pp. 1–70.

Villacorta, P.J., Masegosa, A.D., Castellanos, D., & Lamata, M.T. (2014). A new fuzzy linguistic approach to qualitative Cross Impact Analysis. *Applied Soft Computing Journal*, 24, 19–30. https://doi.org/10.1016/j.asoc.2014.06.025

Villagra, J., Acosta, L., Artuñedo, A., Blanco, R., Clavijo, M., Fernández, C., Godoy, J., Haber, R., Jiménez, F., Martínez, C. et al. (2017). *Automated driving*; Elsevier Inc.: Amsterdam, The Netherlands,. https://doi.org/10.1016/B978-0-12-812800-8.00008-4

Wang, S., & Zhao, J. (2019). Risk preference and adoption of autonomous vehicles. *Transportation Research Part A: Policy and Practice*, 126, 215–229. https://doi.org/10.1016/j.tra.2019.06.007

Wang, Z., Safdar, M., Zhong, S., Liu, J., & Xiao, F. (2021). Public preferences of shared autonomous vehicles in developing countries: A cross-national study of Pakistan and China. *Journal of Advanced Transportation*, 2021, 5141798. https://doi.org/10.1155/2021/5141798

West, D.M. (2016). *Moving Forward: Self-Driving Vehicles in China, Europe, Japan, Korea, and the United States*; Brookings: Washington, DC.

Whittle, C., Whitmarsh, L., Hagger, P., Morgan, P., & Parkhurst, G. (2019). User decision-making in transitions to electrified, autonomous, shared or reduced mobility. *Transportation Research Part D: Transport and Environment*, 71, 302–319. https://doi.org/10.1016/j.trd.2018.12.014

Xu, X., & Fan, C.K. (2019). Autonomous vehicles, risk perceptions and insurance demand: An individual survey in China. *Transportation Research Part A: Policy and Practice*, 124, 549–556. https://doi.org/10.1016/j.tra.2018.04.009

Yigitcanlar, T., & Bulu, M. (2015). Dubaization of Istanbul: Insights from the knowledge-based urban development journey of an emerging local economy. *Environment and Planning A*, 47, 89–107.

Yigitcanlar, T., Degirmenci, K., Butler, L., & Desouza, K. (2022). What are the key factors affecting smart city transformation readiness? Evidence from Australian cities. *Cities*, 120, 103434. https://doi.org/10.1016/j.cities.2021.103434

Yigitcanlar, T., Sabatini-Marques, J., da-Costa, E., Kamruzzaman, M., & Ioppolo, G. (2019). Stimulating technological innovation through incentives: Perceptions of Australian and Brazilian firms. *Technological Forecasting and Social Change*, 146, 403–412.

Yigitcanlar, T., Kankanamge, N., Regona, M., Ruiz Maldonado, A., Rowan, B., Ryu, A., Desouza, K., Corchado, J., Mehmood, R., & Li, R. (2020). Artificial intelligence technologies and related urban planning and development concepts: How are they perceived and utilized in Australia? *Journal of Open Innovation: Technology, Market, and Complexity*, 6, 187.

Zali, N., Zamanipoor, M., Ahmadi, H., & Karami, M. (2018). Analysis of key factors influencing air pollution of metropolises in developing countries by year 2025 (case study: Tehran metropolis, Iran). *Anuário do Instituto de Geociências*, 41, 548–559. https://doi.org/10.11137/2018_3_548_559

Zeadally, S., Hunt, R., Chen, Y.S., Irwin, A., & Hassan, A. (2012). Vehicular ad hoc networks (VANETS): Status, results, and challenges. *Telecommunication Systems*, 50, 217–241. https://doi.org/10.1007/s11235-010-9400-5

Appendices

Items	Details	Number		Percent
Age group	Less than 30 years	9		36
	Over 30 years	16		64
Field of expertise	Transportation planning/traffic engineering	4		16
	Regional planning	1		4
	Urban planning	3		12
	Urban design	1		4
	Automotive engineering	2		8
	Electrical engineering	2		8
	Futures studies	3		12
	Project management	1		4
	Law	3		12
	Applied physics/physics	2		8
	Business management/economics	2		8
	Applied mathematics	1		4
Educational Level	Bachelor	3		12
	Masters	15		60
	PhD/PhD candidate	7		28
Years of experience	less than 5 years	9		36
	Between 5 and 10 years	4		16
	Between 10 and 15 years	7		28
	Over 20 years	5		20
Employment sector	Government agencies	12		48
	Consultants	10		40
	Self- employed	3		12
Sum		25		100

Figure 8A.1 Profile of Delphi experts.

TABLE 8A.1

The Long List of Factors

Drivers	Factors
	Loss of 'freedom' and 'pleasure' of driving
	Loss of driving-based jobs
	Vulnerability of AVs navigation systems against human and natural threats
	Users' trust in the companies responsible for introducing and selling AVs
	Media and marketing
	Resistance to lifestyle change among potential customers
	Residence in larger cities (cosmopolitans)
	Concerns related to cybersecurity and endangering privacy and personal information
	Concerns about AVs accident with other vehicles (autonomous—ordinary)
	Concerns about AVs accidents with pedestrians
	Social acceptance of new technologies
	Cost-effectiveness compared to human-driven vehicles
	Individual readiness
	Awareness of the environmental benefits of using AVs
	Awareness of the social benefits of using AVs
	Legal and technical gaps related to AVs
	Feasibility of using AVs in collective and shared modes
Infrastructure	Quality of streets and roads (quality of markings and signs, quality of asphalt, etc.)
	Assigning a separate route to the movement of AVs
	Revision in the geometric design of roads
	Revision in road paving standards
	Revision in the design and standards of bridges
	Revision in the design of parking spaces
	Backup equipment, data clouding, and maintenance
	Collecting and transmission of online data on road conditions
	Supplying roadside service stations (charging and repair stations)
	5G Internet coverage
	V2X (passing information from a vehicle to any entity)
	Availability of local data centres
	ITS facilities
	Bandwidth
Policy and legislation	Developing transparent supportive laws/regulations for the use of AVs in the field of priority and right of way, insurance, and certification
	Developing legislation in the field of criminals, including responsibility for road accidents, privacy, and cybersecurity
	Supporting R&D activities
	Funding mega projects on AVs
	Providing incentives to attract private sector investment
	Reduction of tariffs for importing AVs vehicles, systems, and technologies
	Encouraging PPT (public and private partnership) in developing AVs systems

TABLE 8A.1 (Continued)

The Long List of Factors

Drivers	Factors
	Supporting skilled and qualified workforce (in the areas of manufacturing, repair, maintenance, and operation)
	Funding support of research/education activities
	Managing the international sanctions for foreign investment
	Establishing a political vision for developing AVs
	Supporting transparency and accountability in the AV market
	Establishing a coordination plan among stakeholders and influential parties
	Refining and developing the legal system in addressing the AVs regulations
Technology and innovation	Tendency to mass production and industrial partnerships
	Number and reputation of companies delivering and supporting AVs
	Number of patents, innovations, and products related to AVs
	The volume of investment in AVs
	Development of high-level cybersecurity protection
	Development of IoT
	Allocation of a share of the car market to electric cars
	Number and quality of research publications/outputs and training/education activities

TABLE 8A.2

Types of Factors Based on the Influence-Dependence Chart in Figure 8.2

Factors	Notes
Input factors (1)	These factors are highly effective and independent. Such factors describe the studied system and the dynamic conditions of the system. If possible, these factors should be prioritized for strategic operational plans.
Intermediate factors (2)	These factors are highly effective and dependent. Thus, they are unstable by nature. Any action on these factors will flow into the rest of the system and deeply affect the dynamics of the system.
Resultant factors (3)	These factors are not effective but are highly dependent. Thus, their status explains the effects of other factors (mainly input and intermediate factors).
Excluded factors (4)	These factors are neither effective nor dependent. Thus, they have little effect on the studied system. As a result, excluding these factors will have few consequences for system analysis.
Clustered factors (5)	These factors are not effective or dependent enough to be included in the previous classifications. No definite conclusions can be drawn about these factors and their effect on the system.

9

Concluding Remarks and Future Directions

9.1 Introduction

This book, the exploration of *Autonomous Urban Mobility: Understanding Adoption Parameters, Perceptions, Perspectives,* underscores the transformative potential of autonomous vehicles (AVs) within our urban environments. As autonomous technologies become a reality, they challenge us to reimagine cities as spaces that are not only technologically advanced but also inclusive, accessible, and sustainable. Through rigorous examination of public perceptions, adoption parameters, and stakeholder perspectives, this book has sought to provide a roadmap for those invested in shaping the future of urban mobility.

This conclusion chapter revisits the central themes of the book while expanding on the implications of AV integration for environmental sustainability, urban equity, and economic resilience. By highlighting successful pilot programmes, regional adaptations, and case studies, this book serves as an essential guide for policymakers, city planners, and industry leaders committed to building AV ecosystems that prioritise public trust, regulatory support, and ethical integration. This chapter, in the following sections, underlines the future directions for a healthy AV adoption in cities.

9.2 Autonomous Vehicles as Solutions for Sustainable Urban Futures

As cities grapple with escalating environmental challenges, AVs offer a unique opportunity to integrate sustainable practices into urban transportation frameworks. Nonetheless, their effectiveness depends on deliberate, city-wide approaches that incorporate environmental goals, inclusive urban

DOI: 10.1201/9781003605676-9

planning, and technological innovation. The potential for AVs to positively impact urban life lies not just in their operation but in how they are integrated within a larger ecosystem of sustainable urban practices. The future efforts concerning AVs as solutions for sustainable urban futures should concentrate on the following strategies.

Reducing Carbon Emissions and Urban Pollution: AVs, particularly when paired with electric powertrains, have the potential to drastically reduce the carbon footprint associated with urban travel. With global trends moving towards decarbonisation, the alignment of AV deployment with clean energy initiatives is paramount. Cities like Oslo and Amsterdam serve as pioneering examples, combining AVs with green energy grids and low-emission zones to promote eco-friendly transportation options. By embedding AVs within these broader sustainability frameworks, cities can foster a cleaner, greener environment, reducing urban pollution levels and contributing to long-term climate goals.

Optimising Traffic Flow and Minimising Congestion: AVs possess the capability to communicate with each other and with urban infrastructure, optimising traffic flow and reducing congestion. Advanced algorithms allow AVs to adjust routes dynamically, anticipate traffic patterns, and avoid bottlenecks, leading to smoother traffic movement. Case studies from Los Angeles and Tokyo have illustrated that such optimised flow can reduce commute times and enhance road safety. Through intelligent traffic systems, AVs can actively contribute to reducing urban congestion—a major cause of air pollution and reduced quality of urban life.

Facilitating Multimodal and Shared Mobility Ecosystems: A critical aspect of sustainable urban mobility is reducing reliance on private vehicles. AVs facilitate shared and multimodal transport networks by seamlessly integrating with public transportation systems, thereby encouraging the adoption of shared mobility. In cities like Vancouver and Zurich, AVs have been introduced as a 'feeder' to existing transit, connecting commuters to larger transit nodes and promoting a culture of shared, multimodal travel. Such an approach not only minimises vehicle use but also fosters a culture of efficient, collaborative urban mobility, transforming how residents engage with transportation infrastructure.

9.3 The Role of Public Awareness and Pilot Programmes

The journey of AV adoption begins with public awareness and familiarity, which are essential in building the foundational trust necessary for mass adoption. AV technology, while highly promising, is still a source of scepticism for many, with common concerns around safety, job displacement, and data privacy. This section reiterates the importance of awareness-building and emphasises successful pilot programmes as stepping stones towards public acceptance. The future efforts reinforcing the role of public awareness and pilot programmes should concentrate on the following strategies.

Addressing Misconceptions through Targeted Educational Campaigns: Public knowledge of AVs remains limited, often shaped by sensational media portrayals that may instil unnecessary fears or misconceptions. For instance, public concerns about job displacement in sectors like delivery and public transportation can be mitigated through targeted communication efforts that clarify AV benefits and address the technology's implications for the workforce. Campaigns that involve clear, relatable explanations of AV technology and its community benefits have proven effective in cities like Seoul and San Francisco, where residents became more receptive after participating in community workshops and educational events.

The Power of Pilot Programmes and Their Psychological Impact: Pilot programmes allow communities to experience AV technology first-hand, creating opportunities for direct interaction and feedback. Successful pilots in cities like Las Vegas and London have shown that such interactions can reduce public scepticism and ease anxieties surrounding AV safety and functionality. In Las Vegas, autonomous shuttles were introduced along popular routes, allowing residents to experience AVs as part of their daily commute. Surveys conducted after these pilots indicated a significant increase in public confidence in AV technology, emphasising the effectiveness of pilots in establishing familiarity and comfort.

Transparency and Public Trust: An often-overlooked aspect of pilot programmes is the transparency with which results are reported. Cities that transparently communicate the successes, challenges, and learnings from AV pilots tend to foster higher levels of public trust. In Helsinki, the city administration published regular updates and public reports on its AV pilot programmes, creating an open dialogue with the community and demonstrating an unwavering commitment to safety. Such transparency helps demystify the technology, bridging the gap between industry advances and public acceptance.

9.4　Bridging Connectivity Gaps: The First and Last Mile Challenge

One of the greatest promises of AVs lies in their potential to address the first and last mile problem, a perennial issue in urban transportation. As the final link between public transit hubs and commuters' destinations, AVs can provide an on-demand, accessible solution that enhances the convenience and appeal of public transit. The future efforts for bridging connectivity gaps by tackling the first and last mile challenge should concentrate on the following strategies.

Enhancing Accessibility for All Users: AVs have the potential to extend public transit access to underserved neighbourhoods, especially in sprawling suburban areas. In a pilot programme in Phoenix, AVs were deployed as shuttle services connecting outer residential areas with major transit hubs, filling a vital gap in the city's public transportation system. Such connectivity promotes public transit use, reduces the dependency on private vehicles, and improves overall urban mobility by offering reliable, on-demand options that reach areas underserved by existing transit routes.

Dynamic and Adaptive Routing for Efficiency: Unlike traditional public transport, AVs can adapt routes based on demand, optimising travel times and increasing user convenience. Such dynamic routing enables AVs to operate as demand-responsive transport, creating a flexible and personalised travel experience that aligns with individual commuter needs. In Singapore, AVs integrated with mobile applications allow users to request shuttle services at specific times, with routing adjusted in real-time. This adaptability has made AVs a popular option for bridging the first and last mile, illustrating their potential to reshape urban travel patterns.

Supporting Reduced Parking Infrastructure and Urban Reclamation: Widespread AV use could lessen the need for extensive parking infrastructure, a factor that influences urban land use. As more people rely on AVs for short commutes, cities can gradually reduce parking facilities, reclaiming land for parks, housing, and other community spaces. For instance, Los Angeles' Mobility Plan 2035 envisions reclaiming portions of parking lots for community projects, creating greener, more liveable urban spaces. This repurposing aligns with broader city goals of improving liveability, enhancing public spaces, and addressing housing shortages.

9.5 Embracing Inclusivity and Diversity in AV Adoption

As AVs become embedded within urban ecosystems, it is essential that adoption strategies reflect the diverse needs of urban populations. The equitable deployment of AVs ensures that all urban residents—regardless of socio-economic status, age, or physical ability—can benefit from the technology. The future efforts concerning embracing inclusivity and diversity in AV adoption should concentrate on the following strategies.

Designing for Accessibility for Meeting the Needs of the Elderly and Disabled: Accessible design is paramount in AV adoption. AVs can be tailored with features like wheelchair access, automated entry systems, and user-friendly interfaces, making them more inclusive for elderly passengers and individuals with disabilities. In Tokyo, AVs equipped with accessible features have been deployed to support the city's ageing population, enabling older adults to maintain their mobility and independence. Such inclusive design reflects a broader commitment to ensuring AV technology serves a wide range of users, supporting the social dimension of urban mobility.

Addressing Economic Disparities through Subsidised AV Services: Economic barriers can impede access to AV technology, particularly for low-income communities. To address this, cities could explore subsidised AV services or offer incentives that make AVs a viable option for underserved populations. Public–private partnerships, such as those seen in Chicago's Low-Income Mobility Program, demonstrate that inclusive AV deployment can mitigate economic barriers, ensuring that all residents can access safe, efficient, and affordable transportation.

Cultural Adaptation and Cross-Demographic Customisation: Understanding that different demographic groups have varied transportation needs is vital. Young, tech-savvy professionals may prioritise speed and connectivity, while elderly passengers may value safety and ease of use. By adapting AV features and services to align with these diverse needs, cities can increase acceptance and ensure AVs serve as an inclusive urban transportation solution.

9.6 Regulatory Frameworks and Stakeholder Collaboration

The path to widespread AV adoption requires cooperation across multiple sectors and the establishment of regulatory frameworks that support innovation while prioritising safety and accountability. The future efforts concerning regulatory frameworks and stakeholder collaboration should concentrate on the following strategies.

Developing Adaptive, Responsive Regulations: Cities around the world are experimenting with regulatory frameworks that enable AV testing while safeguarding public welfare. In places like Dubai and San Francisco, governments have developed adaptive regulations that evolve in response to technological advancements, creating a supportive environment for AV development. Such frameworks often involve phased deployment strategies, allowing regulators to monitor AV impact and make adjustments based on real-world data.

Clarifying Liability and Insurance Models: AV adoption introduces complex questions around liability and insurance, particularly regarding accident accountability. Traditional insurance models may need to evolve to account for the nuanced risks associated with autonomous driving. In Germany, recent legislation has set standards for AV liability, clarifying responsibility for manufacturers and owners. Such policies illustrate the importance of proactively addressing liability issues to ensure that AV integration is legally sound and that users feel protected.

Building Multi-Stakeholder Partnerships for Success: Collaboration among government entities, technology providers, transit agencies, and advocacy groups is crucial for effective AV deployment. Multi-stakeholder partnerships foster a collaborative environment where resources, expertise, and responsibilities are shared, ensuring that AV programmes align with city goals and community needs. New York's Automated Vehicle Initiative exemplifies this approach, involving various stakeholders in planning, pilot testing, and evaluating AV impact.

9.7 Social and Psychological Dimensions of Adoption

As emphasised throughout the book, the success of AVs is contingent on public trust, acceptance, and the thoughtful integration of social and psychological factors. The future efforts concerning social and psychological dimensions of adoption should concentrate on the following strategies.

Building Public Confidence through Transparency and Engagement: Trust is essential for AV adoption. Cities that prioritise transparency, provide clear information about safety protocols, and engage directly with communities often experience higher levels of public acceptance. Transparent data-sharing policies, as seen in cities like Amsterdam, help demystify AV technology, fostering a public understanding rooted in trust and shared knowledge.

Addressing Ethical Concerns and Digital Privacy: Ethical considerations around data privacy and user consent must be carefully addressed. Cities must ensure that AV data collection practices respect individual privacy, and AV providers should commit to stringent data protection standards. Ethical use policies, widely communicated to the public, can alleviate concerns about surveillance and ensure AV deployment aligns with citizens' privacy expectations.

Overcoming Technological Anxiety through Familiarisation Programmes: Public fear of AVs is often rooted in unfamiliarity and lack of understanding. Familiarisation programmes, where residents engage directly with AV technology, can play a significant role in overcoming apprehension. In cities like Boston, community events allow residents to interact with AVs, ask questions, and learn about the technology, promoting a cultural shift towards acceptance.

9.8 A Vision for the Future of Urban Mobility

In conclusion, this book presents AVs not only as a technological solution but as a transformative force in urban development. The insights shared here offer a vision for AVs as cornerstones of a more equitable, efficient, and sustainable future. As cities continue to evolve, the success of AVs will depend on collaborative efforts, adaptive regulations, and a commitment to inclusivity.

By reimagining transportation, cities can create an urban landscape where autonomous mobility enhances quality of life, aligns with sustainability goals, and serves all residents fairly. This book stands as a roadmap, inviting stakeholders to engage with AVs not just as tools but as essential elements of a future that prioritises well-being, resilience, and social responsibility. The journey is ongoing, and the potential for AVs to reshape urban life continues to expand, promising a smarter, safer, and more connected urban world.

Index

For Product Safety Concerns and Information please contact our EU
representative GPSR@taylorandfrancis.com
Taylor & Francis Verlag GmbH, Kaufingerstraße 24, 80331 München, Germany

www.ingramcontent.com/pod-product-compliance
Lightning Source LLC
Chambersburg PA
CBHW060335220326
41598CB00023B/2709